Springer Proceedings in Business and Economics

Springer Proceedings in Business and Economics brings the most current research presented at conferences and workshops to a global readership. The series features volumes (in electronic and print formats) of selected contributions from conferences in all areas of economics, business, management, and finance. In addition to an overall evaluation by the publisher of the topical interest, scientific quality, and timeliness of each volume, each contribution is refereed to standards comparable to those of leading journals, resulting in authoritative contributions to the respective fields. Springer's production and distribution infrastructure ensures rapid publication and wide circulation of the latest developments in the most compelling and promising areas of research today.

The editorial development of volumes may be managed using Springer Nature's innovative EquinOCS, a proven online conference proceedings submission, management and review system. This system is designed to ensure an efficient timeline for your publication, making Springer Proceedings in Business and Economics the premier series to publish your workshop or conference volume.

This book series is indexed in SCOPUS.

Mihail Busu
Editor

Rethinking Business for Sustainable Leadership in a VUCA World

17th International Conference on Business Excellence, ICBE 2023, Bucharest, Romania, March 23–25, 2023

Springer

Editor
Mihail Busu
Bucharest University of Economic Studies
Bucharest, Romania

ISSN 2198-7246 ISSN 2198-7254 (electronic)
Springer Proceedings in Business and Economics
ISBN 978-3-031-50207-1 ISBN 978-3-031-50208-8 (eBook)
https://doi.org/10.1007/978-3-031-50208-8

This Springer imprint is published by the registered company Springer Nature Switzerland AG
The registered company address is: Gewerbestrasse 11, 6330 Cham, Switzerland

Contents

Tackling Climate Change: Effective Measures for Accelerating the Decarbonisation of Maritime Transport Amid a World of Mounting Distress

Maria Judith Kristina Baciu

Abstract Maritime transport is one of the main drivers of economic growth and its role is essential to the consolidation of international trade. At the same time, maritime transport is a major polluter and has been facing increased pressure to lower its carbon footprint. The shipping industry has committed to reducing its greenhouse gas emissions in compliance with the provisions enforced by its global regulator, the International Maritime Organization. Aligned with long-term strategic objectives ratified by the United Nations Framework Convention on Climate Change, the global shipping industry has made significant progress with regard to the strengthening of its overall sustainability measures. Although recent efforts may seem slow and disarrayed, there is strong belief that the shipping industry will achieve its decarbonisation targets by 2050. The current economic conditions are particularly challenging with inflationary pressures continuing to rise, global debt at record highs and skyrocketing energy prices. Beyond doubt, distress is substantial and all industries are seeking ways to mitigate risks and build resilience. Turmoil is pervasive, but the shipping industry is accustomed to extensive threats and high levels of risk, for it is inherently linked to volatility, uncertainty, complexity and ambiguity—against which it has managed to deploy exceptional adaptive capabilities. As the current climate crisis continues to magnify its impact, sustainability-driven ambitions are gaining traction in worldwide shipping. This paper seeks to provide a comprehensive assessment of the progress made towards the decarbonisation of maritime transport. The analysis focuses on benchmarking relevant performance records against short- and long-term formal targets, and further identifies several measures aimed at accelerating the reduction of greenhouse gas emissions. Key findings are outlined in detail and concluding remarks are explicitly set forth.

M. J. K. Baciu (✉)
Faculty of Theoretical and Applied Economics, Bucharest University of Economic Studies, Bucharest, Romania
e-mail: baciumaria22@stud.ase.ro

M. Busu (ed.), *Rethinking Business for Sustainable Leadership in a VUCA World*, Springer Proceedings in Business and Economics,
https://doi.org/10.1007/978-3-031-50208-8_1

Keywords Maritime policy · Environmental compliance · Greenhouse gas emissions · Decarbonisation · Green shipping · Sustainability · Climate change

1 Introduction

'Volatility', 'uncertainty', 'complexity' and 'ambiguity' are popular keywords which characterise today's fast-changing economic conditions. The 'VUCA' acronym was primarily introduced in 1987 by the United States Army War College to map out future trends (Barber 1992), but the concept still finds strong appeals in current contexts underpinned by large-scale, multipolar shifts. Global networks of business and politics are rapidly being reformed as a result of fast-paced social, economic and technological transformations. The world is changing in plain sight and the global economy is experiencing fundamental twists. This growing instability is causing widespread distress and poses serious challenges for both new and mature businesses involved in all industries.

Far-reaching changes have progressively reshaped entire organisations and have cast doubt on the effectiveness of value-creating mechanisms rooted in neoclassical economics theory. Above all else, such transformations have influenced entire business ecosystems and have explicitly remodelled organisational intent, structure and bottom line impact. Under the increasing influence of globalisation, business continues to be carried out in conditions bound by insecurity, heterogeneity, disruption and lack of clarity. As contemporary times remain subject to spontaneous swings and various unknowns, organisations across the globe are actively seeking ways to revive, adapt and upgrade. Environmental dynamism is gaining momentum and so are the stakes that complement it. With growing social concerns over climate change, pollution and waste disposal (Kiron et al. 2013), many executives are shifting away from conservative business practices and are instead pursuing participatory environmental responsibility. This change of focus is largely attributable to the fact that durable economic advantages can no longer be achieved in the absence of sustained environmental stewardship. Alternatively stated, businesses are discarding conventional strategies embedded in macroeconomic postulations of a closed economic system and are, instead, placing increased emphasis on open strategies fostering channelled learning, rapid-cycle innovation and fast adaptation to a global, circular economy model. Most notably, businesses in all parts of the world are actively engaging in targeted actions against climate change and global warming, and future responses are likely to strengthen. In this sense, political cohesion, public awareness and legislative provisions may certainly accelerate this transformation process and ensure its long-haul success.

Climate change is accelerating its spread and there is growing concern about its potentially irreversible impact on our planet. Beyond any shadow of a doubt, global warming is a major cause of widespread distress and its direct effects are in full swing. Extreme weather has recurred in recent years, often causing prolonged periods of torrid heat, severe frost, violent storms, massive floods and intense droughts. These rapid changes in our climate are putting whole ecosystems at great risk and related consequences are further expected to sharpen in the coming decades. Needless to say, our joint efforts to tackle the current climate crisis need to be improved and accelerated. Without firm, coordinated actions, our planet may no longer be able to supply the necessary living conditions for entire species of plants and animals, which could implicitly lead to mass extinctions. Our common future is threatened by unprecedented risks, yet current efforts to combat climate change remain marginal. We have certainly succeeded in raising public awareness of the great dangers that lie ahead, but so far our response to the climate crisis has been slow and inconsistent. Case in point: we have been underperforming in our attempts to address global warming and our future actions must clearly improve. Therefore, we need to renew our commitment to reducing our carbon footprint and start delivering upon our environmental ambitions with greater tact and efficiency. All things considered, it is our duty to mend the harm that we have caused to the natural environment over the years, and there is no doubt that we must handle this issue with utmost caution and promptness. The consistency of our response to climate change is paramount to the survival of all species populating the planet, but our recent efforts to reduce greenhouse gas (GHG) emissions seem to fall short of the explicit targets endorsed by the 2015 Paris Agreement (United Nations, https://unfccc.int/sites/default/files/english_paris_agreement.pdf).

Although we have been slow in responding to the climate crisis, we can still find the time and resources to combat global warming before it is too late. In view of this, it is evident that our future measures will require significant upgrades per both their form and content. Full reliance on current 'mitigation'— and 'adaptation'-driven practices (Intergovernmental Panel on Climate Change: Summary for Policymakers 2014) may only yield modest improvements, but achieving long-term climate goals will likely necessitate the enforcement of fundamental shifts. On such grounds, subsequent actions should therefore go beyond initial measures and seek to deliver radical transformations. As the diffusion of technological innovation continues to gain speed, our permanent decoupling from fossil fuels and our full transition to a net-zero carbon economy can no longer be regarded as objectives rooted in a remote future. However, this unique opportunity will require additional changes to ensure its effectiveness and long-term sustainability. More specifically, we will need to rethink our consumption habits and other relevant behaviours (e.g. land use, waste disposal, etc.) and keep these aligned with the shifting environmental conditions. Against this background, it is mandatory that our future actions on climate change become firmer, swifter and more diverse. Most importantly, further efforts will need to prioritise convergence, cohesion and integration, but their overall effectiveness will continue to depend on the global pace and magnitude of related interventions.

2 Literature Review

Maritime transport lies at the core of international trade and plays a major role in the development of the global economy. Every year, merchant ships around the world facilitate the relocation of more than 10 billion tons of wet and dry cargo (Walker et al. 2019), which altogether account for no less than 90% of the annual global trade volume (Community and Shipowners' Associations, https://ec.europa.eu/info/law/betterregulation/have-your-say/initiatives/13219-Festlegung-des-Kurses-fur-einennachhaltigen-blauen-Planeten-Aktualisierung-der-Agenda-fur-die-internationaleMeerespolitik/F2827345_de). Demand for maritime transport services continues to rise worldwide, but trade volumes remain heavily dependent on developing countries where land transport carries out a minor role towards the consolidation of intraregional trade (United Nations Publications 2022).

The steady growth of population as well as the increase in the global gross domestic product (GDP) and the intensification of trade activities have played a critical part in the rapid development of worldwide shipping (Publishing 2020). Maritime trade growth has also been driven by factors related to the accelerating globalisation, fast pace of industrialisation, rising demand for energy, expansion of free trade agreements and liberalisation of national economies. Considering that the world's population continues to increase and that all emerging economies are intensifying their demand for raw materials, intermediate and final consumer goods, there are strong expectations that the span of seaborne trade will further broaden in the coming decades (International Chamber of Shipping, https://www.ics-shipping.org/shipping-fact/shipping-and-world-trade-world-seabornetrade/). Given its increased reliability, high efficiency and cost-effectiveness, the shipping industry's competitive positioning within the global transport network remains unrivalled, and its future development is generally viewed with great confidence.

Despite its role as most effective means of transportation over long distances, the shipping industry is also known to be a major carbon emitter (UK Research and Innovation Technology, https://www.ukri.org/news/shipping-industry-reduces-carbon-emissionswith-space-technology/). In overall terms, its environmental impact is mainly attributable to causes having to do with:

- Water contamination triggered by fuel and cargo spills or by the discharge of oily bilge ballast water and other similar non-biodegradable solid wastes;
- Atmospheric pollution resulted from vessel power systems;
- Soil contamination induced by land- and water-based infrastructure developments;
- Widespread ecological damages caused by the introduction of novel species of plants and animals, pathogens and non-pathogenic microorganisms (Publishing 1997).

Although all of the above-mentioned threats are equally significant per their negative impact on the natural environment, atmospheric pollution transcends all other concerns. This owes much to the fact that GHGs such as carbon dioxide (CO_2), methane (CH_4), sulphur and nitrogen oxides (SO_x, NO_x) are directly related to global warming and are therefore considered to represent the main threats to the preservation of environmental integrity.

According to the latest available data, maritime transport is responsible for the emission of approximately 940 million tonnes of CO_2 annually, which constitutes around 2.5% of the world's total CO_2 outflows (UK Research and Innovation, https://www.ukri.org/news/shipping-industry-reduces-carbon-emissionswith-space-technology/). The shipping industry is additionally liable for the emission of other GHGs, among which SO_x, NO_x, CH_4, black carbon (BC) and particulate matters (PMs) are classified as most prevalent atmospheric pollutants besides CO_2 (Environment and Agency; Gössling et al. 2021). Furthermore, recent statistics highlight the fact that shipping emissions linked to SO_x and NO_x cover nearly 13% of the annual global air pollution (Print 2018). Although maritime transport ranks lowest per its overall carbon footprint among its comparable, land-based peers (International Chamber of Shipping n.d), there is growing concern around its potentially irreversible impact on the natural environment, especially since the rising demand for shipping is known to have driven significant increases in measured maritime emissions over the years (King 2022).

With the growing importance of environmental sustainability, the global shipping industry has actively tried to control and reduce pollution by strengthening its technologies, operations and strategic emphasis. While most improvements have spurred in reaction to international regulations and government-based legislative provisions, others have been driven by voluntary measures led by shipowners and formal advisers. Many of the innovations carried out to date have targeted the upgrading of structural modules such as propulsion systems, hull designs and communications systems, whereas other developments have focused on the replacement of conventional marine fuels. Some of the most significant improvements include:

- High-performance, anti-fouling paints—such as biocide-free or low-risk biocide, silicon-based, fouling-release coatings used by Teekay, the world's largest operator of mid-size tankers (Teekay Corporation https://www.teekay.com/wpcontent/uploads/2019/10/Teekay-Corporation-2018-Sustainability-Report.pdf)—designed to decrease hydrodynamic drag, reduce fuel consumption, shrink voyage costs and narrow GHG emissions (Hu et al. 2020),
- Duct-type energy-saving devices—such as the Becker Mewis Ducts available for both newbuild and retrofit applications, found on board some of the chemical tankers owned and operated by Odfjell, one of key players in the chemical tankship segment, (Skipsrevyen, https://www.skipsrevyen.no/artikkelarkivbecker-marine-systems-becker-mewis-duct/becker-mewis-duct-for-odfjell/478408)—which improve vessel propulsion and drive significant power savings independent of ship draught and cruise speed levels (Marine and Systems, n.d),
- Motor controllers—such as the variable frequency drives that have been installed on Teekay's tankships (Teekay Corporation 2021)—which boost fuel efficiency and facilitate water cooling and engine room ventilation,

- Ballast water treatment systems—such as Wärtsilä's Aquarius®EC and Aquarius®UV filtration and disinfection plants (Wärtsilä 2023)—aimed at removing and destroying certain types of biological organisms,
- Emission monitoring systems—such as ABB's CEMcaptain (ABBGroup, https://library.e.abb.com/public/dca509aa929f40e5b4b12b40a3e75c3c/PB_CEMcaptain-EN_REV_A.pdf) or Martek Marine's Evolution EMS (Martek Marine Innovative Ship and Solutions, https://www.martek-marine.com/wp-content/uploads/2021/04/EvolutionDatasheet.pdf)—designed to preserve safety on board, ensure full compliance with legislative provisions targeting GHG emissions and deliver real-time tracking and diagnosis,
- Dual-fuel propulsion systems—such as the 2- and 4-stroke main engines produced by Wärtsilä (n.d) and MAN Energy Solutions (MAN Energy and Solutions, https://www.man-es.com/energy-storage/products/dual-fuel-engines)—which enable vessel operation on liquefied natural gas (LNG), liquid biofuels or conventional marine fuels, and reduce exhaust gas emissions (Wärtsila, https://www.wartsila.com/marine/products/engines-and-generating-sets/dual-fuelengines/wartsila-34df; Energy and Solutions, https://www.man-es.com/energy-storage/campaigns/future-proof-engines).

The development and enforcement of environmental guidelines has been essential to the strengthening of decarbonisation initiatives in worldwide shipping. Currently, the International Maritime Organization (IMO) represents the sole institution responsible for setting the statutory requirements targeting the global shipping industry. IMO remains committed to preventing pollution from ships and strongly supports the achievement of all Sustainable Development Goals (SDGs) endorsed by the United Nations (UN) (Lowe-Martin 2015). Over the years, the IMO has instrumented a number of technical and legal frameworks based on which maritime transport has gradually become safer and cleaner. However, the growing environmental issues and their close connection to climate change have raised IMO's concerns and have therefore prompted significant changes in the organisation's policies.

Shortly after its inception in 1948, IMO became intensely involved in the regulation of operations bordering maritime safety and navigational efficiency. The 'International Convention for the Prevention of Pollution from Ships', commonly known as MARPOL, was adopted in 1973 and its provisions entered into force in 1978. Since that time, MARPOL's resolutions have been periodically reviewed and updated, resulting in significant changes per both their structure and emphasis (Lim, https://www.un.org/en/chronicle/article/role-international-maritime-organizationpreventing-pollution-worlds-oceans-ships-and-shipping). Regulations governing the prevention and control of environmental pollution are currently divided into six technical annexes. Their key highlights are summarised below, as follows:

- Annex I (entered into force on the 2nd of October, 1983)—acts on the prevention of pollution by oil;
- Annex II (entered into force on the 2nd of October, 1983)—sets control measures against pollution induced by toxic liquid substances carried in bulk;

- Annex III (entered into force on the 1st of July, 1992)—targets the prevention of pollution caused by hazardous substances in packaged form;
- Annex IV (entered into force on the 27th of September, 2003) —pays regard to the prevention of pollution attributable to vessel-based sewage discharges;
- Annex V (entered into force on the 31st of December, 1988) —focuses on the prevention of pollution triggered by waste disposal;
- Annex VI (entered into force on the 19th of May, 2005)—aids the prevention of air pollution. By setting firm limits on SO_x and NO_x emissions from ship exhausts, Annex VI provides an additional overview of mandatory energy efficiency measures aimed at reducing GHG emissions related to global shipping International Maritime Organization: International Convention for the Prevention of Pollution from Ships (n.d).

Since the introduction of MARPOL Annex VI, decarbonisation initiatives in the shipping industry have gradually improved, but the rate of change has often proved unsteady. Although maritime transport was not specifically addressed in the 2015 Paris Agreement, there have since been considerable efforts to reduce GHG emissions from shipping. The enforcement of stricter caps on sulphur content in marine fuels took effect in early January, 2020, yet strategic focus on the reduction of GHG outflows started to take shape in 2018, when the IMO introduced its explicit aim to decrease shipping emissions by at least 50% by 2050, compared to 2008 levels (European Parliamentary Research Service, https://www.europarl.europa.eu/RegData/etudes/BRIE/2020/659296/EPRS_BRI(2020)659296_EN.pdf).

The 'Fourth IMO Greenhouse Gas Study' found that air pollution from ships was 10% higher in 2018 than in 2012. Additionally, the study stressed that in the absence of tighter regulations, the continued growth of the global fleet could trigger dramatic increases in overall shipping emissions, regardless of the benefits derived from vessel-specific energy efficiency gains (International Maritime Organization, https://greenvoyage2050.imo.org/wp-content/uploads/2021/07/Fourth-IMO-GHG-Study2020-Full-report-and-annexes_compressed.pdf). Furthermore, the study indicated that compared to 2018 levels, GHG emissions from shipping could increase by up to 50% by the year 2050 (International Maritime Organization, https://greenvoyage2050.imo.org/wp-content/uploads/2021/07/Fourth-IMO-GHG-Study2020-Full-report-and-annexes_compressed.pdf), and that the average fleet efficiency could be expected to improve by 25% over the same period (International Maritime Organization, https://greenvoyage2050.imo.org/wp-content/uploads/2021/07/Fourth-IMO-GHG-Study2020-Full-report-and-annexes_compressed.pdf.).

These highlights are particularly suggestive, especially when linked with key trends shaping long-term economic outlooks. Research estimates that the global economy could double by 2050 and further emphasises that its growth prospects will likely be determined by the evolution of emerging markets (PricewaterhouseCoopers, https://www.pwc.com/gx/en/world-2050/assets/pwc-the-world-in-2050-full-report-feb-2017.pdf). Predicted to reach 4.8 billion by 2030, the world population is continuously expanding, thus driving a significant increase in both purchasing power and energy use (European Commission, https://knowledge4policy.ec.europa.eu/growing-consumerism_en). Given the current climate crisis, such evidence signals that the decarbonisation of maritime transport needs to be accelerated in order to ensure full alignment with the IMO's goal to reduce annual air pollution from ships by at least 50% by 2050 compared to 2008 levels (Environment and Division, https://wwwcdn.imo.org/localresources/en/OurWork/Environment/Documents/Presentation%20-%20IPIECA_UCL%20Side%20Event%203%20December%202018.pdf) and the Paris Agreement target committed to "Holding the increase in the global average temperature to well below 2 °C above pre-industrial levels and pursuing efforts to limit the temperature increase to 1.5 °C above pre-industrial levels, recognizing that this would significantly reduce the risks and impacts of climate change" (United Nations, https://unfccc.int/sites/default/files/english_paris_agreement.pdf).

Currently, IMO Resolutions promoting the reduction of GHG emissions in worldwide shipping can be classified into mandatory and voluntary provisions. Mandatory requirements refer to the 'Attained Energy Efficiency Design Index' (EEDI, entered into force on the 15th of July, 2011) (Marine Environment Protection Committee, https://wwwcdn.imo.org/localresources/en/OurWork/Environment/Documents/Technical%20and%20Operational%20Measures/Resolution%20MEPC.203(62).pdf), 'Ship Energy Efficiency Management Plan' (SEEMP, entered into force on the 15th of July, 2011) (Marine Environment Protection Committee, https://wwwcdn.imo.org/localresources/en/OurWork/Environment/Documents/Technical%20and%20Operational%20Measures/Resolution%20MEPC.203(62).pdf), 'Data Collection System for Fuel Oil Consumption of Ships' (DCS, entered into force on the 1st of March, 2018) (Marine Environment Protection Committee, https://wwwcdn.imo.org/localresources/en/KnowledgeCentre/IndexofIMOResolutions/MEPCDocuments/MEPC.278(70).pdf), 'Energy Efficiency Design Index' (EEXI, entered into force on the 1st of November, 2022) (Marine Environment Protection Committee, https://wwwcdn.imo.org/localresources/en/KnowledgeCentre/IndexofIMOResolutions/MEPCDocuments/MEPC.350(78).pdf) and 'Operational Carbon Intensity Indicators' (CII, entered into force on the 1st of November, 2022) (Marine Environmental Protection and Committee, https://wwwcdn.imo.org/localresources/en/KnowledgeCentre/IndexofIMOResolutions/MEPCDocuments/MEPC.352(78).pdf), whereas non-mandatory provisions include the 'Integrated Technical Cooperation Programme' (ITCP, ongoing activity) (International Maritime Organization, https://www.imo.org/en/OurWork/TechnicalCooperation/Pages/ITCP.aspx), 'Voluntary Multi-donor Trust Fund' (GHG TC-Trust Fund, adopted on the 17th of May, 2019) (Marine Environment Protection Committee, https://www.iadc.org/wp-content/uploads/2019/08/MEPC74-18-Add.1-Report-Of-The-Marine-Environment-Protection-CommitteeOn-Its-SeventyFourth-Session-Secretariat-1.pdf) and 'Voluntary National Action Plans to Address GHG Emissions from Ships' (adopted on the 20th of November, 2020) (Marine Environment Protection and Committee, https://wwwcdn.imo.org/localresources/en/OurWork/Environment/Documents/Air%20pollution/Resolution%20MEPC.327(75).pdf).

While this combination of mandatory and voluntary provisions has proved effective, impact to date has been rather moderate and further action is urgently needed in order to strengthen overall results. To this extent, the role of national and regional policies in accelerating the decarbonisation of global shipping remains essential.

The European Union (EU) is considered to be one of the most influential promoters of maritime safety and sustainable shipping. As pressures have built globally to address the escalating climate crisis, the EU has developed a number of frameworks to support the decarbonisation of maritime transport. Prior to the introduction of the 'Fit for 55' legislative proposals, GHG emissions from ships were not regulated in the EU, but were instead tracked through the 'EU Monitoring, Reporting and Verification' system (EU—MRV) that came into effect in 2015 (EUR-Lex, https://eurlex.europa.eu/legal-content/EN/TXT/?uri=CELEX:32015R0757).

The legal proposals included in the 'Fit for 55' package were introduced in 2021 and are intended to facilitate faster alignment with the end objectives endorsed by the 'European Green Deal' committed to driving the EU's transition to climate neutrality by 2050 (European Commission, https://commission.europa.eu/strategy-and-policy/priorities-2019-2024/european-green-dealen). Of these new legislative initiatives, the 'FuelEU Maritime' proposal is considered to be the most representative regulatory mechanism because it advocates the widespread use of low-carbon marine fuels. Compared to the 'EU Emissions Trading System' (EU—ETS) requirements, the 'FuelEU Maritime' proposal focuses on the reduction of GHG emissions from ships and introduces a new compliance framework based on which marine fuels are subject to a specialised qualitative assessment throughout their full life cycle (European Parliamentary Research Service, https://www.europarl.europa.eu/RegData/etudes/BRIE/2021/698808/EPRS_BRI(2021)698808_EN.pdf). Most importantly, the 'FuelEU Maritime' proposal seeks to accelerate transition to sustainable marine fuels and thus takes a broader approach towards decarbonisation by supporting the reduction of GHG emissions across whole supply chains (European Parliamentary Research Service, https://www.europarl.europa.eu/RegData/etudes/BRIE/2021/698808/EPRS_BRI(2021)698808_EN.pdf).

Beyond doubt, the 'Fit for 55' package is an important step in strengthening the EU's climate commitment under the Paris Agreement. With current provisions seeking to deliver impact as early as 2025 (EUR-Lex,; https://eur-lex.europa.eu/legalcontent/EN/TXT/?uri=CELEX%3A52021PC0562&qid=1631197035907; Lim,; https://www.un.org/en/chronicle/article/role-international-maritime-organization preventing-pollution-worlds-oceans-ships-and-shipping), emissions from shipping in the EU are likely to evidence considerable decreases so long as future efforts remain firm and prioritise the adoption of complementary market-based measures (MBMs).

Since 2008, the global shipping industry has made significant progress in reducing its carbon footprint. While the newly introduced EEXI and CII statutory requirements are expected to further improve energy efficiency in worldwide shipping, tighter legislative provisions such as the 'Fit for 55' set of proposals are more likely to prompt the widespread uptake of alternative marine fuels, and thus accelerate the transition to carbon neutrality. There are many aspects that will need to be clarified in regard to the design of future decarbonisation measures, but expectations are definitely high and the shipping industry is keen to take a step forward.

3 Methodology

As formerly claimed, this paper focuses on delineating the current state of progress regarding the decarbonisation of maritime transport. Essentially, the paper seeks to quantify the effects induced by both mandatory requirements and voluntary measures, as evidenced by real-world data. At the same time, the study highlights present levels of performance and benchmarks results against the short- and long-term objectives endorsed by the IMO as part of its Initial GHG Strategy (Marine Environment Protection and Committee, https://wwwcdn.imo.org/localresources/en/KnowledgeCentre/IndexofIMOResolutions/MEPCDocuments/MEPC.352(78).pdf).

Alternatively stated, this study seeks to determine the effectiveness of decarbonisation measures implemented to date and to answer the following question: '*To what extent has the shipping industry met the explicit, IMO-approved, GHG reduction targets?*' Addressing this subject requires a thorough investigation into the relevant factors impacting seaborne trade performance. Accordingly, the paper examines the evolution of maritime trade flows with reference to the influence induced by key economic indicators and includes further provisions for critical information pertaining to fleet-wide highlights. Supplementary data reflecting overall GHG emissions is subsequently linked to interim findings and corresponding results are additionally benchmarked against formal targets.

Appropriate evidence is used to substantiate identified gaps and complementary investigative steps are taken to determine workable solutions that facilitate the achievement of general goals. Acknowledged options are adjusted to considerations involving current resources, leading capabilities and latest industry-driven developments, and related findings are clearly emphasised.

In addition to narrowing down alternative courses of action and forwarding relevant recommendations for future review, the paper proceeds to the final stage of the analysis and answers the following questions: '*What can be done to address the identified gaps?*' and '*What measures should be implemented in order to accelerate the decarbonisation of maritime transport?*'. The foregoing rationale seeks to ensure consistency with scientific foundations and practical purposes, and is therefore committed to supporting related recommendations with arguments involving factual information. Consequently, the analysis employs a mix of descriptive and inferential statistics to outline conclusions and reinforce main findings. For the sake of accuracy, the study strictly uses input acquired from official sources and applies formal techniques to process and interpret data.

Consistent with the above-mentioned considerations, the analysis focuses on:

- Assessing seaborne trade performance (defined by the total weight of goods loaded) given the evolution of relevant economic indicators (measured by GDP per capita, direct primary energy consumption and manufacturing value added);
- Estimating ship-based GHG emissions (comprising of CO_2, SO_x, NO_x, CO, PMs, NMVOC, CH_4, N_2O and BC) and correlating findings with key fleet highlights (size, average age and fuel consumption);

- Benchmarking fleet performance against short- and long-term GHG reduction targets and acknowledging potential gaps;
- Outlining prospective solutions for accelerating transition to carbon neutrality and proposing several measures to support the achievement of overall decarbonisation objectives.

4 Results and Discussions

4.1 Relationship Between Seaborne Trade and Key Economic Indicators

The Analysis Determines that:

- Seaborne trade has been intensifying and continues to display further growth prospects (Fig. 1);
- Since 2015, annual volumes of loaded cargo (based on weight conversions) have more than doubled compared to 1997 levels (Fig. 1);
- Global GDP per capita (Fig. 2), manufacturing value added (Fig. 3) and direct primary energy consumption (Fig. 4) have grown significantly since 1997;
- In overall terms, seaborne trade performance has been strongly influenced by factors such as GDP per capita, direct primary energy consumption and manufacturing value added (Figs. 1, 2, 3 and 4). Results indicate that: World seaborne trade by total goods loaded (metric tons, mil.) = −0.656898 Global GDP per capita ($) + 440.81939 Global manufacturing value added ($, tril.) + 0.12023 Global direct energy consumption (TWh)—6734.103611. Analogously, the analysis proves that there is a very strong correlation between the predicted data and

Fig. 1 Seaborne trade by total goods loaded (metric tons, mil.): Global annual figures, 1997 – 2021 *Source(s)* Author's research, input retrieved from UNCTAD—UNCTADstat

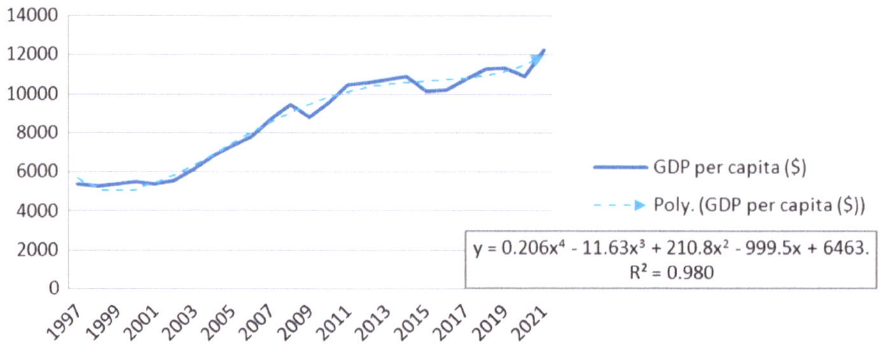

Fig. 2 GDP per Capita ($): 1997–2021: Global Annual Figures, 1997–2021. *Source(s)* Author's research, input retrieved from World Bank Group—DataBank

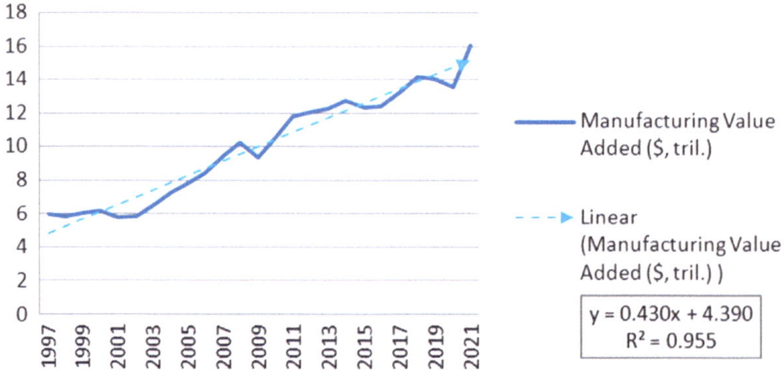

Fig. 3 Manufacturing value added ($, tril.): Global annual figures, 1997–2021. *Source(s)* Author's research, input retrieved from World Bank Group—DataBank

the observed data ($R^M = 0.995451$), and additionally highlights that predictors can explain 99.1% of the variance which corresponds to the observed data ($R^2 = 0.990922$). Furthermore, results acknowledge very strong, positive, linear relationships among variables: r (Seaborne trade, GDP per capita) = 0.965298, r (Seaborne trade, Manufacturing Value Added) = 0.979551, r (Seaborne trade, Direct Primary Energy Consumption) = 0.987547;

- Seaborne trade structure could undergo significant transformations in the wake of future developments in the renewable energy sector which might prompt permanent disconnection from fossil fuels. The phase down of non-renewable energy sources could additionally lead to the softening of historically close ties between GDP, trade and energy demand, which might further increase protectionism and drive deglobalisation.

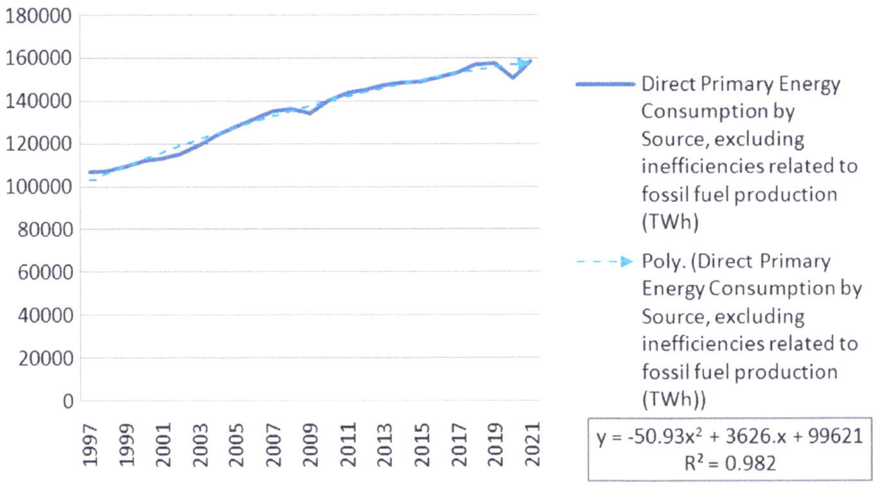

Fig. 4 Direct primary energy consumption (TWh): Global Annual Figures, 1997–2021. *Source(s)* Author's research, input retrieved from Our World in Data

4.2 Link Between Shipping Emissions and Main Fleet Highlights

The Analysis Demonstrates that:

- Between 1997 and 2008, the rise in GHG emissions from global shipping was closely linked to the expansion of seaborne trade (Figs. i and 5);
- Between 2008 and 2014, GHG emissions were gradually decoupled from the growing demand for maritime transport services (Figs. i and 5);

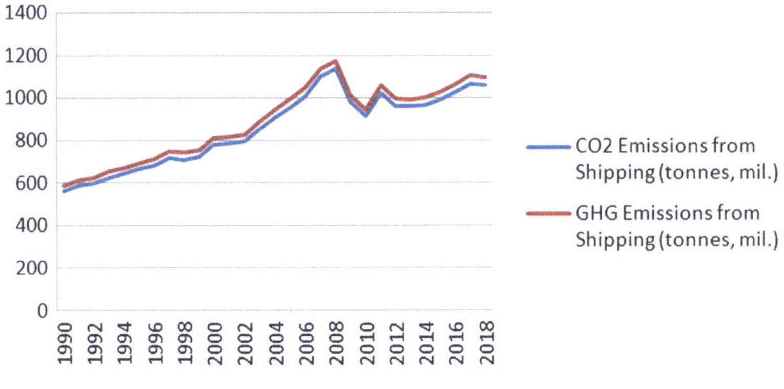

Fig. 5 CO_2 versus GHG emissions from global shipping: Annual figures, 1990–2018. *Source(s)* Author's research, input retrieved from IMO—Second IMO GHG Study 2009, Third IMO GHG Study 2014 and Fourth IMO GHG Study 2020

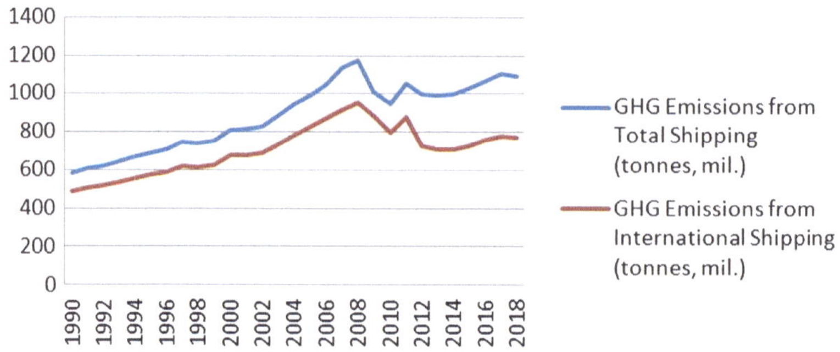

Fig. 6 GHG emissions—Total Shipping versus International shipping: Annual figures, 1990–2018. *Source(s)* Author's research, input retrieved from IMO—Second IMO GHG Study 2009, Third IMO GHG Study 2014 and Fourth IMO GHG Study 2020

- Between 2014 and 2018, GHG emissions evidenced slight increases (Fig. 5);
- Among the aggregate GHG shipping emissions measured between 1990 and 2018, CO_2 outflows ranged between 95.72% and 96.66% (Fig. 5);
- GHG emissions from international shipping were prevalent during the period 1990–2018, with a median of 83.13% (Fig. 6);
- GHG emissions from global shipping can be confidently predicted based on fuel consumption, fleet size and fleet average age (Figs. 7 and 8). Results show that: GHG emissions from global shipping (tonnes, mil.) = 3.232301 Total fuel consumption (tonnes, mil.)—0.0053741 Global fleet size (DWT, mil.)—0.656353 Global fleet average age +23.243579. Analogously, the analysis proves that there is a very strong correlation between the predicted data and the observed data (R^M = 0.999916), and additionally stresses that predictors are able to explain 100% of the variance which corresponds to the observed data (R^2 = 0.999832). Furthermore, results attest the presence of a very strong, positive, linear relationship between GHG emissions and fuel consumption (r = 0.9997, otherwise extremely close to a perfect correlation);
- Compared to GHG emissions, CO_2 outflows from global shipping can only be predicted based on fuel consumption and fleet size (Fig. 9): CO_2 emissions from global shipping (tonnes, mil.) = 3.144996 Total fuel consumption (tonnes, mil.)—0.00825287 Global fleet size (DWT, mil.) +5.012859. The p-value which corresponds to the global fleet average age equals 0.138809, hence the variable has been excluded from the model. At the same time, results substantiate the presence of a very strong, positive, linear relationship between CO_2 emissions and fuel consumption (r = 0.9997);
- Over the period 1990–2008, average annual fuel consumption was 251.75 million tonnes (Fig. 7);
- Between 2008 and 2014, average annual fuel consumption rose to 317.37 million tonnes (Fig. 7);
- Over the period 2014–2018, average annual fuel consumption increased to 327.65 million tonnes (Fig. 7);

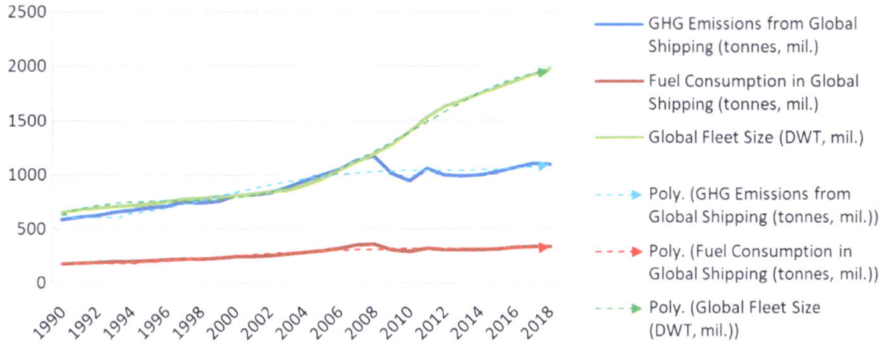

Fig. 7 GHG Emissions from Global shipping versus Fuel consumption and fleet size: Annual figures, 1990–2018. *Source(s)* Author's research, input retrieved from IMO—Second IMO GHG Study 2009, Third IMO GHG Study 2014, Fourth IMO GHG Study 2020 and UNCTAD—Review of maritime transport, 1990–2018. Author's Note Fleet size has been adjusted to E-O-Y counts

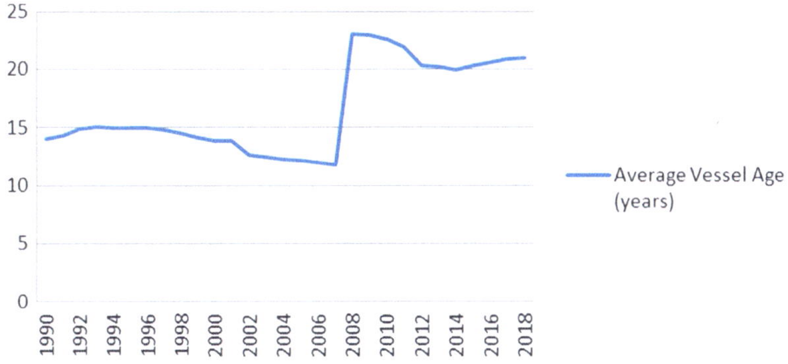

Fig. 8 Average age of the world's merchant fleet: 1990–2018. *Source(s)* Author's research, input retrieved from UNCTAD—Review of Maritime Transport, 1990–2018. Author's Note: Average age has been adjusted to E-O-Y counts

- In recent years, the world's merchant fleet has aged considerably. Among the main vessel types, bulkers are the youngest and general cargo carriers are the oldest (Fig. 10).

4.3 *Progress Carried Out in Regard to the Decarbonisation of Maritime Transport*

The Analysis Confirms that:

- In order to meet all of the explicit GHG reduction targets endorsed by the IMO in 2018, the global shipping industry would have to ensure that:

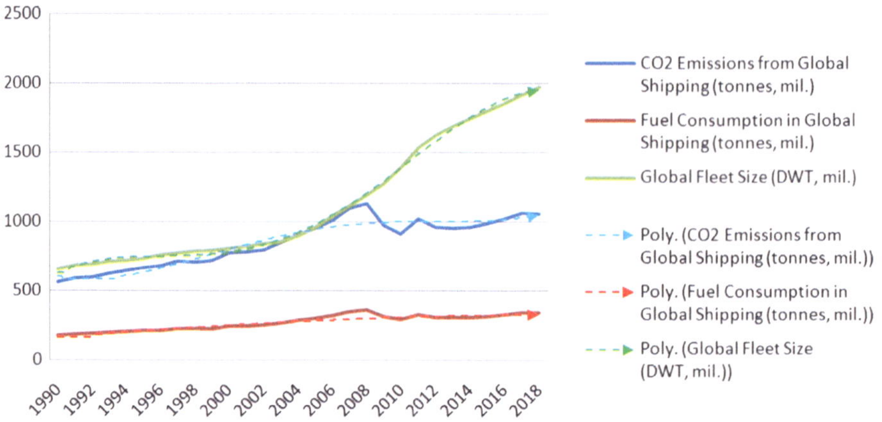

Fig. 9 CO_2 Emissions from global shipping versus Fuel consumption and fleet size: Annual figures, 1990–2018. *Source(s)* Author's research, input retrieved from IMO—Second IMO GHG Study 2009, Third IMO GHG Study 2014, Fourth IMO GHG Study 2020 and UNCTAD—Review of Maritime Transport, 1990–2018. Author's Note Fleet size has been adjusted to E-O-Y counts

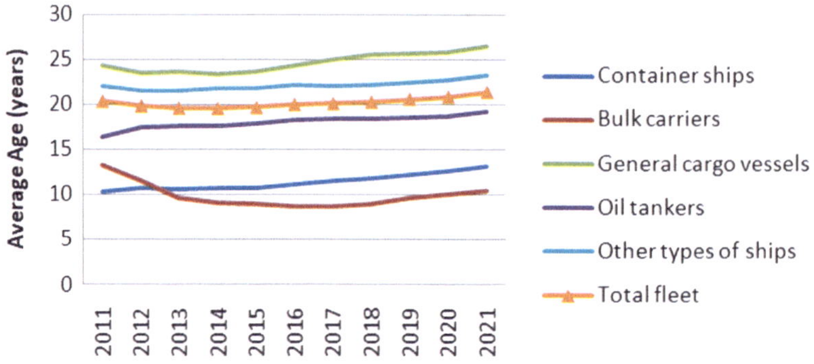

Fig. 10 Average Age of Main Vessel Types, 2011–2021. *Source(s)* Author's research, input retrieved from UNCTAD—UNCTADstat

- Aggregate GHG emissions in 2050 will not exceed 587.18 million tonnes;
- CO_2 emissions in 2050 will be equal to or lower than 340.5 million tonnes;
- CO_2 emissions in 2030 will not rise above 681 million tonnes;

- After three consecutive periods of improved efficiency, CO_2 emissions rose by 4.91% in 2021 compared to 2020 (Fig. 11), as a result of the COVID-19 pandemic and the related port congestion issues which often prompted the extended use of auxiliary engines;
- In 2021, CO_2 emissions from shipping accounted for 833 million tonnes (Fig. 11), but overall GHG emissions were not reported;

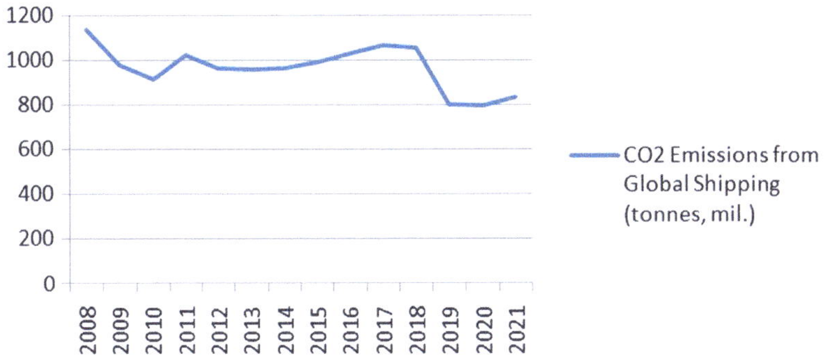

Fig. 11 CO_2 emissions from global shipping: Annual figures, 2008–2021. *Source(s)* Author's research, input retrieved from IMO–Third IMO GHG Study 2014, Fourth IMO GHG Study 2020 and Simpson Spence Young

- Assuming that the central tendency within the dataset was to be kept constant beyond 2018, GHG emissions could have likely reached around 864.46 million tonnes in 2021;
- Over the period 2008–2018, CO_2 outflows accounted for approximately 96.36% (median value) of all GHG emissions (Fig. 5);
- Between 2018 and 2020, CO_2 emissions decreased by 24.24% and 0.75% respectively, but in 2021 CO_2 emissions were 4.91% higher than in 2019 (Fig. 11);
- Achieving the overall decarbonisation objectives endorsed by the IMO would greatly depend on the reduction of CO_2 emissions (Fig. 5);
- Based on the measured shipping emissions to date, a constant reduction rate of minimum 2.21% enforced from 2021 onwards could be able to lower CO_2 outflows to 681 million tonnes by the end of 2030, yet in order to meet all targets, annual reduction rates covering the remaining 20 periods would have to be reconsidered. On such grounds, an additional annual reduction of minimum 3.41% would ensure that CO_2 emissions reach 340.5 million tonnes in 2050. Should the 70% reduction target gets postponed, the enforcement of a further annual reduction rate of minimum 0.92% from 2030 onwards would ensure that by the end of 2050, GHG emissions are limited to 587.18 million tonnes so long as the central tendency within the dataset remains unchanged until at least 2030. Alternatively, provided that the central tendency in CO_2 emissions had stayed unmodified up to 2022, an annual reduction of 2.42% would bring CO_2 outflows down to 681 million tonnes in 2030, but a further annual reduction rate of 3.41% would have to be implemented in order to lower CO_2 emissions to 340.5 million tonnes by the end of 2050. In the event that 70% reduction target gets deferred, an additional annual reduction of at least 0.92% from 2030 onwards could guarantee that GHG emissions would not exceed 587.18 million tonnes in 2050, given that the central tendency within the dataset keeps stable until at least 2030;
- CO_2 and GHG emission levels evidenced significant fluctuations over the period 1990–2018 (Fig. 5) and up to 2021 in the case of CO_2 (Fig. 11), but the analysis

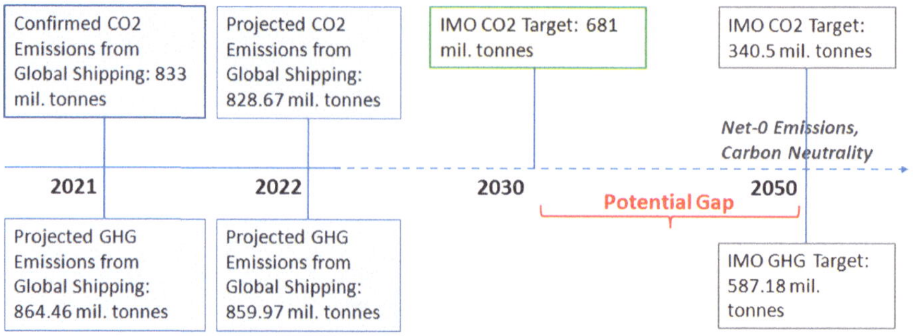

Fig. 12 Current levels of emissions from global shipping versus IMO targets. *Source(s)* Author's elaboration. *Author's Note* Forecasts are based on the aforementioned assumptions

deliberately opted for linear reduction rates in view of the fact that certain vessel types (e.g. general cargo carriers, Fig. 10) are considerably older than others (e.g. containerships and bulkers, Fig. 10). Pursuing the achievement of improved environmental efficiency through accelerated reduction rates covering shipping emissions could put extreme pressures on the ageing fleet, hence the option has been excluded from the analysis carried out on aggregate figures;

'To what extent has the shipping industry met the explicit, imo-approved, GHG reduction targets?'

Results Infer that:

- In respect to the achievement of short-term carbon intensity goals, recent progress appears to be reassuring and the global shipping industry will likely be able to further decrease its CO_2 emissions;
- Fleet environmental performance to date highlights a significant gap between recent progress and long-term objectives (Figs. 11 and 12), and suggests that reaching the 2050 decarbonisation targets will require substantial efforts from various key stakeholders including, but not limited to, governments, regulatory bodies, fuel producers, lenders, investors, insurers and shipowners.

4.4 Prospective Solutions for Accelerating the Achievement of Long-Term Decarbonisation Goals and Fostering Transition to Carbon Neutrality

'What can be done to address the identified gaps?'

Findings emphasise that:

- In order to ensure full alignment with the long-term environmental objectives endorsed by the IMO, future efforts should pursue the extended use of sound

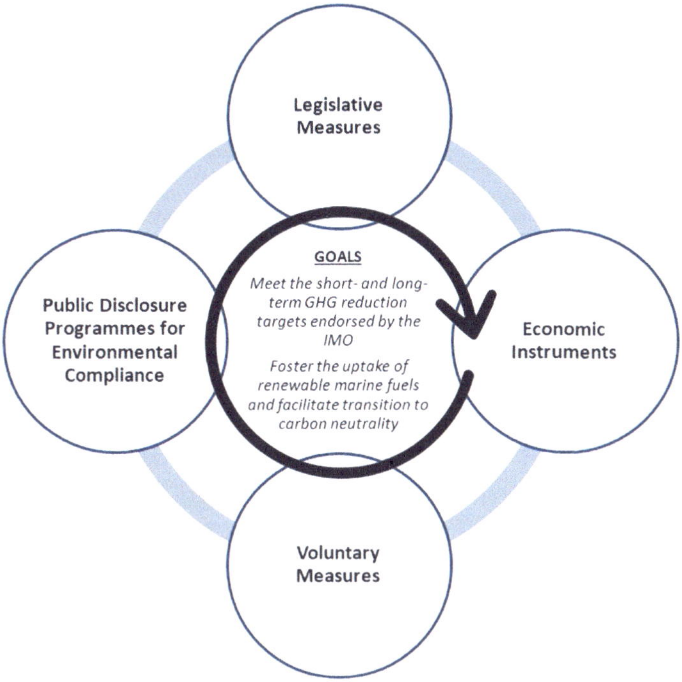

Fig. 13 Key considerations for driving improved environmental performance in worldwide shipping. *Source(s)* Author's elaboration

economic instruments, legislative measures, public disclosure programmes and voluntary actions (Fig. 13);

- Accelerating the path to carbon neutrality would require substantial investments targeting the scale-up of renewable marine fuels, the development of relevant supply infrastructures (port facilities, bunkering capabilities) as well as the upgrading of the world's commercial fleet (vessel replacements and retrofits);

'What measures should be implemented in order to accelerate the decarbonisation of maritime transport?'

The Study Settles that:

- Bilateral and multilateral funding mechanisms (Fig. 14) have strong potential to incentivise first-movers, hasten transition and prevent the occurrence of competitiveness gaps between developed and developing economies;
- Carbon pricing strategies, feebates and cap-and-trade programmes (Fig. 14) could accelerate the uptake of renewable fuels by the late majority and laggards;
- Complementary national and regional binding provisions (Fig. 14) fostering policy convergence and harmonisation could expedite transition to carbon neutrality and close potential competitiveness gaps;

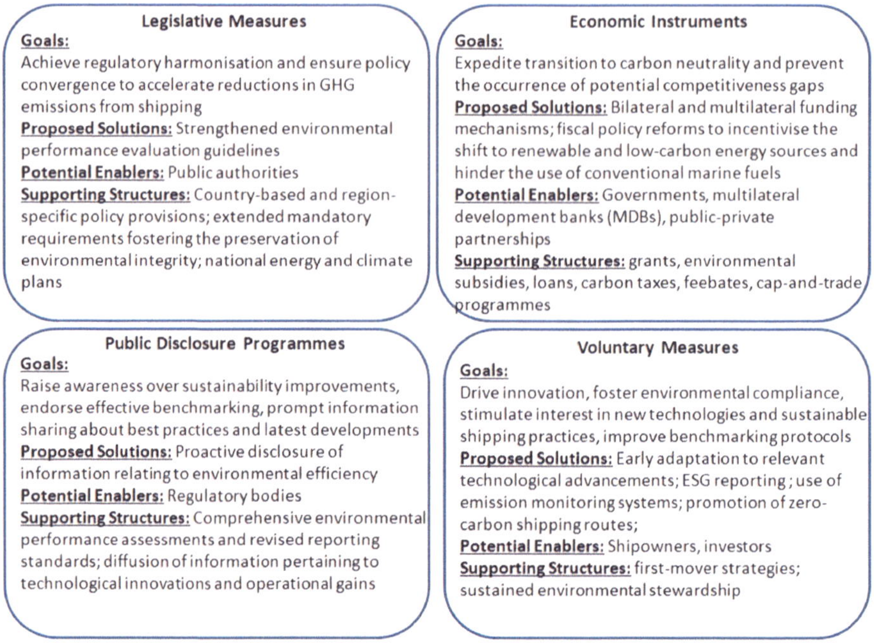

Fig. 14 Tactical measures for accelerating the reduction of GHG emissions from shipping. *Source(s)* Author's elaboration

- Tighter performance assessment guidelines (Fig. 14) could incentivise the uptake of emerging technologies by the early majority;
- Public disclosure programmes (Fig. 14) pertaining to the diffusion of comprehensive information concerning ship-based environmental performance could effectively hasten the adoption of technological innovations, boost operational efficiency and foster sound benchmarking;
- Voluntary measures (Fig. 14) such as proactive ESG reporting could facilitate information sharing about new technologies and their effects, and could therefore induce faster switches to technological advancements;
- The extensive use of emission monitoring systems (Fig. 14) across the global commercial fleet could support real-time GHG tracking and optimise benchmarking processes;
- Zero-carbon shipping routes (Fig. 14) could complement the uptake of renewable marine fuels and trigger the development of 'green' port infrastructures—which could further drive sustainable, resilient and inclusive economic growth.

5 Conclusion

Recognised as one of the main enablers of international trade, the maritime industry is also known to be a major carbon emitter. As global shipping continues to display strong growth prospects, its GHG emissions are likely to intensify, if left unregulated. In response to the escalating climate crisis, the IMO adopted several mandatory requirements designed to reduce GHG emissions from ships and thus strengthen the environmental performance of the world's fleet. The Initial GHG Strategy endorsed by the IMO in 2018 (Marine Environment Protection and Committee) aims to cut total shipping emissions by at least 50% by 2050 compared to 2008 levels and to reduce the average carbon intensity by at least 40% by 2030, and 70% respectively, by the end of 2050. Consistent with the Paris Agreement temperature goals, the IMO has further expressed keen interest in taking additional steps to facilitate transition to carbon neutrality and phase out all ship-related emissions as soon as possible.

In a broad sense, shipping's ability to achieve its full decarbonisation objectives strikes as both simple and difficult. While short-term emissions reduction targets appear to be attainable, long-term goals, on the other hand, seem very challenging. The study demonstrates that in order to reach carbon neutrality by 2050, global shipping should have pursued a constant reduction rate of at least 50.81% in annual GHG emissions from 2021 onwards, provided that CO_2 outflows had remained steady per their contribution to the overall GHG emissions beyond 2018 levels. Based on past information reflecting the emissions performance of ships (Fig. 5), this goal is still too ambitious and unlikely to materialise over the next 27 years.

The study additionally highlights that short-term reduction targets will probably be met by 2030 without significant effort. In support of this conclusion, the analysis shows that a constant annual reduction rate of minimum 2.42% in CO_2 emissions from 2022 onwards could decrease carbon intensity to 681 million tonnes by 2030. With the future widespread availability of dual marine engines suitable for retrofitting and the delivery of new propulsion systems operating on ammonia and methanol (Gas and Diesel 2023); (Energy and Solutions); (TradeWinds 2024) by 2024, the global fleet will most likely be able to reduce its carbon footprint without facing major challenges. Through the judicious deployment of additional technical and operational measures (e.g. optimal structural design features, waste heat recovery systems, slow steaming, higher port efficiency, preventive and corrective maintenance, etc.) geared towards the achievement of improved energy efficiency, the world's commercial fleet stands strong chances to meet the 2030 IMO-approved carbon intensity target without having to undergo significant upgrades involving substantial investments.

On the other hand, findings indicate that meeting the 2050 emissions reduction objectives could be extremely challenging, given the technological uncertainties concerning the widespread availability of future renewable fuels (green ammonia, green hydrogen, bio-methanol or e-methanol). The study further draws attention to the fact that in order to meet the 2050 IMO targets, a constant annual reduction of at least 3.41% in CO_2 emissions from 2030 onwards could lower carbon intensity to 340.5 million tonnes by the end of 2050, while a constant annual reduction of minimum 0.92% in CO_2 emissions starting 2030 could ensure that GHG outflows do not exceed 587.18 million tonnes in 2050, provided that the 70% carbon intensity goal gets deferred. However, given the recent environmental performance of the world's fleet (Fig. 5), it becomes obvious that achieving long-term decarbonisation objectives could prove problematic.

To further align with the 2050 reduction targets, the global shipping industry will likely have to pursue major transformations. Accelerating transition to carbon neutrality will inevitably require the enforcement of complex measures aimed at phasing out conventional fossil fuels and incentivising the widespread uptake of zero-emission alternatives. Findings indicate that there are many potential solutions that could speed up the decarbonisation of maritime transport, but in overall terms, prospective actions should prioritise the use of legislative provisions, economic instruments, public disclosure programmes and voluntary measures.

Regardless of the future efforts pursued, one argument remains firm: the shipping industry has commenced its 'green' voyage and the progressive uptake of renewable fuels will have a profound impact on the manner in which the global commercial fleet will be financed, insured, operated, maintained, regulated and classed. While energy efficiency is known to have a significant effect on asset values (Ådland et al. 2018), it is clear that the transition to sustainable marine fuels will trigger major consequences for the valuation of second-hand vessels. In all probability, older ships will continue to be financed, but lending decisions will most likely hinge on considerations pertaining to vessel-specific environmental credentials (design and operation).

The journey to carbon neutrality is both complex and debatable. Based on the fact that the world's commercial fleet is highly diverse (distinct age brackets and contrasting performances) and that shipping markets are very different from one another, future decarbonisation efforts will need to prioritise the use of information relating to fleet and market highlights.

To optimise deliverables and maximise impact, further research should look into:

- Setting up a comprehensive action plan to back the implementation of tactical measures committed to reducing emissions from ships;
- Determining the extent to which retrofitting existing vessels could work out to be an economically viable option for delivering a swifter transition to carbon neutrality;
- Developing segment-specific strategies to speed up alignment with IMO's long-term decarbonisation objectives.

References

UK Research and Innovation.: Shipping industry reduces carbon emissions with space technology. https://www.ukri.org/news/shipping-industry-reduces-carbon-emissionswith-space-technology/. Last accessed 12 Feb 2023

European Environment Agency.: Air pollution sources. https://www.eea.europa.eu/themes/air/air-pollution-sources-1. Last accessed 14 Feb 2023
International Chamber of Shipping.: Environmental performance. Comparison of CO_2 emissions by different modes of transport. https://www.ics-shipping.org/shipping-fact/environmental-performance-environmental-performance/#:~:text=Shipping%20is%20the%20least%20environmentally,marine%20pollution%20from%20human%20activities. Last accessed 14 Feb 2023
Teekay Corporation.: Sustainability report. pp. 27–28, p. 31. https://www.teekay.com/wpcontent/uploads/2019/10/Teekay-Corporation-2018-Sustainability-Report.pdf. Last accessed 15 Feb 2023
Skipsrevyen: Becker Mewis Duct for Odfjell. https://www.skipsrevyen.no/artikkelarkivbecker-marine-systems-becker-mewis-duct/becker-mewis-duct-for-odfjell/478408. Last accessed 12 Feb 2023
Becker Marine Systems.: Becker Mewis duct. https://www.becker-marinesystems.com/products/product-detail/becker-mewis-duct.html. Last accessed 15 Feb 2023
Teekay Corporation.: Sustainability report, pp. 21–38. https://www.teekay.com/wpcontent/uploads/2022/04/2021-Sustainability-Report.pdf
Wärtsilä.: Wärtsilä encyclopedia of marine and energy technology. AQUARIUS ballast water treatment plants. https://www.wartsila.com/encyclopedia/term/aquarius-ballastwater-treatment-plants. Last accessed 16 Feb 2023
Martek Marine Innovative Ship Solutions.: Evolution EMS: Engine efficiency monitoring system. https://www.martek-marine.com/wp-content/uploads/2021/04/Evolution-Datasheet.pdf
Wärtsilä.: Dual Fuel Engines: Future Proof Dual Fuel Engines. https://www.wartsila.com/marine/products/engines-and-generating-sets/dual-fuel-engines. Last accessed 16 Feb 2023
MAN Energy Solutions.: Providing the flexibility needed for reliable power generation. https://www.man-es.com/energy-storage/products/dual-fuel-engines. Last accessed 16 Feb 2023
Wärtsilä: Wärtsilä 34DF – Reliable Power with High Fuel Efficiency and Low Emissions. https://www.wartsila.com/marine/products/engines-and-generating-sets/dual-fuelengines/wartsila-34df. Last accessed 16 Feb 2023
MAN Energy Solutions.: Now ready for zero. Decarbonize with proven gas and dual fuel engines. https://www.man-es.com/energy-storage/campaigns/future-proof-engines. Last accessed 15 Feb 2023
United Nations: Paris Agreement, p. 4. https://unfccc.int/sites/default/files/english_paris_agreement.pdf
International Maritime Organization: International Convention for the Prevention of Pollution from Ships (MARPOL). https://www.imo.org/en/about/Conventions/Pages/International-Convention-for-the-Prevention-of-Pollution-from-Ships-(MARPOL).aspx. Last accessed 17 Feb 2023
European Parliamentary Research Service: Briefing. Decarbonising maritime transport. The EU Perspective, p. 1. https://www.europarl.europa.eu/RegData/etudes/BRIE/2020/659296/EPRS_BRI(2020)659296_EN.pdf
International Maritime Organization: Fourth IMO greenhouse gas study, p. 26. https://greenvoyage2050.imo.org/wp-content/uploads/2021/07/Fourth-IMO-GHG-Study2020-Full-report-and-annexes_compressed.pdf. Last accessed 17 Feb 2023
International Maritime Organization.: Fourth IMO greenhouse gas study 2020, p. 236. https://greenvoyage2050.imo.org/wp-content/uploads/2021/07/Fourth-IMO-GHG-Study2020-Full-report-and-annexes_compressed.pdf. Last accessed 17 Feb 2023
International Maritime Organization.: Fourth IMO greenhouse gas study, p. 238. https://greenvoyage2050.imo.org/wp-content/uploads/2021/07/Fourth-IMO-GHG-Study2020-Full-report-and-annexes_compressed.pdf. Last accessed 17 Feb 2023
PricewaterhouseCoopers.: The long view. How will the global economic order change by 2050?, p. 5. https://www.pwc.com/gx/en/world-2050/assets/pwc-the-world-in-2050-fullreport-feb-2017.pdf
European Commission.: Knowledge for policy. Growing consumption. https://knowledge4policy.ec.europa.eu/growing-consumerism_en. Last accessed 18 Feb 2023

United Nations.: Paris agreement, p. 3. https://unfccc.int/sites/default/files/english_paris_agreement.pdf
Marine Environment Protection Committee: MEPC 62/24/Add.1. Annex 19, Resolution MEPC.203(62). Amendments to the Annex of the Protocol of 1997 to Amend the International Convention for the Prevention of Pollution from Ships, 1973, Modified by the Protocol of 1978 Relating Thereto, pp. 10–12. https://wwwcdn.imo.org/localresources/en/OurWork/Environment/Documents/Technical%20and%20Operational%20Measures/Resolution%20MEPC.203(62).pdf
Marine Environment Protection Committee: MEPC 62/24/Add.1. Annex 19, Resolution MEPC.203(62). Amendments to the annex of the protocol of 1997 to Amend the international convention for the Prevention of Pollution from Ships, 1973, Modified by the Protocol of 1978 Relating Thereto, p. 13. https://wwwcdn.imo.org/localresources/en/OurWork/Environment/Documents/Technical%20and%20Operational%20Measures/Resolution%20MEPC.203(62).pdf
Marine Environment Protection Committee: Resolution MEPC.278(70). Amendments to the Annex of the Protocol of 1997 to Amend the International Convention for the Prevention of Pollution from Ships, 1973, as Modified by the Protocol of 1978 Relating Thereto, pp. 1–7. https://wwwcdn.imo.org/localresources/en/KnowledgeCentre/IndexofIMOResolutions/MEPCDocuments/MEPC.278(70).pdf
Marine Environment Protection Committee: MEPC 78/17/Add.1. Annex 12. Resolution MEPC.350(78). 2022 Guidelines on the Method of Calculation of the Attained Energy Efficiency Existing Ship Index (EEXI), pp. 1–11. https://wwwcdn.imo.org/localresources/en/KnowledgeCentre/IndexofIMOResolutions/MEPCDocuments/MEPC.350(78).pdf
Marine Environment Protection Committee.: MEPC 78/17/Add.1. Annex 14. Resolution MEPC.352(78). 2022 Guidelines on operational carbon intensity indicators and the calculation methods (CII Guidelines, G1), pp. 1–5. https://wwwcdn.imo.org/localresources/en/KnowledgeCentre/IndexofIMOResolutions/MEPCDocuments/MEPC.352(78).pdf
International Maritime Organization: Integrated Technical Cooperation Programme (ITCP). https://www.imo.org/en/OurWork/TechnicalCooperation/Pages/ITCP.aspx. Last accessed 2023/02/18
Marine Environment Protection Committee: MEPC 74/18/Add.1. Annex 17. Terms of Reference for the Establishment of a Voluntary Multi-Donor Trust Fund to Sustain the Organization's Technical Cooperation and Capacity-Building Activities to Support the Implementation of the Initial IMO Strategy on Reduction of GHG Emissions from Ships ("GHG TC-TRUST FUND"), pp. 1–3. https://www.iadc.org/wp-content/uploads/2019/08/MEPC74-18-Add.1-Report-Of-The-Marine-Environment-Protection-CommitteeOn-Its-SeventyFourth-Session-Secretariat-1.pdf
Marine Environment Protection Committee: MEPC 75/18/Add.1.: Annex 4. Resolution MEPC.327(75). Encouragement of member states to develop and submit voluntary National action plans to address GHG emissions from ships, pp. 1–2. https://wwwcdn.imo.org/localresources/en/OurWork/Environment/Documents/Air%20pollution/Resolution%20MEPC.327(75).pdf
EUR-Lex: Document 32015R0757. Regulation (EU) 2015/757 of the European Parliament and of the Council of 29 April 2015 on the Monitoring, Reporting and Verification of Carbon Dioxide Emissions from Maritime Transport, and Amending Directive 2009/16/EC. Official Journal of the European Union, L 123/55. https://eurlex.europa.eu/legal-content/EN/TXT/?uri=CELEX:32015R0757. Last accessed 18 Feb 2023
European Commission.: A European green deal. Striving to be the first climate-neutral continent. https://commission.europa.eu/strategy-and-policy/priorities-2019-2024/european-green-dealen. Last accessed 19 Feb 2023
European Parliamentary Research Service.: Briefing. Sustainable maritime fuels. 'Fit for 55' package. The FuelEU maritime proposal. PE 698.808, p. 5. https://www.europarl.europa.eu/RegData/etudes/BRIE/2021/698808/EPRS_BRI(2021)698808_EN.pdf
European Parliamentary Research Service: Briefing. Sustainable Maritime Fuels. 'Fit for 55' Package: The FuelEU Maritime Proposal. PE 698.808, p. 4. https://www.europarl.europa.eu/RegData/etudes/BRIE/2021/698808/EPRS_BRI(2021)698808_EN.pdf

EUR-Lex: Document 52021PC0562. Proposal for a regulation of the european parliament and of the council on the use of renewable and low-carbon fuels in maritime transport and amending directive 2009/16/EC. https://eur-lex.europa.eu/legalcontent/EN/TXT/?uri=CELEX%3A52021PC0562&qid=1631197035907. Last accessed 20 Feb 2023

Lim, K.: The role of the international maritime organization in preventing the pollution of the world's oceans from ships and shipping. Our Ocean, Our World, LIV(1 & 2). https://www.un.org/en/chronicle/article/role-international-maritime-organization preventing-pollution-worlds-oceans-ships-and-shipping. Last accessed 17 Feb 2023

European Parliamentary Research Service.: Briefing. sustainable maritime fuels. 'Fit for 55' package: The FuelEU maritime proposal. PE 698.808, p. 5. https://www.europarl.europa.eu/RegData/etudes/BRIE/2021/698808/EPRS_BRI(2021)698808_EN.pdf

Marine Environment Protection Committee: MEPC 72/17/Add.1. Annex 11. Resolution MEPC.304(72). Initial IMO Strategy on Reduction of GHG Emissions from Ships, p. 5. https://wwwcdn.imo.org/localresources/en/KnowledgeCentre/IndexofIMOResolutions/MEPCDocuments/MEPC.304(72).pdf

Marine Environment Protection Committee.: MEPC 72/17/Add.1. Annex 11. Resolution MEPC. 304(72). Initial IMO strategy on reduction of GHG emissions from ships, p. 5. https://wwwcdn.imo.org/localresources/en/KnowledgeCentre/IndexofIMOResolutions/MEPCDocuments/MEPC.304(72).pdf

Winterthur Gas & Diesel.: WinGD and HSD engine collaborate to deliver Methanol fueled two-stroke engines. https://www.wingd.com/en/news-media/pressreleases/wingd-and-hsd-engine-collaborate-to-deliver-methanol-fueled-two-strokeengines/. Last accessed 20 Feb 2023

MAN Energy Solutions.: Four-stroke methanol-fuelled engines. https://www.manes.com/marine/strategic-expertise/future-fuels/methanol. Last accessed 20 Feb 2023

TradeWinds: First Ammonia Engines for Commercial Use are Due Out in 2024. https://www.tradewindsnews.com/insight/first-ammonia-engines-for-commercial-use-aredue-out-in-2024/2-1-824790. Last accessed 20 Feb 2023

European Community Shipowners' Associations. https://ec.europa.eu/info/law/betterregulation/have-your-say/initiatives/13219-Festlegung-des-Kurses-fur-einennachhaltigen-blauen-Planeten-Aktualisierung-der-Agenda-fur-die-internationaleMeerespolitik/F2827345_de. Last accessed 15 Feb 2023

International Chamber of Shipping.: Shipping and world trade: world seaborne trade. https://www.ics-shipping.org/shipping-fact/shipping-and-world-trade-world-seabornetrade/. Last accessed 11 Feb 2023

ABB Group.: ABB measurement & analytics: CEMcaptain Emission Monitoring on your Wavelength, p. 2. https://library.e.abb.com/public/dca509aa929f40e5b4b12b40a3e75c3c/PB_CEMcaptain-EN_REV_A.pdf

Ådland, R., Cariou, P., Wolff, F.-C.: Does energy efficiency affect ship values in the second-hand market? Transp. Res. Part a: Policy Pract. **111**, 347–359 (2018)

Barber, H.F.: Developing strategic leadership: The US army war college experience. J. Manag. Dev. **11**(6), 4–12 (1992)

Intergovernmental Panel on Climate Change: Summary for Policymakers. In: Edenhofer, O., Pichs-Madruga, R., Sokona, Y., Farahani, E., Kadner, S., Seyboth, K., Adler, A., Baum,I., Brunner, S., Eickemeier, P., Kriemann, B., Savolainen, J., Schlömer, S., von Stechow, C., Zwickel, T., Minx, J.C. (eds.). Climate change 2014: Mitigation of climate change. Contribution of working group III to the fifth assessment report of the intergovernmental panel on climate change, p. 4. Cambridge University Press, Cambridge, United Kingdom and New York, NY, USA (2014). https://www.ipcc.ch/site/assets/uploads/2018/02/ipcc_wg3_ar5_summary-for-policymakers.pdf

Marine Environment Division: Initial IMO strategy on reduction of GHG emissions from ships and programme of follow-up actions to 2023. Achieving the IMO GHG reduction objectives. Fossil Fuels, Climate Change and Economic Development. COP 24,

p. 9. https://wwwcdn.imo.org/localresources/en/OurWork/Environment/Documents/Presentation%20-%20IPIECA_UCL%20Side%20Event%203%20December%202018.pdf
Gössling, S., Meyer-Habighorst, C., Humpe, A.: A global review of marine air pollution policies, their scope and effectiveness. Ocean. & Coast. Manag., **212**(105824), 1–7 (2021)
Hu, O., Xie, Q., Ma, C., Zhang, G.: Silicone-based fouling-release coatings for marine antifouling. Langmuir, **36**(9), 2170–2183 (2020)
King, A.: Emissions-free Sailing is Full Steam Ahead for Ocean-going Shipping. Horizon. EU Res. & Innov. Mag., (2022). https://ec.europa.eu/research-andinnovation/en/horizon-magazine/emissions-free-sailing-full-steam-ahead-ocean-goingshipping
Kiron, D., Kruschwitz, N., Rubel, H. Reeves, M. & Fuisz-Kehrbach, S.-K.: Sustainability's Next Frontier. Walking the talk on the sustainability issues that matter most. MIT Sloan Manag. Review (2013). https://sloanreview.mit.edu/projects/sustainabilitys-next-frontier/. Last accessed 10 Feb 2023
Lowe, M.: United Nations: Global sustainable development report, 2015 edn, Advance Unedited Version, p. 41. (2015). https://sustainabledevelopment.un.org/content/documents/1758GSDR%202015%20Advance%20Unedited%20Version.pdf
Global Environment Facility.: United Nations Development Programme, International Maritime Organization & Institute of Marine Engineering, Science and Technology: Ship Emissions Toolkit. Guide No.1: Rapid Assessment of Ship Emissions in the National Context, p. xi. Elephant Print, Lewes, East Sussex (2018). https://gmn.imo.org/wpcontent/uploads/2018/10/ship_emissions_toolkit-g1-online.pdf
International Transport Forum.: Future maritime trade flows. Summary and conclusions. ITF Roundtable Reports, No. 178, p. 8. OECD Publishing, Paris (2020). https://doi.org/10.1787/25198785
Organisation for Economic Co-operation and Development.: The environmental effects of freight, p. 11. OECD Publishing, Paris (1997). https://www.oecd.org/environment/envtrade/2386636.pdf
United Nations Conference on Trade and Development.: Review of maritime transport 2022. Navigating Stormy Waters, p. 6. United Nations Publications, Geneva (2022). https://unctad.org/system/files/official-document/rmt2022_en.pdf
Walker, T.R., Adebambo, O., Del Aguila Feijoo, M.C., Elhaimer, E., Tahazzud, H., Edwards, S. J., Morrison, C. E., Romo, J., Sharma, N., Taylor, S. & Zomorodi, S.: Chapter 27—Environmental effects of marine transportation. In Sheppard, C. (Ed.), World seas: An Environmental evaluation. 2nd ed, vol. 3, pp. 505–530. Academic Press (2019)

Dynamics of Research on Corporate Governance Under Crisis Conditions: A Bibliometric Analysis

Andreea Madalina Bojan and Camelia Iuliana Lungu

Abstract This study aims at identifying the dynamics of research on corporate governance under economic and pandemic crisis conditions. The main objective is to conduct a longitudinal analysis over the last 15 years (2008–2022). Bibliometric analysis research method is combined with frequency, descriptive and statistical analysis using SPSS to obtain relevant results for academic and business environments. A final sample of 1688 scientific papers included in Web of Science database were considered in examining the research questions. First, ordering the studies by year and type, the evolution of publication is analyzed for the field of corporate governance research under crisis conditions. An upward trend is observed, especially for the studies published in journals. Next, the most relevant articles are identified based on the times cited in various databases, and the characteristics measured by the cited reference count and the number of pages, complementing the research. Results emphasize the highest values for studies published in journals compared with books or conference proceedings, as well as significant correlations between variables. Muchmore, a significant difference between groups of publications is noticed. These findings assessed the relevance of corporate governance studies published over the last 15 years, their scientific validation, and accuracy of the results. Finally, the current state of collaboration between authors is displayed based on a network visualization technique implemented in VOSviewer. The co-authorship analysis suggests the intensity of authors' collaboration. This study represents a useful resource for research area development and contributes to capturing the evolution of academic environment.

Keywords Corporate governance · Pandemic crisis · Financial crisis · Bibliometric analysis · Statistical analysis · Co-authorship analysis

---

A. M. Bojan (✉) · C. I. Lungu
Bucharest University of Economic Studies, Bucharest, Romania
e-mail: bojanandreea17@stud.ase.ro

C. I. Lungu
e-mail: camelia.lungu@cig.ase.ro

M. Busu (ed.), *Rethinking Business for Sustainable Leadership in a VUCA World*, Springer Proceedings in Business and Economics,
https://doi.org/10.1007/978-3-031-50208-8_2

1 Introduction

Economics is included in the social sciences, which justifies the decisive influence of demographic, environmental, political, technological, and cultural factors. The repercussions of the important roles played by these factors reverberate in both academic and business context. Under current conditions, companies focus on the perception of how internal and external factors impact the activities they carry out. Moreover, among the objectives pursued by economic entities, the achievement of equilibrium from a non-financial perspective, along with the financial one is more and more discussed. In terms of non-financial reporting and performance, an important role is attributed to corporate governance.

Corporate governance is characterized through the mechanisms applied by companies to ensure leadership and control and emphasize the importance of governance practices in maximizing the value created by entities and stabilizing relationships with stakeholders. Research does not expose corporate governance individually, but in an extended sense, along with multiple economic and social dimensions. Dumay et al. (2016) examine governance as a constituent part of integrated reporting, emphasizing the need for co-existence of financial and non-financial data disclosure, Albu and Mateescu (2014) include in the analysis characteristics regarding corporate entrepreneurship, while Shui et al. (2022) correlate corporate governance characteristics with the innovation of environmental factors.

Ruan and Liu (2021) approach governance as an integrated part, along with environmental and social factors, highlighting the importance that stakeholders such as investors, regulatory bodies, or state institutions give to sustainable reporting. Strengthening environmental, social, and governance factors (ESG) is justified in the context of the coronavirus-triggered pandemic, with an impact on transparency in corporate reporting. For example, in the European Union, the *Apply or Explain* principle operates, while the Nasdaq Stock Exchange in the United States issued in 2019, *ESG Reporting Guide 2.0.*

This study aims to identify the evolution of research on the issue of corporate governance, given both the conditions of the economic crisis and the health crisis. Using bibliometric analysis as the main research method, the purpose is to analyze publication patterns in the period 2008–2022. In this regard, based on the research conducted by Baker et al. (2020), complemented by Donthu et al. (2021), and Lungu et al. (2020), the following research questions are addressed: RQ1: *What is the trend for publication on corporate governance under crisis conditions?*, RQ2: *Which are the most relevant papers on corporate governance under crisis conditions?*, RQ3: *Which are the characteristics of papers published on corporate governance under crisis conditions?*, RQ4: *What is the state of collaboration involving research on corporate governance under crisis conditions?*

Given that the purpose of this study is to identify the dynamics and the main areas of corporate governance under crisis as compared with non-crisis conditions, the time interval for which data were collected being between 2008 and 2022. This is a reference period, characterized by the fact that it captures the way in which the academics relate to corporate governance in conditions of risk and uncertainty generated by various crisis' types. This period analyses the research context for both the 2008 financial crisis and 2020 health crisis. Hence, the entire course of the crisis, from the onset to paroxysm, is examined from the research point of view, through published studies. The research methodology combines bibliometric analysis with frequency, descriptive and statistical analysis. Findings are discussed by comparing with the results of previous research, while limitations are presented to open the debate on future research. The relevance of this research and the main contribution reside in the analysis of an extended time interval, 2008–2022, marked both by the effects of the financial recession and those of the pandemic conditions contributing to globalized economic crises.

The rest of the paper is structured as follows. Literature review covers an overview of relevant studies which approached corporate governance under crisis conditions area of research. Methodology presents details on data collection and filter, describes the final sample and the techniques used for bibliometric analysis and the statistical tests. The main findings debated in the context of previous research are captured in Results and discussions. Conclusions summarize the results, limitations, and future directions for research.

2 Literature Review

The global financial crisis started in 2007 and its negative influences are felt in all countries and business sectors of the world. Among the economic instability causes is counted significant stock market capitalization reduction, which creditors, investors, and shareholders faced (Clarke 2010). Nowadays, the consequences of the COVID-19 pandemic have extended their effects, materializing in the form of the global economic crisis. Another unfavorable aspect at the global level is found in the case of capital markets, whose volatility registers significant increases. The crisis caused the unemployment rate to increase, adversely influenced investment activity, and affected trade relations between states. Even though some of the fields of activity, such as on-line commerce, the medical/pharmaceutical field, have enjoyed the rise, the influences at the economic level are severe and characterized as unfavorable. In such conditions, it is essential to understand the changes that have left their mark on the entrepreneurial environment and on the factors with which it coexists (Kells 2020).

The importance of reporting quality information under crisis conditions is supported by Hsu and Yang (2022), who found that during the health crisis caused by the pandemic, financial reporting is inadequate. Despite the winding itinerary traveled by the companies, the authors identify that during crisis periods, the quality of financial reporting can be ensured in the conditions where the board is composed by a larger number of members, they comply with the criteria of independence, and the position of president is not simultaneously held by the executive director.

Baker et al. (2020) performs a bibliometric analysis of board diversity, examining 579 studies published between 1999 and 2019. The most influential articles are identified, based on location, publications, and citations. The analysis is based on issues such as total studies published in every year, top publishing countries and journals, top publishing institutions and authors, top articles classified by citations, and top keywords. The results emphasize a lack of collaboration according to the authors' countries, especially between authors from developed countries and those from developing ones.

Through a systematic literature review, Hossain et al. (2022) approach the subject of small and medium enterprises, emphasizing the multiple difficulties that have had their effects during the pandemic. They identify the main areas affected, and the usefulness of the study is given by the possibility of formulating sustainable strategies, as a response to the unstable context in an emerging economy.

Governance in the context of crisis is advanced in an extended sense, with interesting associations with dimensions aimed at digitization, innovation, aspects related to the social sphere, environment, and sustainable development, organizational climate, and financial factors.

Naeem et al. (2022) analyze corporate governance correlated with gender diversity, considering the influences exerted by the pandemic. The main results reveal that corporate governance practices significantly influence the activity carried out by companies. For example, companies active in the financial field stand out for the relevance of the results, compared to those of non-financial companies. The situation is due to the activity of regulatory authorities, which have the responsibility of monitoring the activity of companies and play a supporting role in the application of corporate governance mechanisms.

Despite these advancements, a clear picture of differences in research interest on corporate governance issues during crisis as compared with non-crisis periods remains unclear. The present study bridges this gap by synthesizing the research papers validated in the literature, identifying the publication trend, the most relevant papers, research characteristics, state of collaboration between authors, and providing insights for future research on corporate governance.

3 Methodology

To address the evolution of research on corporate governance under crisis conditions, the methodology is based on bibliometric analysis used to provide a review of scientifically recognized publications. Fundamental research is addressed through literature review and consists of approaching theoretical valences of corporate governance and crisis periods.

Bibliometric analysis implies several steps: setting up the database depending on the research topic, collecting data, processing (cleaning and filtering) the secondary data, and analyzing the final database (Donthu et al. 2021). Muchmore, the analysis of publications' performance is combined with science mapping of the relationships between authors, as techniques of bibliometric analysis. Regarding the performance analysis, the advantages are represented by creating an overview of the interest that the academic environment grants to corporate governance under crisis conditions research. In addition, the contributions of research constituents are complemented by relevant interactions and structural connections among them, provided by science mapping and included in the results, discussions, and limitations, crucial for future research perspectives.

Similar to Baker et al. (2020), in this study data are collected from Web of Science (Clarivate Analytics) database, for the 2008–2022 reference period. A topic search using the keywords *corporate governance* and *crisis* was conducted to filter the data in such a way as to allow the collection of relevant results for research purposes. The search strategy included a combination of title, abstract, authors keywords, and keywords. The originality of the study is insured by selecting specific WOS field and category (*Business* and *Economics*), and a precise analysis period, to include both the effects of the economic crisis caused by financial recession in 2008 and those of the health crisis caused by the pandemic in 2020.

Through bibliometric combined with frequency, descriptive, statistical, and network analysis, the evolution and the relevance of research are identified as main explanatory aspects for properly answer the four research questions addressed in the Introduction. The *evolution of the research* is determined by the trend for publication and the most relevant papers, while *the relevance of research* could be based both on the publications' characteristics and the state of collaboration.

Initially, the database consisted of 2605 articles, but those that had missing data on publication type and year of publication were eliminated. In the same way, articles to be published in 2023, duplicates, and those with lacking information about authors are eliminated. The number of studies has been reduced to 1688. Relevant indicators for performance analysis were included: publication year, publication type, cited reference count, times cited in all databases. Data collection and secondary processing were facilitated by using *Microsoft Excel* software, while *VOSviewer* was used for co-authorship network visualization. Furthermore, for statistical tests, and the calculation of the main statistical indicators, *SPSS Statistics, Version 23* was used.

4 Results and Discussions

4.1 Research Trend on Corporate Governance Under Crisis Conditions

To answer the first research question (*RQ1: What is the trend for publication on corporate governance under crisis conditions?*), the publications are analyzed by type and year. The results are based on bibliographic data collected from the Web of Science database.

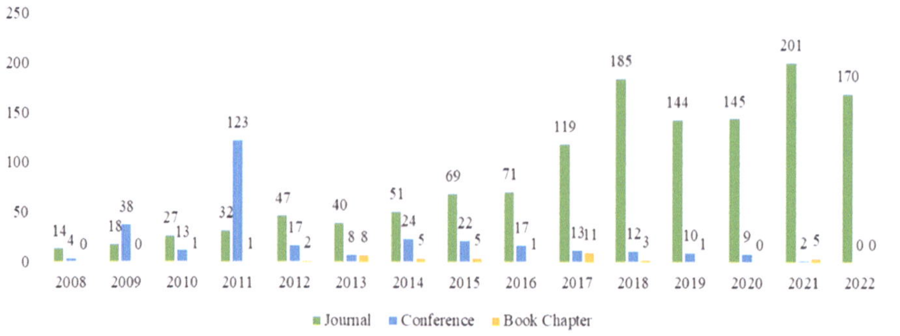

Fig. 1 Evolution of corporate governance under crisis conditions publications. *Source* Authors' own research

Figure 1 presents the evolution of research on corporate governance under crisis conditions between 2008 and 2022, for the 1688 studies included in the sample, according to the type of publication and the publication year retrieved from the database. Results show that that most of the papers (1333 studies or 78.97%) are articles published in journals. Conference papers are ranked in the second position, with a total of 312 studies (18.48%). The smallest number of studies is accounted for books or book chapters category. For the period analyzed, only 43 studies have been published in books.

The explanation of these trends may rely on the fact that many of the authors are interested in leading journal publishing, for several reasons. First, the studies in these journals represent an added value to academic activity and a response to the performance research criteria recognized by universities (Lungu et al. 2020). Second, journals inspire more confidence due to recognition of the authors' scientific contribution and the veracity of the disseminated results. In addition, the journals' stages of review are rigorous, ensuring the quality of research work and of the results achieved.

From year to year, the distribution of studies is heterogeneous, but it may be observed that the trend for studies published in journals is upward, except for 2013 and 2022. Overall, the smallest number of studies published in 2008 (only 18) may be attributed to the fact that the financial crisis was at its beginning when the effects of the imbalance caused by the global economic crisis were difficult to predict. Instability in capital markets, bankruptcies, creditors, and insurance companies' bankruptcy are among the effects of the crisis, which created confusion at the social, professional, and academic levels.

During 2009–2010, the subject of corporate governance under crisis conditions became of interest for the academic environment, resulting in an increased number of journal publications. In times of crisis, company representatives show increased awareness in corporate governance. According to Clarke (2010), corporate governance continues to identify responses to the dynamics that characterize the business world and continuously provides solutions for adverse events. This is an additional argument that justifies the importance of the research on the evolution of corporate governance under crisis conditions.

After overlapping the period of the economic crisis, with all the phenomena and unfavorable effects, the health crisis caused by the COVID pandemic emerged in 2019, posing a new challenge. Analyzing this period, this study suggests an increased interest of researchers in corporate governance under crisis conditions with a lag of 2 years, the largest number of publications being in 2021 (208 studies overall, of which 201 in journals). The year 2021 could be considered one of the referenced years, because researchers are fully aware of both the evolution and the uncertain conditions determined by the crisis.

The years that also stand out are 2011, 2018, 2019, 2020, and 2022, and the number of publications justifies the academic interest in capturing the challenges and opportunities faced by corporate governance in periods of economic and social instability. Based on the analysis, it may be observed that most of the research activity is validated through conference volumes and leading journals publications.

4.2 Research Relevance on Corporate Governance Under Crisis Conditions

The second research question (*RQ2: Which are the most relevant papers on corporate governance under crisis conditions?*) aims to identify the characteristics regarding the relevance of the studies, measured by two indicators showing the number of times a paper was cited in Web of Science Core and in All databases, respectively.

Table 1 presents the main statistical indicators calculated for the two variables extracted from Clarivate Analytics database. The studies included in the sample are grouped by publication type, and the results were generated with *SPSS Statistics, V. 23*.

Table 1 Descriptive statistics based on times cited

Relevance characteristics	Publication type	Valid N (listwise)	Min	Max	Mean	Std. Deviation	One-way ANOVA
Times cited (Web of Science Core)	Book Chapter	43	0	39	4.70	8.082	26.752**
	Conference	312	0	43	0.89	3.394	
	Journal	1333	0	648	18.23	43.549	
Times cited (All Databases)	Book Chapter	43	0	39	4.79	8.178	27.164**
	Conference	312	0	45	0.91	3.518	
	Journal	1333	0	650	18.65	44.219	

** Correlation is significant at the 0.01 level (2-tailed)
Source Authors' own research

For all three publication types, the minimum number of times cited is 0, while maximum limits vary between 39 and 650. The most cited studies are represented by publications in journals, with almost 19 citations on average, followed by studies published in books, with less than 5 citations, and then by conferences not exceeding 1 citation on average. The high difference by publication type represents an argument for validating the research question as the characteristics regarding relevance of research are homogeneous within the same type of publications, but very different between the types. One-way ANOVA tests validate the findings, demonstrating that there is a significant difference between the three groups of publications.

Next, the correlation analysis is performed to discuss the relevance of the studies based on the number of citations. *Pearson* and *Spearman correlation coefficients* are calculated, in order to interpret the strengths and the direction of the link between the variables. The results are summarized in Table 2.

The results suggest that there exists a direct and statistically significant correlation between the analyzed variables. Publication type is directly correlated both with the times cited according to the Web of Science Core and times cited in all databases, and statistically significant, at the 1% level.

Furthermore, after processing the sample data, a classification is achieved for the most relevant 20 publications, the studies being grouped by publication type. The results of the most relevant studies and their source titles (publication name) are displayed in Table 5. It may be noticed that none of the studies is cited less than 137 times, the maximum limit being 650 times. The most frequent source is Journal of Financial Economics, followed by Journal of Business Ethics, and Journal of Banking and Finance. The Socio-Economic Review stands out by the maximum number of times cited, while the minimum number is associated with the Journal of Financial Economics.

Table 2 Relevance characteristics' correlation

	Publication type	Times cited (Web of Science Core)	Times cited (All Databases)
Publication type	1	0.162**	0.163**
Times cited (Web of Science Core)	0.513**	1	1.000**
Times cited (All Databases)	0.513**	0.999**	1

Note Statistics are computed based on Pearson Correlations (upper side of the table) and Spearman Correlations (lower part of the table), applied listwise for the full sample of 1688 papers included in analysis

** Correlation is significant at the 0.01 level (2-tailed)

Source Authors' own research

4.3 Research Characteristics on Corporate Governance Under Crisis Conditions

For the third research question (*RQ3: Which are the characteristics of papers published on corporate governance under crisis conditions?*), the results are presented in Table 3. First, the statistical analysis is applied to research characteristics referring to *cited reference count* and *number of pages*. The cited reference count is understood to be the total number of bibliographic references that the authors cite in their research. The number of references represents an important element in assessing the influence of studies and authors. Kivinen et al. (2017) suggest that a study could be more relevant as more cited references it includes. In this context, the number of references cited denotes thorough documentation, detailed knowledge of the researched subject, validation, and relevance of the results. From this point of view, the studies may represent the basis for the development of future research perspectives.

The results show that studies in the field of corporate governance under crisis conditions with the highest number of references (264) and the larger number of pages (71) are published in scientific journals, closely followed by books. The characteristics of papers published in conference proceedings registered 141 references, and 45 pages, the most. Differences are noticed for the mean values also, smaller between papers published in journals and in books, and higher as compared with conference proceedings, however significant for all three groups, based on one-way ANOVA results (Table 3).

In addition, the correlations between the papers' characteristics were analysed and the results are presented in Table 4.

Direct correlations, statistically significant at the 1% level, are noticed between all three variables. However, the strongest correlation is recorded for the number of

Table 3 Descriptive statistics based on papers' characteristics

Relevance characteristics	Publication type	Valid N (listwise)	Min	Max	Mean	Std. Deviation	One-way ANOVA
Cited reference count	Book Chapter	43	4	217	53.23	39.461	286.698**
	Conference	312	1	141	21.19	19.000	
	Journal	1333	0	264	69.71	34.418	
Number of pages	Book Chapter	43	6	67	24.93	11.124	271.576**
	Conference	312	2	45	9.45	5.010	
	Journal	1333	4	71	21.08	8.571	

** Correlation is significant at the 0.01 level (2-tailed)
Source Authors' own research

Table 4 Studies characteristics' correlation

	Publication type	Cited reference count	Number of pages
Publication type	1	0.423**	0.328**
Cited reference count	0.541**	1	0.547**
Number of pages	0.477**	0.603**	1

Note Statistics are computed based on Pearson Correlations (upper side of the table) and Spearman Correlations (lower part of the table), applied listwise for the full sample of 1688 papers included in analysis
** Correlation is significant at the 0.01 level (2-tailed)
Source Authors' own research

sources included in the reference list and the number of pages of the paper. Furthermore, in Table 6 a top of the first 20 studies that have the most cited references, grouped by publication type, was made.

To increase relevance, information about the authors and publication year was added. Only one of the studies falls into book chapter category, the reaming 19 being journals publications. The study with the highest number of cited references (264) was published in 2018 and the study with the lowest references (168) was published in 2009. According to classification, the publication year is between 2009 and 2020. No study of the top 20 is published in 2008, 2010–2015 or in a conference volume, and no author appears on the list twice.

4.4 Co-authorship Analysis of Research on Corporate Governance Under Crisis Conditions

The fourth research question (*RQ4*: *What is the state of collaboration involving corporate governance under crisis conditions?*) aims to identify collaboration between authors in the corporate governance and crisis conditions research field. To answer the question, *VOSviewer* software is used through the co-authorship network visualization. In Fig. 2 can be observed the network graph constructed.

Based on bibliographic data, *VOSviewer* is used to create a co-authorship map. In this way, it is possible to realize the network graph, which allows one to approach the strength network.

In Fig. 2 it can be observed that each of the ten clusters created is highlighted with a color. For example, cluster 1 is red and contains 6 authors. The second cluster is green and contains 6 authors. The color of the third cluster is blue, and the number of authors included in it is 5. The fourth cluster is yellow, and its total number of authors is 5. For the fifth cluster, total authors are 5 and the corresponding color is purple. The sixth cluster is composed of 4 authors and its corresponding color is turquoise. The color of the seventh cluster is orange, and its authors are 4. The eighth cluster is brown and contains 4 authors. The total number of authors for the ninth cluster is

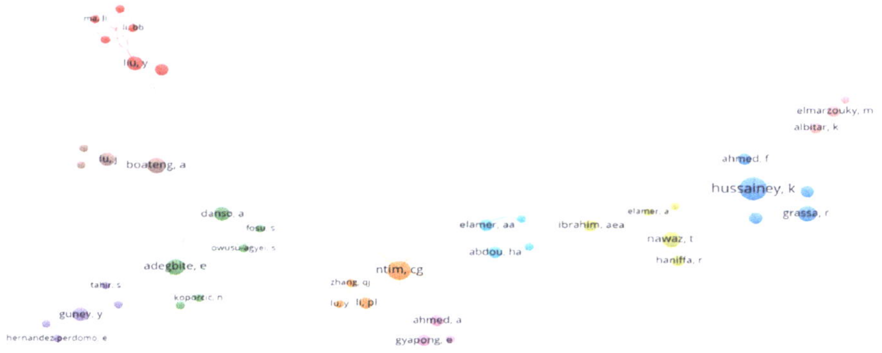

Fig. 2 Co-authorship network visualization. *Source* Authors' own research processed with VOSviewer

3 and its color is pink. For the last cluster, the corresponding authors are 3 and the color is light orange.

Studies with more than 25 authors are not included in the network analysis. The minimum number of documents and authors must be at least 1. The size of the nodes corresponding to each author indicates his relevance in research on corporate governance under crisis conditions. The thickness of the curves and the distance between the nodes signify the collaboration between the authors. The ten generated clusters have in common the links established between the authors who publish studies on the topic of corporate governance and the context of crises, in the sense of scientific collaboration between them. Groups are formed according to the intensity of the collaboration link. The clusters are presented in relevance order.

5 Conclusion

This study aims at providing an overview of corporate governance under crisis conditions research considering its evolution between 2008 and 2022, and the relevance based on specific characteristics. The classification according to the type of publication suggests that the largest number of studies are those published in leading journals (1333 studies from 1688), followed by the conference proceedings (312 papers), and books with only 43 papers. The evolution of publications by year provides the answer for the first research question. Through frequency analysis, it has been observed that the years that stand out through a considerable number of publications are 2011, 2018 and 2021.

Addressing the second research question, the most relevant papers published on corporate governance under crisis conditions are identified through descriptive and correlation analysis. Furthermore, the 20 most relevant studies are displayed, considering the times cited in the Web of Science Core and times cited in all research databases. All 20 studies are published in leading journals and the times cited vary between 137 and 650. Additionally, a direct correlation between the variables analyzed related to the entire sample is found.

In response to the third research question, characteristics as the number of references and the number of pages are analyzed. Findings support a direct and statistically significant correlation between the type of publication, the cited references count, and the number of pages. Additional analysis shows that 19 of the top 20 studies with the most cited references (the reference count varies between 264 and 168) are published in journals, and only one study is included in the book chapters category.

The fourth research question referring to the state of collaboration between authors and the analysis is made based on co-authorship network visualization generated using *VOSviewer* software.

This study makes several contributions to the field of research on corporate governance under crisis conditions because the evolution of research is examined from multiple perspectives. First, the years with the most publications were identified, together with studies classified according to publication type. Second, the most relevant studies were identified by the times cited. Third, the defining characteristics of the studies were identified and analyzed, based on cited reference count and number of pages, Fourth, through network visualization of co-authorship analysis, it is allowed to observe the state of collaboration between researchers.

Additionally, some research limitations and perspectives for future studies could be identified. The limitations consist of the database size and the long-analyzed period, which would not have allowed a more detailed analysis of each article to examine the abstract, methodological details, research methods used, and the main results, conclusions, and discussions. Based on limitations, future studies may use small databases in the study, combining descriptive analysis with content and frequency analysis. Moreover, future studies may combine bibliometric analysis with in-depth statistical analysis, complete with regression equations. In the same way, studies of some specific professional fields could be relevant because they allow for comparability.

Acknowledgement This paper was co-financed by The Bucharest University of Economic Studies during the PhD program.

Appendices

see Tables 5 and 6

Table 5 The journals publishing the top 20 relevant papers based on the number of citations

Publication Type	Source title	Times cited (All Databases)	Times cited (Web of Science Core)
Journal	Socio-Economic Review	650	648
Journal	Journal of Financial Economics	625	622
Journal	Journal of Financial Economics	556	551
Journal	Journal of Corporate Finance	444	436
Journal	Journal of Banking and Finance	392	384
Journal	Journal of Business Ethics	391	381
Journal	Journal of Financial Economics	378	377
Journal	Journal of Corporate Finance	285	280
Journal	Journal of Banking and Finance	255	250
Journal	Journal of Financial Intermediation	248	245
Journal	European Journal of Operational Research	211	200
Journal	Journal of Business Ethics	171	165
Journal	International Review of Financial Analysis	157	143
Journal	Journal of Business Ethics	157	154
Journal	Pacific-Basis Finance Journal	155	150
Journal	Corporate Governance—An International Review	153	148
Journal	Journal of Business Ethics	141	137
Journal	Journal of Management Studies	141	141
Journal	Emerging Markets Review	141	140
Journal	Journal of Financial Economics	140	137

Source Author's own research

Table 6 Top 20 studies by Cited reference count

Publication type	Authors	Cited reference count
Journal	Hassan and Aliyu (2018)	264
Journal	Lu et al. (2022)	264
Journal	Hazaa et al. (2021)	249
Journal	Trung (2022)	241
Journal	Daluwathumullagamage (2022)	223
Book. Chapter	Puaschunder (2017)	217
Journal	de Haan and Vlahu (2016)	200
Journal	Geontarek and Belghitar (2021)	197
Journal	Wu et al. (2021)	196
Journal	Usai et al. (2019)	190
Journal	Kalyvas and Mamatzakis (2014)	187
Journal	Mutlu et al. (2018)	186
Journal	Mercader et al. (2021)	181
Journal	Balachandran and Williams (2018)	179
Journal	Fernandes et al. (2018)	178
Journal	Chahed (2021)	177
Journal	Mehrpouya and Salles-Djelic (2019)	175
Journal	Cunha et al. (2021)	175
Journal	Carnegie and O'Connell (2014)	169
Journal	Morck and Yeung (2009)	168

Source Author's own research

References

Albu, N., Mateescu, R.A.: The relationship between entrepreneurship and corporate governance. The case of Romanian listed companies. Amfiteatru Econ. **17**(38), 44–61 (2014).

Baker, H.K., Pandey, N., Kumar, S., Haldar, A.: A bibliometric analysis of board diversity: Current status, development, and future research directions. J. Bus. Res. **108**, 232–246 (2020)

Clarke, T.: Recurring crises in Anglo-American corporate governance. Contrib. Polit. Econ. **29**(1), 9–32 (2010)

Donthu, N., Kumar, S., Mukherjee, D., Pandey, N., Lim, W.M.: How to conduct a bibliometric analysis: An overview and guidelines. J. Bus. Res. **133**, 285–296 (2021)

Dumay, J., Bernardi, C., Guthriea, C., Demartini, P.: Integrated reporting: A structured literature review. Accounting Forum **40**(3), 166–185 (2016)

Hossain, M.R., Akhter, F., Sultana, M.M.: SMEs in Covid-19 Crisis and Combating Strategies: A systematic literature review (SLR) and a case from emerging economy. Oper. Res. Perspect. **9**, 1–13 (2022)

Hsu, Y.L., Yang, I.C.: Corporate governance and financial reporting quality during the COVID-19 pandemic. Financ. Res. Lett. **47**(Part. B), 1–13 (2022)

Kells, S.: Impacts of COVID-19 on corporate governance and assurance, international finance and economics, and non-fiction book publishing: some personal reflections. J. Account. Organ. Chang. **16**(4), 629–635 (2020)

Kivinen, O., Hedman, J., Artukka, K.: Scientific publishing and global university rankings. How well are top publishing universities recognized? Scientometrics **112**, 679–695 (2017)
Lungu, C.I., Mititean, P., Caraiani, C., Constantinescu, D.: A structured literature review of corporate governance and performance research within an emerging country setting. J. Account. Manag. Inf. Syst. **19**(4), 707–733 (2020)
Naeem, M.A., Karim, S., Mohd Nor, S., Ismail, R.: Sustainable corporate governance and gender diversity on corporate boards: evidence from COVID-19. Econ. Res.-Ekon. Istraživanja **35**(1), 5824–5842 (2022)
Ruan, L., Liu, H.: Environmental, social, governance activities and firm performance: Evidence from China. Sustainability **13**(2), 1–16 (2021)
Shui, X., Zgang, M., Smart, P., Ye, F.: Sustainable corporate governance for environmental innovation: A configurational analysis on board capital, CEO power and ownership structure. J. Bus. Res. **149**, 786–794 (2022)
Stubbs, W., Higgins, C.: Integrated Reporting and internal mechanisms of change. Account., Audit. & Account. J. **27**(7), 1068–1089 (2014)

A Study into the Implementation of Robotic Process Automation in Different Business Processes

Andreea-Izabela Bostan and Oana Virgolici

Abstract The main focus point of this paper has been chosen based on today's trends in the Information Technology field, as companies' interest in automating manual processes is a continuous investment. Specifically, this article focuses on the Robotic Process Automation concept and the way this technology is perceived among companies at present. As part of the scientific research, a questionnaire has been addressed to people working with Robotic Process Automation in different fields, in order to bring to attention different behaviors that Robotic Process Automation teams express today. The objective of this paper is to analyze the responses and interpret the results according to theoretical concepts, chosen as point of reference. The questionnaire was designed to reflect preferences among Robotic Process Automation providers, areas with automation potential, as well as possible drawbacks of the technology, that could reside in potentially areas of improvement. As highlights of the research, it has been identified what are employees' preferences in terms of tools used for project implementation, challenges that employees who work with Robotic Process Automation every day are facing, as well as a general perspective on how this technology is currently affecting internal business processes and it is expected to evolve in the near future.

Keywords Robotic process automation · Digitalization · Digital transformation

1 Introduction

Artificial Intelligence (AI) automation is divided into a wide range of technologies such as Natural Language Processing, Computer Vision, Robotic Process Automation, and Machine Learning. The development of algorithms has led to the automation of cognitive and manual tasks hence replacing human labor (Yarlagadda 2018).

A.-I. Bostan (✉) · O. Virgolici
Bucharest University of Economic Studies, Bucharest, Romania
e-mail: Izabela.bostan@csie.ase.ro

M. Busu (ed.), *Rethinking Business for Sustainable Leadership in a VUCA World*, Springer Proceedings in Business and Economics,
https://doi.org/10.1007/978-3-031-50208-8_3

A foundational question for many BISE (Business and Information Systems Engineering) authors and readers is "What should be automated and what should be done by humans?" (Aalst et al. 2018).

Robotic process automation (RPA) is a software technology that makes it easy to build, deploy, and manage software robots that emulate human actions interacting with digital systems and software. Just like people, software robots can do things like understand what is on a screen, complete the right keystrokes, navigate systems, identify, and extract data, and perform a wide range of defined actions (UiPath 2023).

RPA aims to replace people by automation done in an "outside-in" manner. This differs from the classical "inside-out" approach to improve information systems. Unlike traditional workflow technology, the information system remains unchanged (Aalst et al. 2018). RPA tools aim to reduce the burden of repetitive, simple tasks on employees and its impact is visible in multiple industries.

A Forrester's recently published study forecasts that 25% of Europe-5 countries (France, Germany, Italy, Spain, and the UK) jobs are at risk from automation and 12 million jobs will be lost by 2040. The study shows that information worker jobs have grown nine times faster than the European jobs market over the past 10 years, and over the next 20 years, up to 9 million new green jobs will be created (O'Grady 2040).

RPA is driving new efficiencies and freeing people from repetitive tedium across a broad swath of industries and processes. Enterprises in industries ranging from financial services to healthcare, to manufacturing, the public sector, retail and far beyond have implemented RPA in areas as diverse as finance, compliance, legal, customer service, operations, and IT.

2 Literature Review

As UiPath states in the most recent report ("Automation Trends 2023: What's new, What's next, and your path to real value"): "As technology improves, as adoption increases, automation doesn't just change the way people work - it changes the very work you're capable of doing" (UiPath 2023).

Before analyzing Robotic Process Automation instruments' applicability and results, it is important to make the distinction on different types of robots, based on their role. We can identify four types as follows:

(a) Data entry robots are used to copy inventory load documents to a repository. It also gathers information from inventory documents and cleans and transforms information according to different pre-established business rules. Finally, data entry robots transfer data into an electronic system.
(b) Verification and validation robots: can make calls to a third party and internal systems to verify information effectively. If there is a verification or validation error, these robots tend to pass the human co-worker's issue to handle it the best

way possible. Robots can verify a customer's identity before any transaction is carried out to determine any risk.

(c) System integration robots are used to bring together systems such as core systems of different companies when they merge. System integration helps an organization to integrate difficult invoices regardless of the location. Software robots help to interface systems perfectly without special coding.

(d) Scheduled or trigger robots: Scheduled robots are used to take simple tasks that cannot be executed unless a specific event occurs. These robots remind humans of a certain event that is supposed to happen and the specific date (Table 1).

Top RPA companies, as it reveals in an article (Maguire 2023), are: UIPath (best overall), Automation Everywhere (best for Cloud RPA), Blue Prism (best for integration), WorkFusion (best for Finance Industry), Hyland RPA (best for Third-Party Bots) and the list continues.

According to a Grand View Research study (Grand View Research 2023), the global Robotic Process Automation market size was estimated at USD 2322,9 million in 2022 and it is expected to reach USD 2942,7 million in 2023. At the same time, the global Robotic Process Automation market is expected to grow at a compound annual growth rate of 39,9% from 2023 to 2030 to reach USD 30850 million by 2030. By analyzing these numbers, it is clear that RPA is a continuously evolving technology, and this dominance over competitors is determined by the overall advantages, in a few words: because of RPA's ability to quickly and precisely perform a vast array of key essential tasks.

In terms of efficiency, it is a better use of staff resources: human staff is freed from many tedious, low-value tasks. Companies are choosing this technology for its impressive Return on Investment (ROI). "The cost of licensing a robot is between 3000 and 8000 EUR annually, compared to 30000 EUR for a full-time employee" (Anagnoste 2017). Since a robot could replace the work done by up to three full-time employees, not to mention the continuous work cycle (without breaks, weekends, or vacations), it is understandable why RPA is chosen at the expense of the human workforce. Manyika et al. concluded that by 2030 "between 400 and 800 million individuals could be displaced by automation and need to find new jobs" (Manyika et al. 2017). Still focusing on the human component, but on the customer side this time, with many low-level customer requests handled automatically, the customer experience tends to increase, which leads to great loyalty.

Also, the overall costs of the process are reduced compared to the manual one, although an RPA solution has upfront costs (buying the application, installing it and maintenance costs). In most cases, businesses have been able to get more tasks completed with fewer workers using RPA. In terms of scalability, a group of Automated Robotic Processes, for instance, can be duplicated and programmed to accomplish a similar but slightly different set of procedures.

Additionally, RPA bots can increasingly incorporate AI and ML into their automation, which offers yet another way to scale. For example, Human Resources (HR) is one department that is successfully implementing RPA for conducting day to day activities. Different existing implementations are proving the obvious benefits

Table 1 RPA Tools comparison between vendors. *Source* https://www.keymarkinc.com/rpa-tools-comparison/

Vendor	UIPath	Blue Prism	Power automate	NICE	Automation anywhere
Focus	RPA	RPA	RPA	RPA	RDA
Ease of customization/integration	Moderate	Simple	Difficult	Difficult	Requires third-party for customization
IT involvement	Some	Minimal	Some	Heavy	Heavy
Security	Average	Built-in	Average	Average	Recommended security add-ons
Productivity	Innovative productivity tools including task capture for documentation	Object-oriented. One server runs many bots	Desktop automations for Windows users and compatibility with Microsoft Azure apps; limited elsewhere	Effective automation tools for improving customer service and customer centers. Limited capabilities elsewhere	Script oriented, desktop automation only. Reduced client productivity
Scalability	Moderate	Moderate	Limited	Robust	Limited
Minimum entry cost	$9000	$14,950	Free with Windows. Price/per bot model for full version	Subscription-based, one-time-purchase, or use-case/transaction-based pricing models	$995.00/one-time/user
Features					
Analytics	Yes	Yes	Yes	Yes	Yes
Attended automation	Yes	Yes	Yes	Yes	Yes
Code free development	Yes	Yes	Yes	No	No
Image recognition	Yes	Yes	No	No	No
Optical character recognition	Yes	Yes	No	Yes	No

(continued)

Table 1 (continued)

Vendor	UIPath	Blue Prism	Power automate	NICE	Automation anywhere
Process builder	Yes	Yes	Yes	Yes	No
Unattended automation	Yes	Yes	Yes	Yes	No
Highly detailed immutable execution and change logs	Yes	Yes	N/A	No	No
Single SKU (one product to purchase)	No	Yes	No	N/A	No
Free comprehensive on-line training	Yes	Yes	Yes	No	No

resulting as part of RPA and other AI integrations. A 2017 use case applied in a 700 employees company proved how a Chatbot and RPA solution can save 2300 EUR annually, by simply covering 11% of the activity an HR Business Partner has to do, in order to enroll each employee in six mandatory trainings per year (Anagnoste 2018).

Other than HR, the variety of business processes with automation potential is confirmed by the diversity of industries that are using this technology, with or without other integrations. In cybersecurity, RPA bots can automate security threat detection through privileged data management, by detecting malware in emails by specific keywords and by running cyber threat hunts, regular diagnostics, and fault remediation (conversational AI-enabled RPA bots can leverage responses from command catalogs to troubleshoot users' problems). In education, RPA can be found in course registration, exam grading, booking advising sessions, shortlisting admission candidates, attendance management. In healthcare, RPA can be very useful for patient appointment scheduling, or enhancing patient care by supporting analytics (Dilmegani 2023).

Another advantage is the improved analytics for workflow management. A clear grid of activities and timelines can be established, and an RPA tool's effectiveness can be tracked down to the most minor task and the smallest time increment. Because RPA uses AI and ML, analytics will become more complex. Also, advanced governance can be considered a plus, since an RPA system complies with a clear and precise governance system and RPA software can govern larger systems as well.

On the other side, when discussing the disadvantages of RPA, one of the items identified is the financial expense for implementation and maintenance of Robotic Process Automation (Yarlagadda 2018). Small organizations with limited financial resources cannot purchase, install, and maintain robots because of the high costs.

Lack of technical ability can also be considered, as researchers believe that for robotic systems to flourish, the end-user must possess significant technical skills (Adami 2015). If this element is ignored, therefore, the robotic process will lack the technical skills to perform various complex tasks. Another disadvantage would be the major change that it involves, depending on the company's history with using AI tools for its systems. However, this is something that cannot be felt when the right tool is used (Yarlagadda 2018).

Finally, the risk of a data breach can also be included as a disadvantage that should be taken into consideration. Replacing human labor with software applications can lead to data breaches since this system's information is prone to cybersecurity issues. Businesses that automate applications without security concerns are more prone to the risk of a data breach.

While RPA may automate processes faster and easier, governance structures may become more challenging and complex. Thus, it strongly depends on every organization whether it regards RPA as only a temporary solution or if it plans to engage RPA as part of its strategic capability. Choosing the best project methodology for RPA implementations can be a challenge and lead to a pitfall. "The agile deployment of RPA projects brings about the necessity to consider short-term and long-term influences as well as necessary changes to strategic organization designs. Researchers should seek to determine what kind of strategic approach is suitable, given different

RPA implementation goals" (Hofmann 2020). It is necessary to employ an agile delivery methodology that, depending on complexity, should enable the development of automated processes in cycles of three to five weeks, in order to avoid unnecessarily extending the delivery time. At the same time, organizations need to decide if they want to apply RPA using their own resources or by engaging outsourcing providers (Hofmann 2020). Decision-makers should consider two general KPI groups for measuring RPA's business impacts. The first KPI group should focus on software robots' influences on internal factors, such as employee productivity enhancement, job satisfaction, process acceleration, or cost savings. The second KPI group should consider software robots' influences on external factors, such as customer satisfaction, cooperation with partners and suppliers, or stock market value (Hofmann 2020).

3 Methodology

For this study, results were mostly assessed using descriptive statistics, with the aim to summarize and describe the characteristics of the data set. The main objective was to define trends related to using RPA technology in order to automate manual processes, place RPA teams at the center of IT decision-making and identify best practice recommendations for developing high-quality and efficient implementations.

In order to determine the extent of Robotic Process Automation in companies belonging to different fields, a questionnaire was addressed to 105 respondents, in February 2023. The questionnaire method has been chosen as the best instrument to gather a large set of data from people with different levels of expertise, working in different environments, but actively using Robotic Process Automation, at the same moment in time. It offered the benefits of multiple reviews and simultaneous perspectives.

The questionnaire was created using Google Forms and has been shared online, mostly by email and on a professional social media platform. Results were later interpreted by displaying the data in a visual format such as pie charts, bar charts or tree maps. While bar charts are only highlighting the number of responses and differences between answers are easier to remark, pie charts are representing each result as a percentage of the whole sample. At the same time, tree maps have been chosen in order to better visualize the hierarchical data and distribution of responses. For multiple choice responses, only bar charts have been used for an accurate representation.

The criteria based on which an individual was chosen as a potential respondent were to have been working with RPA at least for the last two years, regardless of the company or role. The study was based on a total of 22 questions, divided into three sections. The main categories were the following: the history of the respondents, a theoretical approach of the RPA technology on a conceptual level and applying these concepts in real life projects using different RPA vendors. The questionnaire was designed to reflect preferences among Robotic Process Automation providers, areas with automation potential, as well as possible drawbacks of the technology, that could reside in potentially areas of improvement.

Responses gathering did not imply any personal data collection as the profile of the respondents was built as part of the first section of the questionnaire. This one has been defined by questions analyzing the following topics: the industry to which their company belongs, the department they are working in, their location, experience with the RPA technology, role and the number of RPA projects they have experienced in the last 12 months. All this information was later correlated with the results observed on the other questions. All questions were mandatory, with multiple choices but single answer and the option to add a different response, if needed for the company industry, department, or role.

Respondents are mainly located in Romania (95%) and have been working with RPA technologies for different periods of time: 2–5 years (59%), less than 2 years (29%) or more than 5 years (12%) (Figs. 1 and 2).

The study was conducted among different companies and as a result, 14 different industries were addressed, as part of the respondents' involvement. Among the most recurrent options are: Professional Services and Technology, with a majority of 29,52%, Consulting (22,86%), Enterprise automation software vendor (26,67%), while some of the other industries addressed are: Banks and Financial Services, Retail, Pharmacy and Cosmetics. Most of the respondents are working in the RPA department of their company (74%) and have developed or took part in the development of more than five robots during the last 12 months (60%) (Figs. 3, 4 and 5).

Fig. 1 Respondents' location. *Source* Authors' own research results

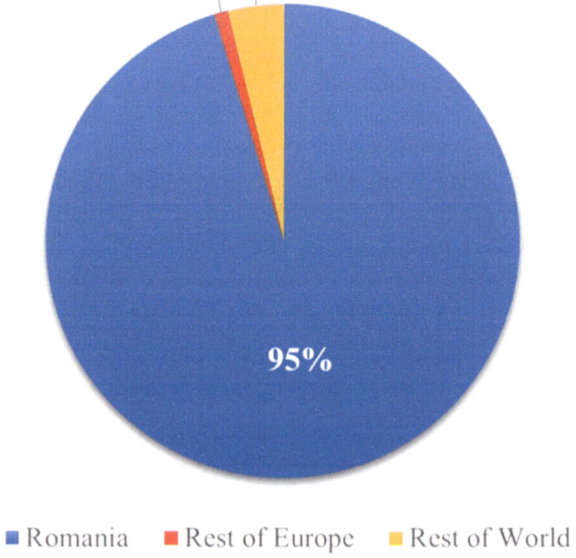

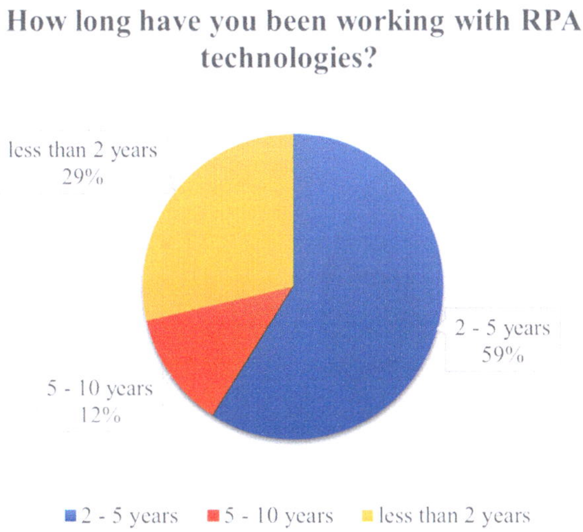

Fig. 2 Respondents' experience with RPA technologies. *Source* Authors' own research results

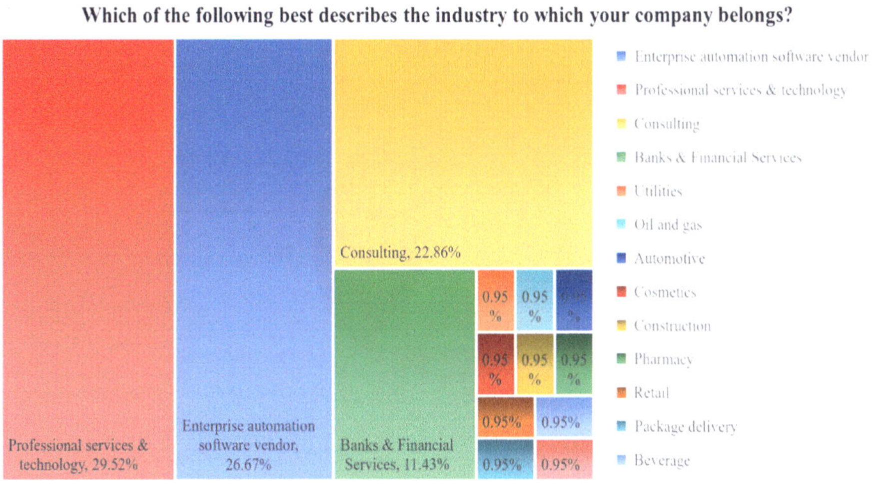

Fig. 3 Industries the respondents' companies belong to. *Source* Authors' own research results

The second section of the questionnaire focuses on the hands-on experience that respondents have with RPA, questions being related to: RPA vendors, advantages of RPA tools, possible improvements for this kind of platforms, types of robots developed, criteria for identifying manual processes with automation impact, fields with potential for RPA implementations, technologies that can be integrated, and advantages and disadvantages of software robots. Another topic of this section was organizing the actual implementation, with all the: challenges experienced, adoption

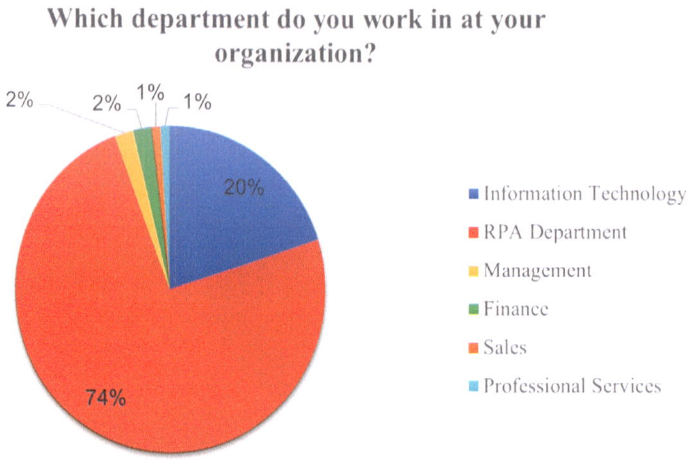

Fig. 4 Departments in which respondents are working. *Source* Authors' own research results

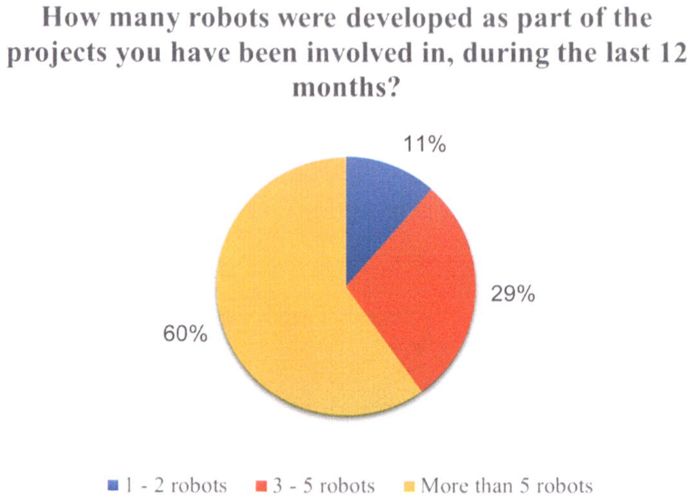

Fig. 5 Number of robots developed by respondents in the last 12 months. *Source* Authors' own research results

of project methodologies, project phases with most risks, and RPA perspective of clients. In this second section, questions are mandatory, and respondents had the option to select multiple responses for some of the applicable items (e.g.: project development challenges, advantages). Most of them also allowed for own responses, in order to prevent respondents from being limited by a specific predefined list.

Finally, the third and last section focuses on the RPA evolution: starting with the past two years, and continuing with prospects, based on each respondent's opinion.

The last two years' time frame has been chosen in correlation with the digitalization and the many changes that companies have faced. All responses in this section are mandatory and single choice, for a more exact conclusion.

4 Results and Discussions

Considering the respondents profile, in terms of automation criteria, the most important attributes that a manual process should have in order to be considered appropriate for automation are considered to be: predictability (that comes with the many repetitive tasks that determine steps in a process; to be easy to anticipate) and availability of input data in a fixed structure exclusively. A low execution time, implementation effort and reduced run frequency, although might reduce the complexity of the implementation, are not considered to be essential when deciding what can or cannot be automated with RPA.

As one of the purposes of the questionnaire was to determine the areas with the greatest potential for automation, the results confirm the hypothesis stating that one of the advantages of RPA is the diversity of fields in which it can be applied. Although Finance & Accounting was mentioned in most of the responses, Reporting and Payroll/HR are the other two most frequent responses. This can confirm the idea concluded in the previous topic, that regardless of the field, as long as data is provided into a structured input file, there will be a great potential to instruct the robot how to manipulate those inputs. This might be the case for a domain that handles a huge amount of numbers and has different regulations and formulas that must be followed for different calculations or statistics, which is one thing that all three fields have in common.

Customer Service, Audit, and different Operational departments are also considered having potential for automation; however, the less predictable a process within a field is (in terms of actions and/or input data), the more is required to opt for another Artificial Intelligence (AI) integration, that will maximize the success rate of the implementation. This being said, it is remarkable that the majority of respondents consider that an RPA tool can be used to its maximum efficiency if it allows multiple AI integrations. The allowance of different simultaneous connections is also one of the top answers, as this is one method that supports teamwork and collaboration, which are the main objectives pursued by project management methodologies, regardless of their type (Figs. 6, 7, and 8).

In what RPA vendors are concerned, 90 respondents recognized UiPath as the main provider for software robot's implementation. This comes in correlation with the main functionalities that RPA teams are looking forward to, when choosing the right tool: ease of use, ability to integrate with other tools, and coding experience. Other advantages mentioned are constant updates, rapid deployment, stability/robustness of the tool, scalability, ability to reduce workload and to automate complex actions. The ability to create attended and unattended robots is another advantage of RPA

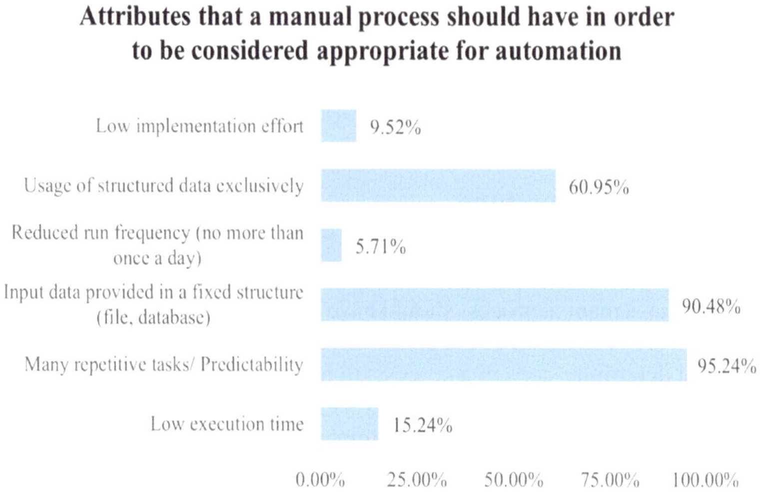

Fig. 6 Attributes that a manual process should have in order to be considered appropriate for automation. *Source* Authors' own research results

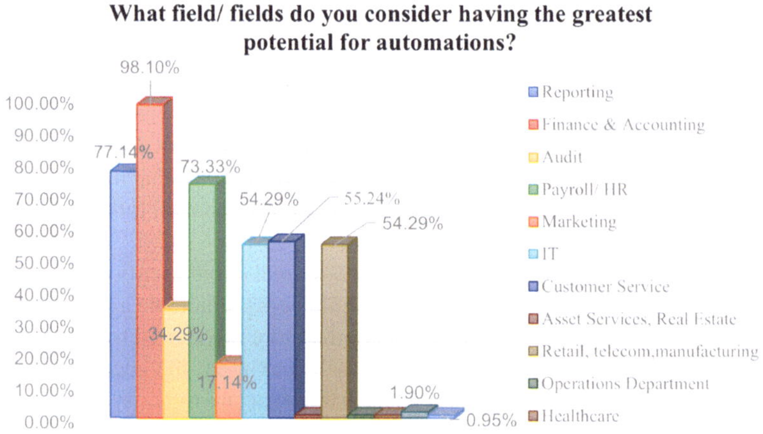

Fig. 7 Fields that allow RPA implementation. *Source* Authors' own research results

tools, as 76% of respondents are using both types, and 20% are using unattended robots exclusively (Figs. 9, 10, and 11).

Another topic addressed as part of the study was the usual RPA project lifecycle, with a focus on the project phases in which teams are often facing difficulties. A majority of 30%, most of the respondents are considering the Discovery phase to be the most error prone. Solution Design and Development phases are the next preferred options, with a 20% choice rate each. By analyzing these numbers, it is also important to mention that most of the respondents are working as a Developer/Architect (74%), the difference between the other existing roles being remarkable: Business Analyst/

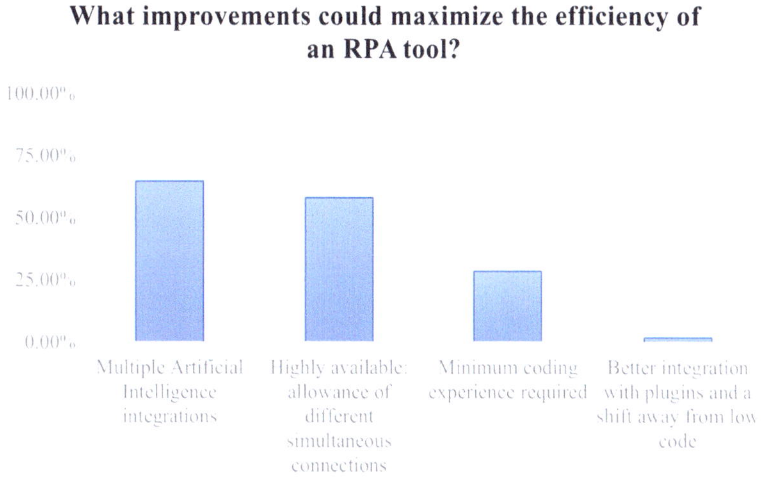

Fig. 8 Improvements that could maximize the efficiency of an RPA tool. *Source* Authors' own research results

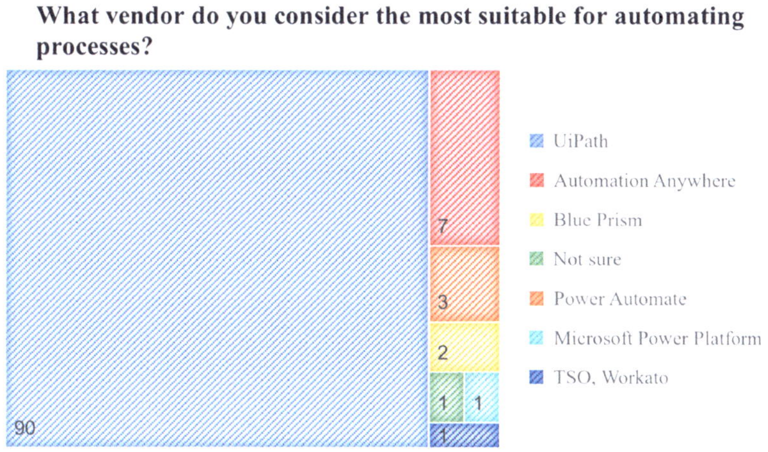

Fig. 9 Respondents' RPA vendor preferences. *Source* Authors' own research results

Product Owner (13%), Management positions (8%), Project Manager/Scrum Master (2%). In this context it is interesting to notice to what extent the initial phases of a project (Discovery, Design) are considered to be the most challenging in terms of the number of risks that could potentially affect the whole project implementation. Going even more into this subject, the fact that the initial phases of a project are usually the most demanding is confirmed by responses received to a different question: "What are the challenges that may be experienced during a project development?" Respondents were allowed to select multiple choices, and the results show that the top three most difficult aspects are faced when: the initial manual process is not clear enough or is

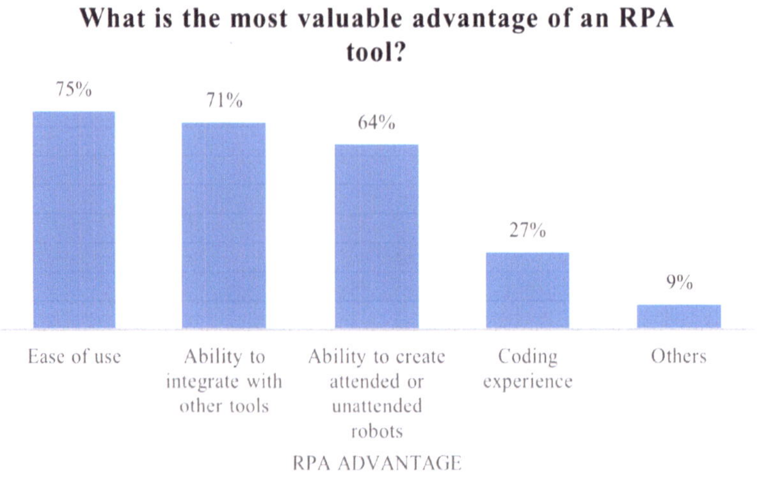

Fig. 10 Most valuable advantages of RPA tool. *Source* Authors' own research results

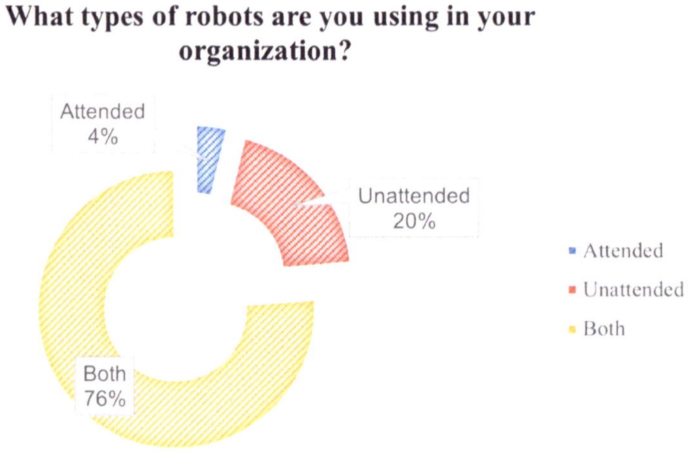

Fig. 11 Types of robots developed, based on the manual input requirement. *Source* Authors' own research results

lacking adequate documentation (with a reference to the first phases of the project implementation), customers are expressing reluctance or adding complexity to the implementation during the process and when edge cases are not covered by the software robot (which, again, can be referenced to the same Analyze and Design phases) (Figs. 12 and 13).

Among responses, 25,71% of the results are targeting the adoption of an unsuited project methodology. Even if it can be considered that automation and methodology are mutually exclusive, as there are no limitations when having to decide on a methodology for an RPA specific project, like any other project in the IT field, organizing

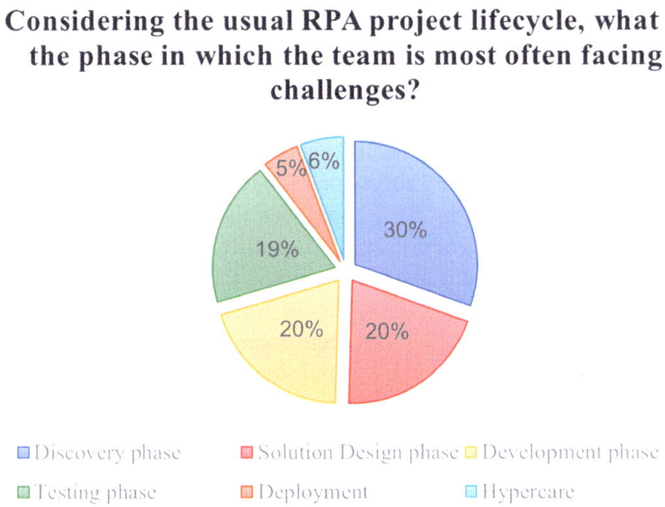

Fig. 12 Project phases where teams are having most of the difficulties. *Source* Authors' own research results

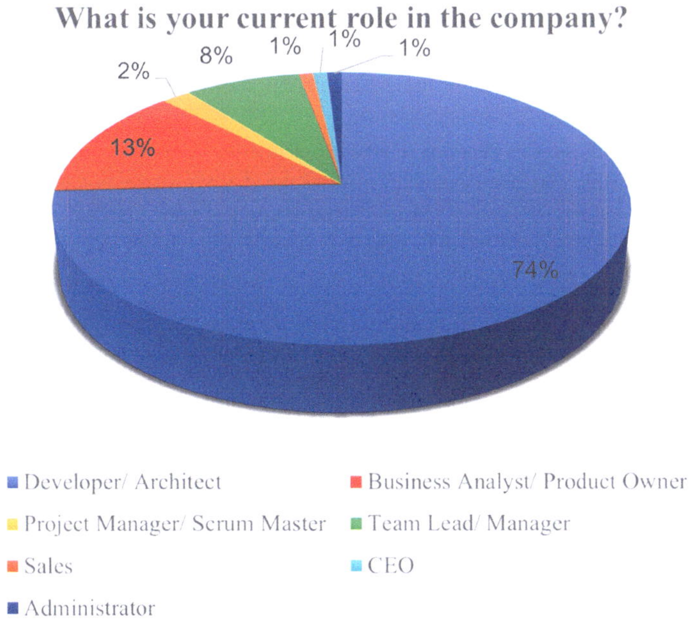

Fig. 13 Respondents' role in their company. *Source* Authors' own research results

the workflow and resources based on a specific methodology is the element that adds value to the implementation and may conclude in the success or failure of the project. Software development methodologies, from traditional to agile approaches, are available for organizations to choose from. Even if some are taking a hybrid approach, their methodology preference is related to specific organizational projects and team traits. The "Pulse of the Profession" (2018) study conducted by PMI (the Project Management Institute) in 2018 concludes that 71% of organizations were choosing the Agile approach, even if applied to different methods, based on each team's needs. This number was a result of a survey that highlighted feedback from 5702 Project Managers and other management roles, from a range of industries, including Information Technology. "More and more organizations are recognizing that agility—the capability to quickly sense and adapt to external and internal changes to deliver relevant results in a productive and cost-effective manner—is helping them stay competitive" (Project Management Institute 2023). The same conclusion can be highlighted considering the study that made the subject of this paper, as 72% of respondents considered that Agile methodologies can best be associated with automation processes (Figs. 14 and 15).

Regarding advantages of RPA implementations, 60% of respondents state that reducing the manual work is the most valuable input of this technology. Other responses include: the improvement of the process quality (19%), overall cost reduction of the process that is being addressed (14%), less errors observed in the automated process in comparison to the manual one (4%). On the other hand, when discussing the potential disadvantages, only 23,81% believe that RPA automations are not implying any disadvantages. There are a number of items mentioned as being less favorable to the end users, and the main one (53,3% of responses) is financial implications, more exactly as initial investment costs. From a social perspective, it has also been mentioned the need to invest in reskilling training or the cost of hiring new staff (33,33%), both options being related to the support or maintenance of the

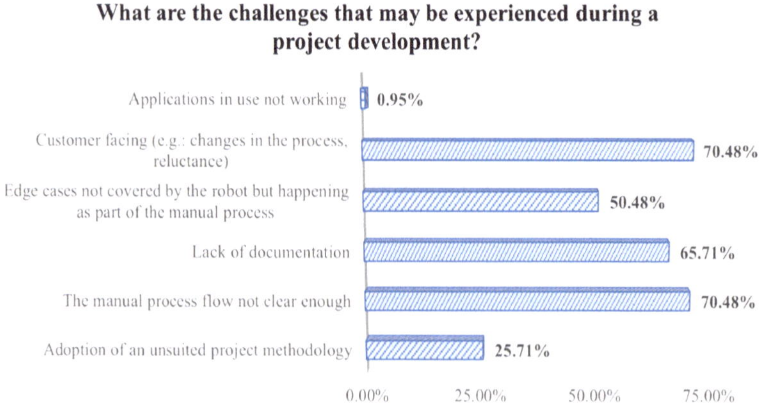

Fig. 14 Challenges experienced during an RPA project development. *Source* Authors' own research results

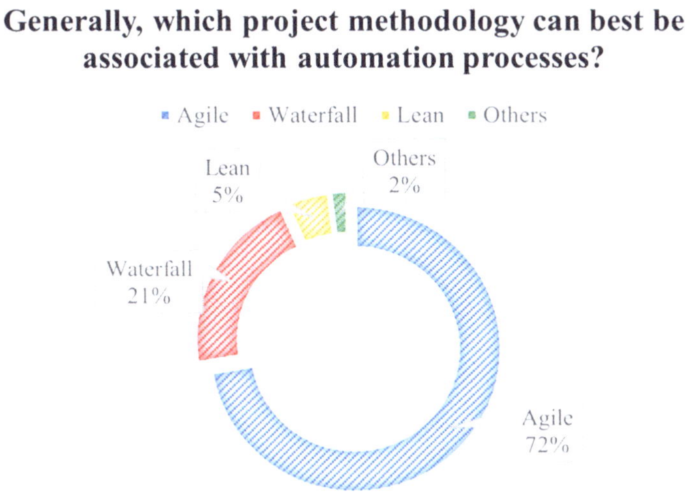

Fig. 15 Project methodologies most used for RPA implementation. *Source* Authors' own research results

robots. The potential job loss is prioritized by only 13,33% of respondents. This confirms that, indeed, RPA implies future transformations of jobs, but investing in the continuous development of employee skills can be seen as part of the progress and will prevent decrease in manual labor as long as companies are investing in reskilling. Other disadvantages of RPA implementations considered are: the need for some outputs to be checked by a human user (in case of processes that cannot be covered completely by software and will also inquire the human input) and potential automation failures that are negatively perceived by end users (Figs. 16 and 17).

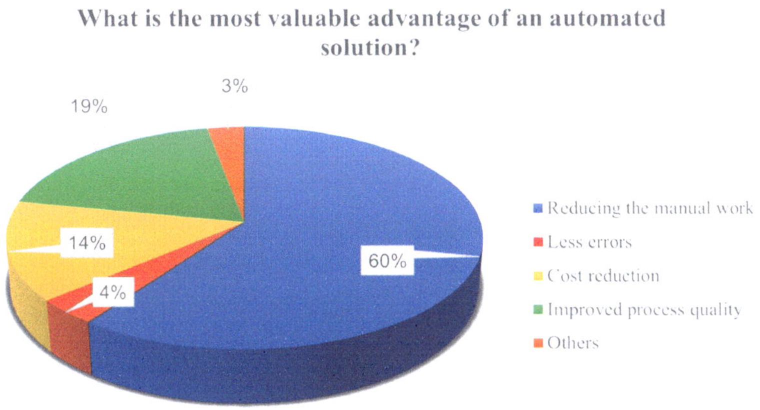

Fig. 16 Advantages of an automated solution. *Source* Authors' own research results

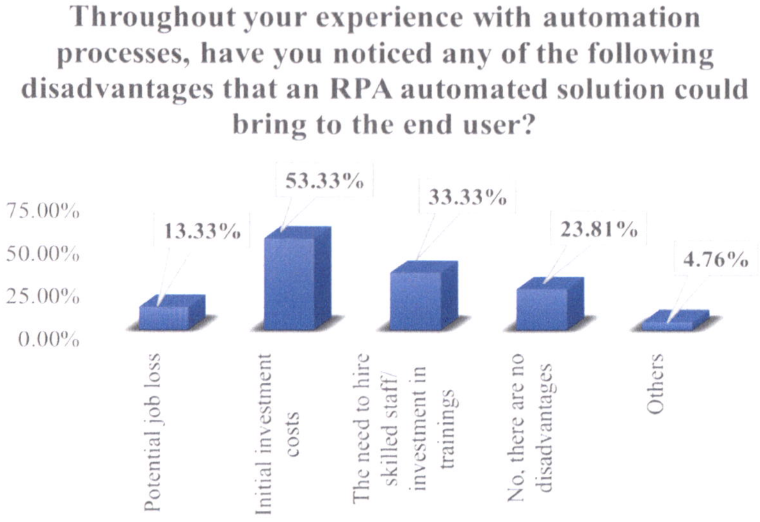

Fig. 17 Disadvantages of an automated solution. *Source* Authors' own research results

Regarding the customer relationship involved in RPA processes, 76,19% of people that are often facing the end user's reluctance are connecting this with the lack of knowledge that the customer may have towards this technology. They also reported the idea that RPA robots could take over their tasks, or they don't believe in minimizing or excluding manual work. High costs of the implementation can be one of the justifications as well, but not to the same extent as the other options (only 28,57%) (Fig. 18).

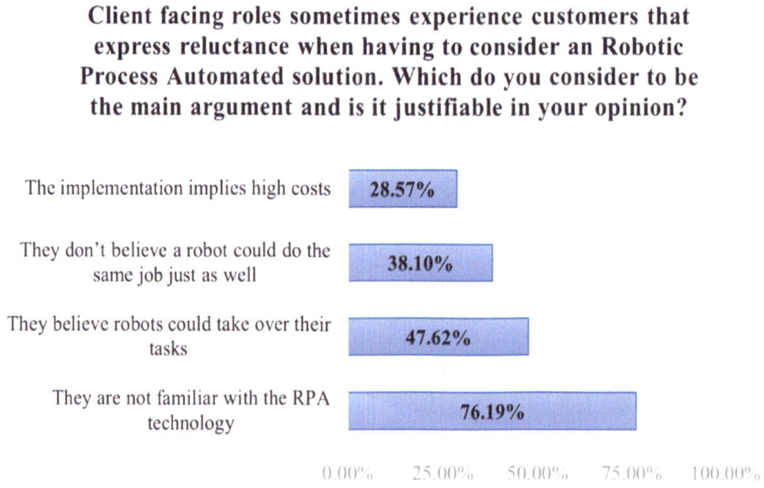

Fig. 18 Client facing challenges in RPA implementations. *Source* Authors' own research results

In terms of the evolution of RPA implementations, it is visible that the automation necessity has increased in the last 2 years, considering the need to adapt to a digital environment, where human interactions are minimized. The first two options have a percentage close to one another: 39% of respondents observed an increase in the number of RPA projects between 25 and 50%, while 34% of respondents reported more than 50% project implementations. At the same time, the opinion of the majority (58%) is that RPA impacts the productivity of a company to a large extent, which doesn't come as a surprise considering the variety of advantages that RPA brings in a process, as detailed in the previous section. However, even if the number of RPA implementations has increased, and this technology is considered to bring a plus of value to companies, 52% of respondents consider that Robotic Process Automation will not replace manual work completely or to a great extent in the next five years. Even if in this case the problem of people's reluctance cannot be raised, considering that the respondents are actively involved in the RPA field and are admitting its plus of value, the percentage can be justified when analyzing this fact from a different perspective. First of all, working exclusively with automated processes would involve a very abrupt change of direction, and even with the whole progress of technology, this is a step that might be done over a longer period of time. Secondly, RPA alone might not bring the desired result, this being a reason why people are tending to opt for tools that are also facilitating other forms of Artificial Intelligence. Optical Character Recognition (OCR), Chatbots and Machine Learning are the other Intelligent Automation technologies that people are using the most as integrations with software robots (Figs. 19, 20, 21 and 22).

5 Conclusion

As confirmed by the results of the questionnaire, which has addressed a range of domains, decision making in what automated processes are concerned, like any other important choice in a company, requires a strategic approach. RPA is a powerful tool, with a great impact on companies' productivity, as the Grand View Research forecast (2023) and this survey, which represented the focus of this paper, are both showing.

It is essential to analyze the design of human–robot interaction patterns, as well as the potential impact of RPA on employees and their perception of software robots. Employee engagement and their skills development should be considered in all strategic implementation initiatives. Companies must reconsider reskilling employees, given the changing areas of responsibility. According to Manyika et al. (2017), foreseeing millions of jobs displaced by automation by 2030 is less likely to be confirmed since 52% of respondents consider that RPA cannot replace manual work to a great extent in the next years of development. Even if replacing manual processes to a great extent might be a long run initiative, as can be concluded from the above study, the digital workforce should start taking steps into this direction, as technology will only continue to develop and cover even more areas. In these conditions, job loss cannot be considered a risk to the human workforce; however

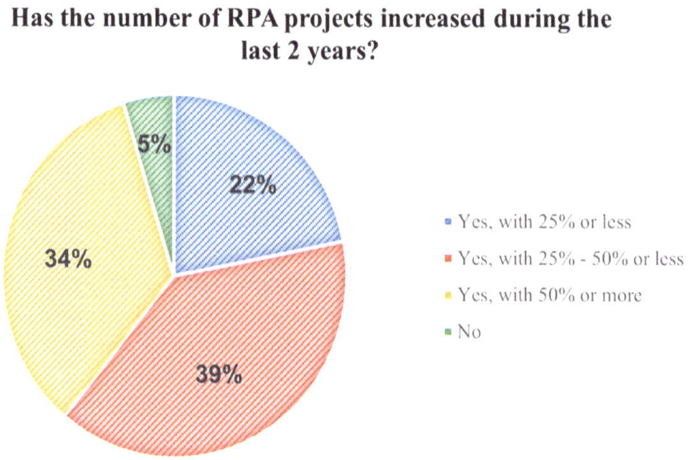

Fig. 19 Increase in the number of RPA projects. *Source* Authors' own research results

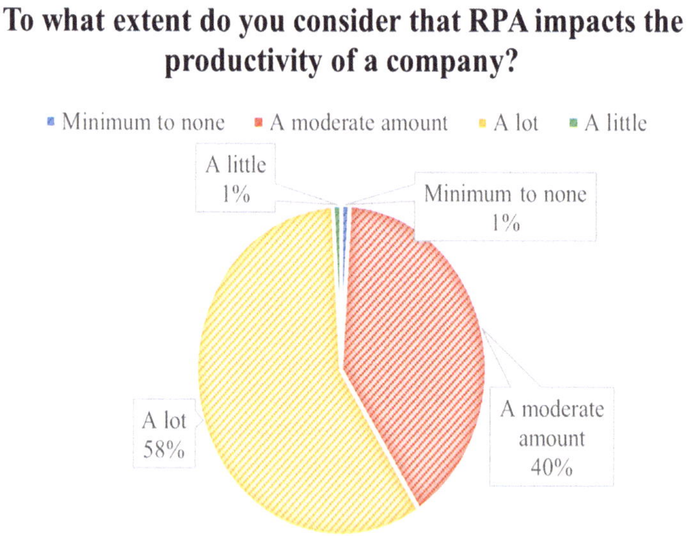

Fig. 20 Impact of RPA in the company's productivity. *Source* Authors' own research results

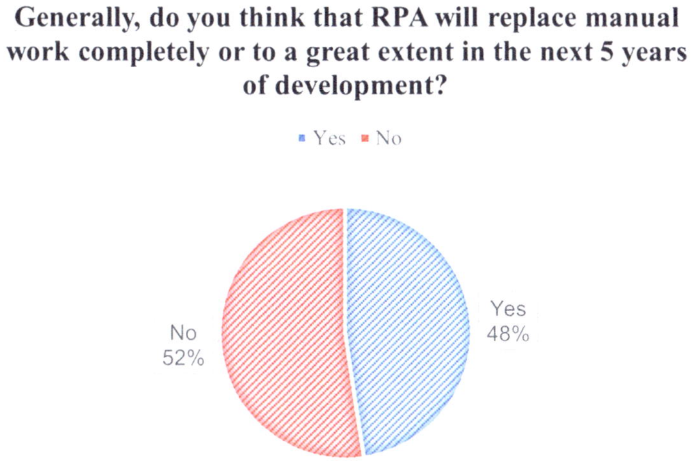

Fig. 21 RPA development in the next 5 years. *Source* Authors' own research results

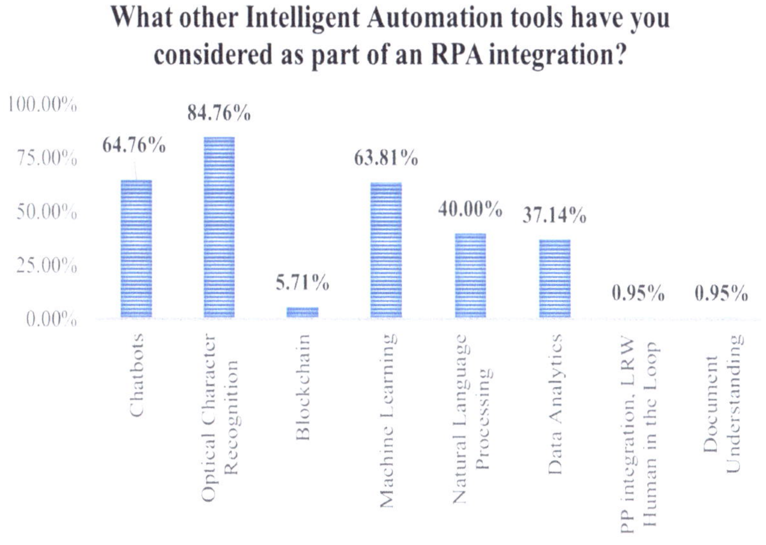

Fig. 22 Artificial intelligence integrations for RPA. *Source* Authors' own research results

it is understandable that the employees will have to develop new skills, even if it is in order to implement better and more efficient software robots, or to learn to work along with them. "The most common fear among humans is that RPA might replace human labor, yet robots' main aim is to give a supporting hand to humans at the workplace" (Adami 2015).

Robotic Process Automation applicability will continue to expand in different fields, as RPA advantages are becoming more and more prominent and adopted by an increasing number of companies. Even if the market today offers different options for RPA vendors, there is a preference for one specific tool that offers the majority of functionalities developers are looking for (ease of use being the most selected option provided in the survey). UiPath was the RPA company chosen by most of the respondents, which approves the overall comparison done by Maguire (2023). One improvement that RPA vendors could address, however, is the value of initial costs. Yarlagadda (2018) mentioned this as one of the disadvantages that this technology has among implementations, specifically for small organizations. And it seems that even nowadays, the initial investment cost is seen as the main drawback, as stated by the majority of respondents.

As it can be deduced from the statistics of the research, processes with the most automation potential will always have in common the predictability of the actions and the availability of data in a structured format; however, the ability to integrate multiple Intelligent Automation instruments is extending the capabilities. This latter resulted to be another one of the most appreciated attributes of a tool, this way confirming the "Robotic Automation Process—The operating system for the digital enterprise" study (Anagnoste 2018), where the two use cases were presented to highlight the Chatbots and Intelligent Optical Character Recognition integrations as Artificial Intelligence tools that maximize RPA results. Chatbots and IOCR are, at the same time, the technologies respondents are using the most, in RPA implementations.

In terms of project methodology and project development phases, the research identified that teams are tending to adopt an Agile approach, which fosters close collaboration and communication. In connection to that, the study shows, at the same time, that one improvement that would maximize the efficiency of RPA tools is the ability to allow simultaneous connections.

As a future direction of research, one proposal would approach potential solutions that could minimize the number of challenges that RPA teams are facing during the Discovery and Solution Design phases. Considering the results of this study, these two phases, part of the project lifecycle, are connected with most of the risks in a project implementation and a potential solution would minimize the errors that are experienced during an RPA project implementation.

References

Adami, C.: Robots with instincts. Nature **521**, 426–427 (2015)

Anagnoste, S.: Robotic automation process-the next major revolution in terms of back office operations improvement. Proc. Int. Conf. Bus. Excellence **11**(1), 676–686) (2017)

Anagnoste, S.: Robotic automation process–the operating system for the digital enterprise. Proc. Int. Conf. Bus. Excellence **12**(1), 54–69) (2018)

Dilmegani, C.: 43 Back Office Automation Examples in 2023: RPA, WLA, AI/ML. https://research.aimultiple.com/back-office-automation/#back-office-automation-examples (2023). Last accessed 2023/03/01

Grand View Research, Robotic Process Automation Market Size, Share & Trends Analysis Report By Type, By Application, By Deployment, By Organization, By Region, And Segment

Forecasts, 2023–2030. https://www.grandviewresearch.com/industry-analysis/robotic-process-automation-rpa-market (2023). Last accessed 2023/03/01

Hofmann, P.: Robotic process automation. Electron. Mark. **30**(1), 99–106 (2020)

Maguire, J.: Top 15 Robotic Process Automation (RPA) Companies. https://www.datamation.com/artificial-intelligence/top-15-robotic-process-automation-rpa-companies/ (2023). Last accessed 2023/03/01

Manyika, J., Lund, S., Chui, M., Bughin, J., Woetzel, J., Batra, P., et al.: Jobs lost, jobs gained: what the future of work will mean for jobs, skills, and wages (2017)

O'Grady, M.: Twenty-Five Percent Of Europe-5 Jobs Are At Risk From Automation; 12 Million Jobs Will Be Lost By 2040. https://www.forrester.com/blogs/twenty-five-percent-of-europe-5-jobs-are-at-risk-from-automation-12-million-jobs-will-be-lost-by-2040/ (2022). Last accessed 2023/03/01

Project Management Institute, "Pulse of the Profession". https://www.pmi.org/-/media/pmi/documents/public/pdf/learning/thought-leadership/pulse/pulse-of-the-profession-2018.pdf (2018). Last accessed 2023/03/01

UiPath, Automation Trends 2023: What's new, What's next, and your path to real value. https://www.uipath.com/hubfs/2023_Automation_Trends_Ebook_UiPath.pdf?mkt_tok=OTk1LVhMVC04ODYAAAGKLnl4kVHwFnaQYBAcdhLklgtqTAtT_nqAH8dNMJtY79HL9oD0beHZAvEGGWOkxz8mcSRE6mhZJpatQBCJfrYnrv5wSW2HR8kaT_i1HWyUKrKUO2k (2023). Last accessed 2023/03/01

Van der Aalst, W.M., Bichler, M., Heinzl, A.: Robotic process automation. Bus. Inf. Syst. Eng. **60**, 269–272 (2018)

Yarlagadda, R.T.: The RPA and AI automation. Int. J. Creative Res. Thoughts (IJCRT) (2018). ISSN 2320-2882

Looking at Competitive Intelligence as an Emergent Practice. Recent Developments and Key Takeaways from Romania

Irena Chiru and Mireille Rădoi

Abstract The paper aims to provide a framework for understanding and analyzing the practices associated with competitive intelligence in young democracies. It starts from the premise that in the last decade we have witnessed a gradual but slight reduction of the gap between the consistent development of competitive intelligence—in the Anglosphere and Western Europe—and the South-Eastern Europe. In this sense, Romania represents a good example as it provides an intimate view on not only the geographical impacting variables, but also on the historical, cultural, and economic specificities. From a methodological perspective, in order to explore the key characteristics, meanings, and implications of the Romanian case, the paper will apply a qualitative case study research design. The results can be further on use for the understanding of the impact on national variables on the competitive intelligence function as well as for future comparison surveys seeking to shed light on the development of private intelligence practices.

Keywords Competitive intelligence · Business intelligence · Intelligence consulting · Training

1 Introduction: The Development of Competitive Intelligence

As the practice of collecting data in order to inform military and political decisions, intelligence has a long history, one that overlaps with the history of waging wars and outclassing enemies. The transparent, assumed, institutionalized, and professionalized orientation towards business, industries and markets happened more recently

I. Chiru (✉)
"Mihai Viteazul" National Intelligence Academy, Bucharest, Romania
e-mail: chiru.irena@animv.eu

M. Rădoi
Bucharest University of Economic Studies, Bucharest, Romania

M. Busu (ed.), *Rethinking Business for Sustainable Leadership in a VUCA World*, Springer Proceedings in Business and Economics,
https://doi.org/10.1007/978-3-031-50208-8_4

and it is related to the progressive development of what initially were ad hoc and informal processes within some (mainly US) firms in the '70 s. These processes involved circumstantial gathering and indexing data on firms' competitors as well as on the composition and dynamics of the industry in which they were performing.

Beyond practical needs, awareness on data-driven business decisions was raised by several significant theoretical contributions coming from Stevan Dedijer who began thinking of how the principles of intelligence might be applied to business. Also named the "father of business intelligence" (Marren 2005), Dedijer brought together the ideas of strategy and information, discussed the emergence of "social intelligence" and pleaded for shifting from stealing secrets towards intelligent analysis and "open sources". In time and with the conceptual and applied support from research and analysis centers providing not just business-oriented intelligence products but also training and scholarship (e.g. the Futures Group of Jan Herring or Fuld and Company), competitive intelligence emerged as a new discipline. To this, a rich contribution was brought by the works of Michael Porter on strategy (as related to the notion of competitive advantage) and by his systematization of arguments for creating a system for analyzing competition (Porter 1980). Hence, by capitalizing on a long tradition of gathering data in order to take better decisions, competitive intelligence started to be acknowledged as a per se field of practice, mainly build upon the transformation of secret intelligence into a formal and visible process conducted by dedicated professionals. To this transformation and to the crystallization of competitive intelligence, the massive growth of data and the underpinning technological development has had a very strong input.

Along these decades of refinement, what we call today competitive intelligence has intertwined and seldom been used as "business intelligence", "market research", "corporate information", and "knowledge management". However, although names like "business intelligence" or "corporate intelligence" are preferable as they more accurately capture the essence of this activity (Murphy 1980), "competitive intelligence" (CI) is the most commonly used term. Definitions coming from the *Society for Competitive Intelligence Professionals* (SCIP) describing CI as a systematic and ethical program to collect, analyze and manage external information that can affect your company's plans, decisions and operations or from authors as Cris West (2001), Ben Gillad (2008a) or Fleisher and Bensoussan (2012) depict CI as a process enabling organizations to know both their competition and the space in which they operates, hence be use this data in the decision-making process offering a perspective on facts.

As one can easily imagine, from its beginnings, CI has undergone significant changes given by the accumulation of tacit and explicit knowledge, lessons learnt, professionalization and specialization of CI experts and, last but not least, by the gradual but decisive technologization that proved to be a crucial element in the development of the field. For example, if in the past human source collection was the backbone of the CI profession delineating competitive intelligence forcefully and clearly from market research and library services (Gillad 2008b), today CI works with software that make data gathering and structuring automatic processed streaming in time data to executives about the state of the business. In other words, the field benefits from a doctrine providing theoretical and practical oriented frameworks, fora for exchanging ideas and building upon best practices (e.g. SCIP) and also from tailor made solutions designed to facilitate the data emersion processes of intelligence analysts (e.g. Maltego).

Against this encouraging picture of a more that five decades systematic maturing, the landscape beyond the Anglosphere does not echo the same thriving. According to a SCIP (2005–2006) report on *State of the Art: Competitive Intelligence* aiming to prove a snapshot of competitive intelligence professionals, their practices, and clients, 77% of the respondents, all SCIP members, came from North America with only 62 of the respondents coming from Europe (Fehringer and Hohhof 2006). Another survey, conducted at a 10 year difference (2016) indicates a raise of the European number to 117; same raise in the case of Africa which had 12 respondents in 2006 and 76 in 2016. There are several limitations that does not allow formulating a general conclusion on a raise of interest in Europe based on these two reports (e.g., the 2016 survey does not provide the total number of subjects and in both cases the pool of respondents was given by SCIP members leaving aside other potential respondents involving in CI but without being member in the SCIP). In addition, the most recent report issued by SCIP in 2022 on *State of Competitive Intelligence* (2022) emphasizes an extraordinary amount of progress (but no region/countries data are provided). Despite these limitations, existing data indicate the global, but yet unbalanced development of CI practices.

Alternately, the small number of respondents can be used as a strong indicator for assessing the existence or absence of "organized competitive intelligence" (Gillad 1989) as a systematic alternative to informal intelligence activities which are performed in organizations on a daily basis; when in place, organized CI should be to a certain extend connected to the professional network of peers (which in this case is undoubtedly SCIP). This observation on the geographically unbalanced dispersion of CI is reinforced by the evaluation made by S. Wright in her seven European countries analysis (Wright 2005). Looking into the state of affairs in Finland, Germany, Lithuania, Russia, Spain, Sweden, and the UK, she highlights the existence of varying levels of competitive intelligence knowledge, practice, and longevity between countries. Such a diversity that does not allow a "side-by-side" comparison. From the multitude of concepts by which CI practices are called (competitor, competitive, business, economic intelligence) to the different organizational functions to which CI is associated (e.g. marketing in Germany, market research, customer management, web searching or environmental scanning in UK) and different level of use (e.g. tactical in Finland and strategic in Sweden), the European landscape as illustrated by the seven case studies proves to be diverse, and variable.

Against this background of diversity it is worth mentioning the particular example of France illustrating the footprint of the national strategic culture. Here, CI is characterized by an interconnection between defense, government, and business. It is illustrated not only in and by the backgrounds of the reputed figures who have contributed since the' 80 s to the development of the discipline, but also in the construction of what it is called "intelligence économique" combining both public and private sector (Baumard 2003). Also, in 2016 France created "Service de l'Information Stratégique et Sécurité Économiques" seeking to "raise the awareness of economic actors about economic security" (Le Service de l'information stratégique et de la sécurité économiques 2022). Hence, CI in France has its multidimensionality given by the involvement of levels of government, numerous support organizations from the private and the public sectors as well as public private partnerships and quasi-governmental organizations. *Le Rapport Martre* (1994) on national competitiveness illustratively identified three pillars of CI for France: 1. Encourage the practice of CI at the company level; 2. Optimize the transfer of information between the private and public sectors; 3. Construct data banks in light of the user needs (Smith and Kossou 2008a).

This progressive advancement was reinforced by a more recent study that looked for the specificities of the European competitive intelligence practices (Calof et al. 2018). Based on 156 responses, more European firms appear to have formalized intelligence functions and are using these intelligence functions for a broader range of intelligence objectives (range of projects and targets). To sum up, the European landscape can be characterized as being: (1) emergent—progress is visible, (2) dependent on the national strategic culture and the engagements of the political level—integration in the EU was a trigger, (3) highly receptive to existing legacy of scholars—in the case of Sweden, the hundreds of articles signed by Stevan Dedijer shaping the understanding of intelligence impact on strategy.

One may reasonably ask what is beyond the case studies briefly discussed above and more at length in the literature of the field? What is the state of the art in countries that have not been in the radar of intelligence studies so far? Starting from these legitimate questions and the absence of most countries from Eastern Europe in the surveys conducted so far (except for Russia which out of obvious reasons constitutes a special case), the present article will investigate the case study of Romania. Presumably, this will bring a change of direction in the investigation of CI practices beyond well-trodden paths.

2 The Methodological Framework in Assessing Competitive Intelligence Developments

Most of the surveys that have collected empirical data to map the landscape of competitive intelligence in the last two decades embraced a quantitative approach using questionnaires applied to SCIP members (Fehringer et al. 2006; Wright and Calouf 2006; Calof et al. 2018). Alternatively, other papers used a historical approach based on the identification and discussion of national variables illustrating a certain path of development (Millán and Comai 2004; Smith and Kossou 2008b; Xie and Xuehui 2018).

The current research will also employ a historical approach but as part of a single case study methodology. From this perspective, an exploratory case study was considered appropriate as it allows us to look at contemporary phenomena understood as a broad notion of studying the present but without excluding the recent past, hence where direct observations can still be made (Yin 2014). By looking at a case that have so far fallen under the radar of traditional surveys that are published, we tried to see to what extend and how has the scholarship of CI been developed in post-communist Romania as well as to explore the main benchmarks of this development. Indirectly, the results will be used for inferring on the status of CI practice, its context-driven facilitating/limitation factors and its strengths/weaknesses.

Case study evidence came from four sources: documents/publications, interviews, and personal observation referring to developments after 2010 when the incipient debates about practicing intelligence in private organizations started. However, given the invisible role that past circumstances play in tailoring the subsequent events, the discussion extended to the post- communist era.

3 Romania: A Mystery Caged in a Puzzle

A valuable discussion about intelligence practices in Romania cannot be made without the reference to the variables shaping post-communist Romania. By bleak contrast to events elsewhere in central and eastern Europe during the course of 1989, the collapse of the Ceaușescu regime in Romania was extremely violent with the execution of the country's Communist dictator, killing of civilians and street fighting between armed forces, members of the Securitate and so-called 'terrorists'. As opposed to "the swift, bloodless and essentially peaceful overthrow of the Communist regime in the then Czechoslovakia", "Romania's 'revolution' in December 1989 was far from being 'velvet'. The same is true of Romania's subsequent attempts to transform its economic and political systems" (Light and Phinnemore xxxx). Furthermore, the 1989 Revolution images about the Securitate and its abdominal acts have prolonged by feeding collective perceptions of intelligence as being corrupt, subversive, and against human rights.

One of the first observation that can be made and which has been time-spanning since the very beginnings of discussing intelligence until now is about terminology. Generally speaking, lack of clarity in defining the core concepts of a field of theory and practice leads to lack consistence in strategies and subsequent actions. In the case of Romania, the transition towards democracy was also imprinted with a conceptual passage from the terminology that had been traditionally used by intelligence services to terms that echoed the academic scholarship on one hand, but also the language of Western intelligence services on the other. Hence, the concept of "intelligence" has started to be progressively used as a more eloquent alternative of the classic "informații" (Romanian word for information). After two decades since the 1989 and the Romanian revolution, and assisted by initiatives such as the *Romanian Intelligence Studies Review* (*Revista Română de Studii de Intelligence*), the *National Institute for Intelligence Studies (Institutul Național de Studii de Intelligence*) or the master's program in *Intelligence Analysis* (*Analiza de Intelligence*), *intelligence* has started to make its way in particular in the academic environment but also with reverberations in mass media.

Despite these significant steps that marked the beginnings of intelligence studies in Romania and opened a long conversation on its uses (Light and Philammore 2001), two interrelated conceptual streams took form: one of the institutional language remaining loyal to the Romanian "informații" (also reflected in the formal name of the pertaining institutions), and one of the academic debate. On one hand, the use of "intelligence" per se in Romanian and its progressive assimilation was due to the explosion to the Western intelligence studies consistent literature and to understanding the imperfections of the Romanian correspondent. On the other hand, it proved to be a very efficient facilitator in Romanian scholars' making proof of using a common acknowledged language and hence enhancing the dialogue with well-reputed international academics. This latter observation, although it may show a very practical orientation, can be also considered illustrative for the manner in which the whole spectrum of intelligence studies (including academic events and publications, study programs, international consortia) have developed in Romania making it a systematically discussed best practice (Sebe 2010; Gill and Phythian 2012). What we have witnessed in Romania is relatively similar to the condition of intelligence and intelligence scholarship in the majority of countries (Spain, France) that looked for their own model of strategic intelligence as an alternative for importing already existing models or parts of these (e.g. the intelligence doctrine of Canada, US or UK). However, such a construction requires certain determination, time and prowess in combining different knowledge resources (political science, sociology, military science), and hence in achieving the right conceptual accuracy as a precondition of a national school of thought in intelligence.

Relatively similar conceptual struggles has featured the avenue of competitive intelligence. Its first formalized use goes back to 2012 when the *Intelligence for Businessman Association* (*Asociația Specialiștilor în Informații pentru afaceri—ASIA*) (Puyvelde et al. 2020) and its journal were created with the involvement of several former intelligence officers. Both provide a very good example of the interconnectivity between generally acknowledged concepts (such as intelligence and competitive intelligence) and the local and national variations. In this particular case, both the Romanian concept of "*informații pentru afaceri*" and its English counterpart—competitive intelligence are combined in an attempt to promoting the mission of"fostering the competitiveness of the Romanian business environment"[1]. Although the activities of the association ended the same year (with the opening in June and last posted activity in October 2012), this initiative has been continued by organizing periodical sessions of training dedicated to competitive intelligence—*Business and Competitive Intelligence* program along with other programs dedicated to elicitation, OSINT or competitive analysis.

A consistent debate engaging both the scholars from the governmental intelligence schools' sector (pertaining to the ministry of interior, internal intelligence service) and also the experts from the very emergent intelligence private sector was initiated about the competencies that intelligence analysts needed jointly with the attempt to formalize a competencies framework. The letter led to the creation of a new occupational standard including the skills, knowledge, values, and interests needed and also clarifying the professional contexts and the activities specific to experts in what was called "private intelligence" (www.asia.org.ro, last accessed 2023).

On an academic level, two master's programs were created by the National Intelligence Academy: *Competitive Intelligence* and *Competitive Intelligence for Public Administration*. While initiated at the same time (October 2014), they had a different, however complementary, focus: the former was mainly dedicated to developing the knowledge, skills, abilities, and experience needed to stay ahead of one's business or marketplace competitors, while the letter sought to instill the formal practices of intelligence in public organizations that could this way systematically and deliberately organize themselves, and develop insights about the competitive environment in which they performed.

This debut was rapidly followed by other related initiatives. During the same time, the *Competitiveintelligence.ro* (https://codcor.ro/2422-specialisti-in-domeniul-politicilor-administrative#expert-informatii 2023) first newsletter was issued. Also characterized by a sporadic appearance (3 issues in 2015 and one in 2016), the newsletter aimed to facilitate the understanding of competitive intelligence benefits for small and medium firms. By introducing international (SCIP) and national CI activities (ASIA), presenting the bios of intelligence experts (most of them former governmental intelligence analysts), and retrieving several job offers from LinkedIn, this initiative somehow confirmed the already existing status quo, indirectly illustrating the lack of progress in Romania.

Currently, there are several small entities that officially declare their involvement in collecting and analyzing data about markets and competitors. They seem to embrace a similar portfolio of activities ranging from consulting to training, both in Romania and in the South- Eastern Europe, with a main declared focus on competitive intelligence or to related activities (e.g. risk analysis, due diligence, financial analysis, computer security and forensics, surveillance and counter-surveillance):

- *IAdvice*—offers consultancy in competitive intelligence, market intelligence, counterintelligence, and also tailor made competitive intelligence training programs in Romania (https://competitiveintelligence.ro/ 2023);
- *Iceu Group*—a "competitive intelligence company" dedicated to "consultancy and training" https://i-advise.ro/servicii-de-consultanta/ 2023) which combines using HUMINT and OSINT with profiling and simulated behavior detection. Its products are destined various sectors of activity: HR, IT, retail, selling, banking, insurance, politics, legal, education, private security (https://iceu.ro/ 2023);
- *Corporate Intelligence Agency*—"help companies and corporations limit industries specific risk and vulnerabilities, while enhancing fraud awareness and building a bullet proof risk management system that allows growth and business development" (https://i-advise.ro/i-advise-meeting-of-the-minds/ 2023);
- KWLG—based on a "40 years previous experience in governmental intelligence", focuses on competitive and business intelligence, market intelligence, media intelligence and offers a "pre-academic intelligence consulting" pretesting future intelligence operatives and analysts abilities (https://corporateintelligence.ro/en/);

- *ISCO INTELLIS*—group of companies specialized in custom-made security services, business intelligence and private investigations (https://kwlg.ro/ 2023);
- *Total Trust*—provides support for corporate and private security interests. We deliver intelligence insights through a range of projects and formats, with investigation, net assessment and traditional risk analysis as the pillars of our unique approach (https://isco-intellis.ro/en/services/ 2023);
- *Interdiligence*—corporate investigations and risk consulting"operating in Eastern and Central Europe" (www.total-trust.com 2023).

Although at a glance most of these initiatives present themselves as being unique on the Romanian market ("the only company" or "the only program"), they seem to focus on similar areas namely providing: (1) consulting for business development based on data about industry and competitors, (2) training for completive intelligence expertise, and (3) related services such as targeted private investigations, investigative due diligence, physical protection. Many display a combined experience in both governmental and private area, usually advancing prior expertise in governmental intelligence as a calling card for their professionalism and credibility in the field. Hence, IC profession seems to have been built on an extrapolation of competencies from intelligence services towards private organizations. This reflects a more general *modus operandi* specific to the development of IC in the US and Western Europe which in itself is not problematic unless gradually the profession starts building upon its own scholarship, and a coherent and cohesive relation between education, research and practice.

A more recent initiative (2022), the *Master of Business Intelligence* granted by the Bucharest University of Economic Studies seeks to build competencies in data collection, analysis and communication that will help improve products, services, and strategies in constantly changing markets and realities (https://intell.fabiz.ase.ro 2023). Although not a new concept at a global scale, for Romania the program is unprecedented as it combines the approach of two universities (one in economics and one in intelligence), it blends the expertise of academics and practitioners from Romania and from abroad teaching in tandem, and it offers an integrated platform for internships and on field visits augmenting opportunities for developing professional connections and offering hands-on experience in industry.

In a nutshell, at the moment the landscape of CI in Romania shows a certain dynamic, but looks rather fragmented and solely concentrated on the practice and tradecraft of CI. What is evidently lacking is a literature as a reflection of advancement in the knowledge about and in CI as few existing writings are mere reflections of previous non-autochthonous concepts and theories of CI without including a scientific research-driven systematization of best practices and lessons learnt. Having a literature would not only structure existing knowledge, but would also allow overcoming the risk in-silos developments, lack of self-reflection and of awareness of own strengths and challenges or of examples worth following.

The model which places practices at the center of CI profession is not a sustainable one. Certainly, working in the field provides the context for "having a story to tell", however not the sole condition for "having a say" over an ethical development of CI or over its interconnections to other areas of study and practice. Such an observation draws upon the viable advancement of CI initially in the US and later on in several Western European countries in which a considerable effort has been paid and still is on first understanding, structuring and only then applying the existing knowledge on how to collect and analyze data. Presumably, the effort for doctrinal and conceptual progress would also encourage other currently missing puzzle pieces to form and prove effectual. One example are the professional associations and networks providing key professional support, the right environment to grow correctly, while safeguarding the CI profession and occupation. Related are the set of regulations setting general standards, requirements, restrictions and guidelines and professional deontology code of conduct serving as a platform for expressing professionals' public commitments, enforcing norms, values, and ethics.

4 Conclusions and Way Ahead

In the last two decades, Romania has witnessed an unceasing process of developing the practices of intelligence, in both the private and the governmental sector. This process, which has indirectly but frequently connected private and governmental intelligence, was initially hampered by a terminological exploration ended with a progressive, however not autochthonous appropriation of "intelligence" as a concept. Despite its use in the professional language of intelligence entities for more than two decades, "intelligence" is still inexistent in the Romanian language dictionary. Certainly this has not and does not impede continuing to use it as a synthetic formula expressing the whole philosophy, theory and practice of what intelligence is. Once assumed, the lack of a Romanian similar or at least very proximate concept can be reframed not as a permanent battlefield for why using an indigenous term but rather for accommodating it to the national variables and making it a very useful agent of international collaboration and local conceptual adjustment.

Similar to other Eastern European countries, in Romania scholars and professionals need to devote themselves to empirically investigating the state of play of competitive intelligence by systematically raising questions (e.g. about the role of CI in organizations, the skills required in the field, the benefits of CI, the interconnections of CI with other organizational functions), while moving beyond the simple postulating or recursive invoking of intelligence as an imported practice. Such a research-driven endeavor requires coordination between academia, business and government bodies stimulating collaborative formats for intelligence professionals, trainers, and scholars. Just as importantly, since Romania along with several other countries in South and Eastern Europe share a common post-communist destiny, a future direction of study would require investigating multiple national case studies that may allow comparative perspectives and a structured approach on differences and communalities. Ideally, this would diminish the peril of developing CI in silos, so much likely given the very self-protective nature of intelligence, and inspire a more open dialogue on professionalizing CI.

References

Baumard, P.: Entretien paru. Lettre Sentinel, **7**, (2003). http://www.infosentinel.com/info/article_07_baumard.php. Last accessed 12 Dec 2022
Calof, J., Arcos, R., Sewdass, N.: Competitive intelligence practices of European firms. Technol. Anal. Strat. Manag. **30**(6), 658–671 (2018)
Fehringer, D., Hohhof, B., Johnson, T.: State of the art: competitive intelligence, Research report 2005–2006. CI Foundation, (2006). https://static.canalblog.com/storagev1/vtech.canalblog.com/docs/f_060608_stateofart_sum.pdf. Last accessed 03 Mar 2023
Fleisher, C. S., Bensoussan, B.: Analysis without paralysis: 12 tools to make better strategic decisions. FT Press (2012)
Gill, P., Phythian, M.: Intelligence studies: some thoughts on the state of the art. Ann. Univ. Buchar., Polit. Sci. Ser., **14**(1), 5–17 (2012). https://nbn-resolving.org/urn:nbn:de:0168-ssoar-389493
Gillad, B.: The role of organized competitive intelligence in corporate strategy. C. J. World Bus., Winter
Gillad, B.: The future of competitive intelligence. J. Compet. Intell. Manag. **11**(5), (2008). www.scip.org. Last accessed 03 Mar 2023
https://codcor.ro/2422-specialisti-in-domeniul-politicilor-administrative#expert-informatii. Last accessed 03 Mar 2023
https://competitiveintelligence.ro/. Last accessed 03 Mar 2023
https://corporateintelligence.ro/en/. Last accessed 03 Mar 2023
https://i-advise.ro/i-advise-meeting-of-the-minds/. Last accessed 03 Mar 2023
https://i-advise.ro/servicii-de-consultanta/. Last accessed 03 Mar 2023
https://iceu.ro/. Last accessed 03 Mar 2023
https://isco-intellis.ro/en/services/. Last accessed 03 Mar 2023
https://kwlg.ro/. Last accessed 03 Mar 2023
https://www.interdiligence.com/. Last accessed 09 Feb 2023
https://www.total-trust.com/. Last accessed 02 Mar 2023
https://intell.fabiz.ase.ro/. Last accessed 07 Mar 2023
Le Service de l'information stratégique et de la sécurité économiques. https://sisse.entreprises.gouv.fr/fr/qui-sommes-nous. Last accessed 10 Dec 2022
Light, D., Phinnemore, D.: Post-communist romania. coming to terms with transition. Palgrave Publishers, (2001)
Marren, P.: The father of business intelligence. J. Bus. Strat. **26** (1) (2005).
Millán, J. T., Comai, A.: Competitive intelligence in Spain: A situational appraisal. J. Compet. Intell. Manag., **2**(3), (2004)
Murphy, C.: Competitive intelligence. Gathering, analysing and putting it to work. Routledge, (1980)
Porter, M.E.: Competitive strategy: Techniques for analyzing industries and competitors. The Free Press (1980)
Van Puyvelde, D., Wirtz, J. J., Holeindre, J. V., Oudet, B., Bar-Joseph, U., Kotani, K., Matei, F. C., Díaz Fernández, A. M.: Comparing national approaches to the study of intelligence. Int. Stud. Perspect. **21**(3), 298–33 (2020)
Sebe, M.: Despre intelligence (II). Rom. Intell. Stud. Rev. **3**, 115–127 (2010)
Smith, J., Kossou, L.: The emergence and uniqueness of competitive intelligence in France. J. Compet. Intell. Manag. **4**(3), (2008)
Smith, J., Kossou, L.: The emergence and uniqueness of competitive intelligence in France. J. Compet. Intell. Manag. **4**(3) (2008)
State of Competitive Intelligence.: (2022). https://www.scip.org/page/2021StateofCompetitiveIntelligence. Last accessed 03 Mar 2023
West, C.: Competitive intelligence. Palgrave Basingstoke, UK, New York (2001)
Wright, S.: Seven European nations. A profile of current CI practice. J. Compet. Intell. Manag., (2005)

Wright, S., Calouf, J.L.: The quest for competitive, business and marketing intelligence: A country comparison of current practices. Eur. J. Mark. **40**(5–6), 453–465 (2006)

www.asia.org.ro. Last accessed 03 Mar 2023

Xie, X., Xuehui, J.: The evolution of competitive intelligence in China. J. Intell. Stud. Bus. **1**, 61–75 (2018)

Yin, R.K.: Case study research design and methods, 5th edition. Sage, (2014)

Romania's Cluster Potential. Integrating Intelligence Systems to Spur Competitiveness in the Agriculture Industry

Cristiana-Ioana Coman and Valentin Cojanu

Abstract A clusters' role in stimulating, creating, and expanding a region's competitiveness is axiomatic. The research questions that have steered our study are: which are the Romanian industries with the highest potential for cluster development? How did the European top-ranked agricultural clusters integrate intelligence systems into their operations? How could we adapt those intelligent systems to develop Romania's agro-clusters? The results were obtained by employing a two-fold methodology: (1) to evaluate Romania's cluster potential, we have used the Cluster Index of Sternberg and Litzenberger, and (2) to design sustainable policies meant to spur the sector's competitiveness, we have used the benchmarking for 14 agricultural clusters awarded with Gold Label by the European Cluster Excellence Initiative. We learned that over half of the Romanian counties have the highest potential to develop clusters in the primary industry—the agriculture, forestry and fishing, and mining and quarrying sectors. The counties that registered the most significant Cluster Index on the agriculture sector are Bihor, Covasna, Harghita, Botoșani, Neamț, and Vaslui.

Concerning the stimulation of the cluster's competitiveness, we concluded that Romania needs a robust investment in knowledge and innovation, implementation of intelligent systems among the value chain, and facilitation of national and international collaboration. Our conclusions could represent a starting point for regional policymakers, cluster directors, and governments to create new strategies to support Romania's agricultural industry and clusters.

Keywords Cluster · Competitiveness · Cluster Index

C.-I. Coman (✉) · V. Cojanu
Bucharest University of Economic Studies, Bucharest, Romania

M. Busu (ed.), *Rethinking Business for Sustainable Leadership in a VUCA World*, Springer Proceedings in Business and Economics,
https://doi.org/10.1007/978-3-031-50208-8_5

1 Introduction

Globalization, urbanization, income and life expected growth, and the swap in technological paradigm disrupt traditional value chains and, implicitly, the agricultural industry. As a result, customers are changing their shopping habits and searching for alternative, more natural, sustainable products.

In consequence, "the new customer" is fueling a demand for high-quality agricultural products whose production process requires no pesticides (or a limited amount) and has a friendly approach to the environment and the other actors involved (e.g., the labor force). These changes, along with the projected population growth, climate change, and food safety, are driving companies to redesign their strategies to become competitive.

Agricultural clusters must reinvent themselves through intensive knowledge, technology, and innovation to respond to these challenges. Today, more traditional agricultural clusters are transitioning towards a technology cluster. In consequence, agricultural clusters are showing interest in implementing intelligent systems using big data, artificial intelligence, the Internet of things (IoT), and other digital technologies to enhance their service (1) at the farm level (e.g., crop & water monitoring, fertilization, irrigation, risk assessment, farm management); (2) at the operation process level (e.g., process improvement, quality assurance, monitoring, transport), and (3) at the customer interaction level (e.g., innovative packaging, quick delivery). These new types of clusters integrate an input of services to each segment of the value chain using concepts such as "smart food systems" or "intelligent agricultural systems".

This paper has two primary research purposes. Firstly, we evaluate Romania's cluster potential by industry and by county. Secondly, based on the results obtained, we shed light on how the agriculture sector's cluster potential could be enhanced by implementing intelligent systems and not limiting them. The inspiration for the policy recommendations is based on a consistent study of the best practices applied by the European agricultural clusters awarded with Gold Label by the European Cluster Excellence Initiative (ECEI) identification.

2 Literature Review

Amidst the current globalized environment, companies take advantage of the free circulation of capital, services, goods, information, and technology. This shift led to an easiness in the flow of intermediaries and implicitly significantly lowered the importance of location in defining the competitive advantage. Thus, the traditional approach to how companies and nations compete must be revamped.

Nowadays, "the economic map of the world is dominated by what I call clusters (…) of unusual competitive success in particular fields". (Porter 2023) Cluster definition has evolved over time. In 1990, Porter described the cluster as a productive sectorial system where actors are linked by similar or related activities in a geographical concentration. (Porter 1990; Delgado et al. 2015) defined the cluster as a geographic concentration of companies (or industries) that form in the proximity of skills, input, demand, and other competitiveness variables. (Delgado et al. 2015; Sosnovskikh 2017) named clusters according to proximity, value creation, and business environment. This perspective is in line with Porter's theory, suggesting that a cluster structure stimulates both the firm and region growth, producing a competitive advantage (Sosnovskikh 2017).

The factors contributing to the cluster's competitiveness were studied more in-depth by a vast number of researchers: (Porter 1990; Nallari and Griffith 2013; Gajšek and Kovač 2016; Chen et al. 2022), while Martínez-Marín et al. (2020) presented how intelligent systems impacts a cluster's competitiveness (Martínez-Marín et al. 2020).

The impact of intelligent systems and technology at a large scale on agricultural cluster competitiveness has been studied by several researchers. Xiong and Qiao (1999) concluded that intelligent systems modify the performance in fertilization, plant protection, and for the cultivation of cotton, rape, tea, wheat, corn, rice, tobacco, sugar cane, orange, and cabbage. (Xiong and Qiao 1999; Kolhe and Gupta (2014); Essien (2019), and Marques (2022) approached a macro-perspective in analyzing the impact of intelligent systems on the agriculture industry. (Kolhe and Gupta 2014; Essien 2019; Marques 2022).

More into-depth, Chowdhury and Tawsifur Rahman (2021), Li et al. (2021), Abioye (2022), and Lykhovyd (2022) examined the precision of water management, irrigation systems and leaf disease detection, while Chen and Hu (2022) investigated the implementation of an adaptive algorithm for the intelligent control method of agricultural greenhouses. (Chowdhury and Tawsifur Rahman 2021; Li et al. 2021; Abioye 2022; Lykhovyd 2022; Chen and Hu 2022).

In regards to the value chain operations that can be enhanced through intelligent systems, Barenji et al. (2019) mentioned the transformation of logistics centers, and Allal-Chérif et al. (2021) have put forward the impact of Artificial intelligence (AI) in redefining the purchasing functions. (Barenji et al. 2019; Allal-Chérif et al. 2021).

3 Methodology

Competitiveness assessment is a preeminent catalyst for a nation's progress. The research questions igniting our study were: which are the Romanian industries with the highest potential for cluster development? Which are the most notable intelligence systems applied among the top-ranked agricultural clusters in Europe? How could we adapt those intelligent systems to develop Romania's agro-clusters? We have used a two-fold methodology approach to evaluate Romania's cluster potential (the cluster index) and to design sustainable policies meant to spur the sector's competitiveness (benchmarking).

The first approach—the Cluster Index of Sternberg and Litzenberger (2004)—measures the concentration (absolute concentration), and it correlates the relative

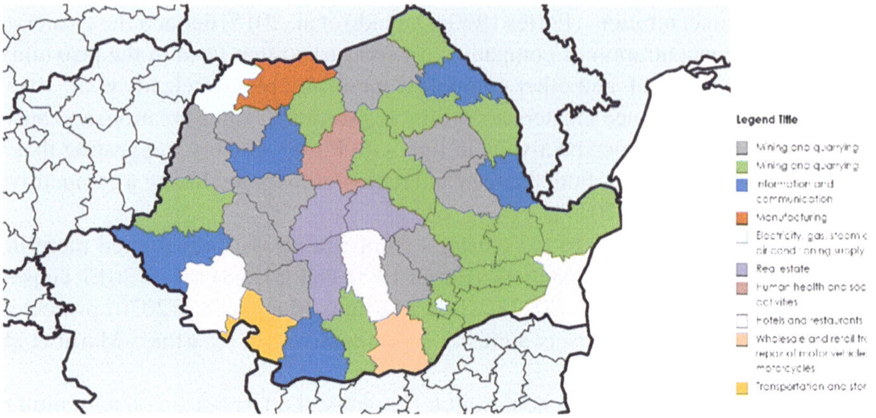

Fig. 1 Romania's specialization map by CI value (2019). *Source* Authors' own research

enterprise density (ID), the relative enterprise status (UB), and the relative company size (BG) (Sternberg and Litzenberger 2004).

Data were retrieved from The National Institute of Statistics for 2019—the latest available period, covering 18 economic activities—CAEN Rev. 2 Code. Regarding the spatial concentration, we employed the NUTS 3 classification—including the Romanian 42 counties. Figure 1 illustrates the results considering the cluster potential by county, respectively industry. As the results show, 24 out of 42 counties recorded the highest value in cluster potential for the primary industry—including agriculture, forestry, and fishing and mining and quarrying sectors (Fig. 2).

$$\mathrm{CI}_{ij} = \mathrm{ID}_{ij} \times \mathrm{UB}_{ij} \times \frac{1}{\mathrm{BG}_{ij}}; \mathrm{ID}_{ij} = \frac{\frac{e_{ij}}{\sum_{i=1}^{n} e_{ij}}}{\frac{i_i}{\sum_{i=1}^{n} i_i}}, \mathrm{UB}_{ij} = \frac{b_{ij}}{\sum_{i=1}^{n} b_{ij}}; \mathrm{BG}_{ij} = \frac{a_i}{\sum_{i=1}^{n} a_i}$$

where,

(j) is the sector and.

(i) is the region,

(eij) is the number of employees,

(bij) the number of enterprises,

(ai) is the size of the region and.

(ii) is the number of inhabitants in the region.

The second approach reviews and compares the most significant initiatives implemented to spur intelligent systems in agriculture. Competitive benchmark means to compare how well (or poorly) an organization (in our case—the agricultural clusters) is doing with respect to the leading competition. Consequently, we have selected the European agricultural clusters awarded with Gold Label by the European Cluster Excellence Initiative for evaluation. A total number of 14 clusters were analyzed

Fig. 2 Recommendations to enhance Romania's agricultural clusters competitiveness. *Source* Authors' own elaboration (2023)

originating from Belgium (1), France (4), Denmark (1), Norway (1), Portugal (2), Romania (1), and Spain (4).

In line with our research, Carmen Păuna, mentioned the Romanian cluster potential by domains targeting the manufacturing industry. The author reviewed the indicators used by the European Secretariat for Cluster Analysis by employing the Herfindhal index, the Gini coefficient, and Sternberg & Litzenberger Index (Păuna 2015). Additionally, a more in-depth analysis of the AgroTransilvania cluster was conducted by Manukyan Izabella, who used the Cluster Initiative Performance Model technique to measure how the cluster's members use cooperation and competition to achieve remarkable results (Manukyan 2021).

4 Results and Discussions

The function of clusters in stimulating, creating, and expanding regional competitiveness is axiomatic. Figure 1 illustrates Romania's specialization map using the cluster-index value. It emphasizes Romania's cluster potential in *primary industries* since almost half of the counties recorded the highest score in one of these sectors. The *agriculture, forestry, and fishing industry* cluster have the highest potential to develop in Bihor, Covasna, Harghita, Botoșani, Neamț, and Vaslui, while, as for the *mining and quarrying industry*, Bistrița- Năsăud, Sălaj, Alba, Bacău, Suceava, and Dâmbovița, are the counties with the highest score. Other industries which registered significant values in CI were registered as follows: *the information and communication sector* (Cluj, Iași, Galați, and Timiș), the *real estate sector* (Brașov, Vâlcea, and Sibiu), *hotels and restaurants sector* (Argeș), and *electricity, gas, steam, and air conditioning supply* (Municipiul București).

In 2009, Goschin et al. analyzed the regional specialization and geographic concentration of Romanian industries using the Herfindahl and Krugman Dissimilarity Indexes. The results concluded a higher degree of concentration of agriculture (hunting and forestry, fishery and fish-farming) and construction, among others (Goschin et al. 2009).

The Ministry of Economy confirmed the agro-food potential clusters in the South-West region, Bucharest-Ilfov, South Region, and North-East Region. In the North-West, precisely in Bihor, a potential for a Geothermal cluster was identified (Ministry of Economy 2022). Clipa et al. (2012) acknowledged Romania's potential to develop competitive clusters, mentioning eight regions meeting all requirements for cluster formation (Clipa et al. 2012). Correspondingly, Duduian (2011) reviewed the innovative clusters in Romania, mentioning several clusters in the agro-food industry (Duduian 2011).

Transitioning to the analysis of the top-ranked European agriculture clusters, Table 1 reviews the agro-clusters awarded with the Golden label by the European Cluster Excellence Initiative's assessment.

According to the above data, Asincar is the cluster with over 41 years of experience, while, at the opposite pole, Food & Bio Cluster has one year of activity. Regarding the largest number of members, three clusters from France could be acknowledged: Vitagora—650, Vegepolys—480, Agi Sud—Ouest Innovation—412. When examining the domain sector and the technology field, a trend towards a cross-industry approach was recognized among all clusters. The selected food clusters are involved in each step of the value chain, from the procurement of raw materials (e.g., water treatment) to the delivery of the finite product and the customer experience (e.g., taste, quality, and package). In order to maintain their competitive position, these clusters place technology at the base of their operations, recognizing themselves as *technology clusters*.

Further on, we reviewed the main initiatives related to technology and innovation, implemented up to the moment by the selected clusters (Table 2).

The selected initiatives are knowledge-intensive and driven to transfer the power of technology in all the clusters' activity fields. Some projects include a collaboration between several clusters, while others include global cooperation between different institutions and companies, strengthening the power of cooperation over competition. For instance, the New Frontiers in Food Fast Forward project brings to the same table 5 food clusters, namely Clusaga, Food Valley, Valorial, Vitagora, and Wagralim. Also, it stimulates the international collaboration and business development of 25 European agrifood SMEs in 4 foreign target markets (Brazil, Canada, China, and the USA). In consequence, cluster members operating within a healthy and stimulating environment gain access to a competitive advantage that is almost impossible for sole companies to compete with.

Table 1 Cluster Identification: 14 clusters in the agriculture sector awarded with ECEI GOLD Label

Cluster	Country/Region	Year of origin	No. of members*	Domain sector/ technology field
Agri sud-ouest innovation Porter (2023)	France/ Nouvelle-Aquitaine and Occitanie	2007	Total: 412 SME: 258 Large enterprises:61 R&D: 53	Water, climate, soil; Sustainable production; Farm and factory performance; Biomass valorization and bioeconomy; Better nutrition; Innovative food services and usages
Agro Transilvania Cluster Porter (1990)	Romania/ Transilvania	2013	Total: 78 SME: 48 Large enterprises: n/ a R&D:8	Research, development, innovation and support center in the field of bioeconomy
Asincar Delgado et al. (2015)	Spain/Asturia	1982	Total: 100 SME: > 95% of total Large enterprises: n/ a R&D: n/a	Food preservation treatments; Packaging systems; Bioeconomy and development of new products; Applied biotechnology for food security
The Galician Food Cluster (Clusaga) Sosnovskikh (2017)	Spain/Galicia	2010	Total: 119 SME: 85 Large enterprises: 19 R&D: 6	Food and beverage sector; Advanced packaging
Flanders' Food Nallari and Griffith (2013)	Belgium/Flanders	2005	Total: 310 SME: n/a Large enterprises: n/ a R&D: n/a	Resilient and sustainable agri-food systems; Personalized food and healthy diets; New and shifting resources
Food & Bio Cluster Denmark Gajšek and Kovač (2016)	Denmark/ Central Jutland	2020	Total: 258 SME: 157 Large enterprises: 66 R&D: 15	Food and bioresources; Foods or foodstuffs; their treatment, not covered by other classes

(continued)

Table 1 (continued)

Cluster	Country/Region	Year of origin	No. of members*	Domain sector/ technology field
Foodservice Cluster of Catalonia Chen et al. (2022)	Spain/Catalonia	2014	Total: 106 SME: 84 Large enterprises: 13 R&D: 8	Manufacture of food products; Food and beverage service activities; Foods or foodstuffs; their treatment, not covered by other classes
Catalan cluster of meat and alternative protein (INNOVACC) Martínez-Marín et al. (2020)	Spain/North of Catalonia	2008	Total: 104 SME: 72 Large enterprises: 18 R&D: 14	Meat and alternative protein sector, pig meat sector; Processing and preserving of meat; Water collection, treatment and supply; Butchering; meat treatment; processing poultry or fish; Treatment of water, waste water, sewage, or sludge
INOVCLUSTER Xiong and Qiao (1999)	Portugal/ Castelo Branco	2009	Total: 175 SME: 140 Large enterprises: 1 R&D: 12	Foods or foodstuffs; their treatment, not covered by other classes; Technologies or applications for mitigation or adaptation against climate change
NCE Heidner Biocluster Kolhe and Gupta (2014)	Norway/Eastern N	2012	Total: 54 SME: 30 Large enterprises: 12 R&D:6	Research, innovation and commercialization in breeding, biotechnology, feed development and residual raw materials; Genetics and biotechnology; Sustainable feedstock; Biorefinery

(continued)

Table 1 (continued)

Cluster	Country/Region	Year of origin	No. of members*	Domain sector/ technology field
Pôle valorial Essien (2019)	France/West: Brittany, Pays de la Loire and Normandy	2006	Total: 390 SME: 154 Large enterprises: 106 R&D: 55	Water, climate, soil; operational excellence of farms & plants; food supply for better eating; sustainable plant & animal production; integral valorization of biomass; innovative food uses & services
Portugal foods Marques (2022)	Portugal/ Northern Portugal	2009	Total: 163 SME: 128 Large enterprises: 17 R&D: 12	Innovation and technology transfer; Digitalization of the industry; Identification and training of suppliers; Internationalization, circular economy, and energy efficiency

(continued)

Table 1 (continued)

Cluster	Country/Region	Year of origin	No. of members*	Domain sector/ technology field
Vegepolys Chowdhury and Tawsifur Rahman (2021)	France/ Pays de la Loire, Auvergne Rhône Alpes, Bretagne, Centre-Val de Loire	2005	Total: 480 SME: 380 Large enterprises: 26 R&D: 40	Varietal innovation and seeds and plants performance; Plant health; New technologies and practices for plant production systems; Nutrition health prevention, well-being, cosmetics; Biobased Materials and bioprocessing; Urban farming & Plants in the city
Vitagora Li et al. (2021)	France/ Burgundy-Franche-Comté	2016	Total: 650 SME: 271 Large enterprises: 83 R&D: 112	Foods or foodstuffs; advanced packaging; alternative proteins, functional food, malnutrition, ageing well, traceability and food safety; food waste management, AgTech, input reduction, bio innovation, packaging solutions; delivery, 3d printing, new ingredients & flavors, personalized food

Source Author's own research based on data extracted from the European Cluster Collaboration Platform (2023)
**Note* SME: Small and medium Enterprises, R&D: Research and Development

4.1 Recommendations

According to our analysis, Romania needs a revamped approach to stimulate and increase the agricultural sector's competitiveness. Several agricultural clusters (e.g., Agrifood Covasna Bionest Cluster, BioDanubius Cluster) are actively exploring new opportunities and support to improve their performance. Although the good practices cannot be copied per se from one cluster to another, based on the actual research and conclusions, several recommendations could be formulated to support the Romanian agro-industry.

A cluster success is defined by several factors acting as individual pieces of an interconnected puzzle. Vertical and horizontal links between the actors of the clusters

Table 2 Intelligence systems implemented by the ECEI GOLD Label agricultural clusters

Cluster	Intelligence systems initiatives
Agri Sud-Ouest innovation	• Diva Project: related to digital marketplaces, big data, cloud, robotics & artificial intelligence, cyber-physical systems, IoT, and digital/ electronic /photonic components • Agro y Aigua: bring together the agrifood and water management sectors, or "AgroTIC Hub," to support the digital transformation of agrifood companies • Food Log Proximity: identify and support companies with digital solutions for short food supply chains An active member in European innovation hubs such as s Plant InterCluster and S3 platforms (Smart Sensors for Agrifood; Traceability and Big Data)
Agro transilvania cluster	• PANACEA: facilitate an exchange between research, industry, and the farming community so that precisely practical solutions can be disseminated • Food in Eco Network-internationalization and global competitiveness of European SMSs in the Food and Eco Logistics Sector: develop a value chain model that ensures cooperation between SMEs in the food group based on the transfer of knowledge and their exploitation, with an emphasis on their practical collection and use • Track: create a favorable inter-regional 4.0 ecosystem to stimulate the matching of new innovative solutions with the specific requirements of the agrifood chain • Systemic Innovations Towards a Zero Food Waste Supply Chain: create a solution to the challenges raised by food waste Other projects: Livada din Crisana, 8 X S3; Operational Programme Competitiveness 2014–2020
Asincar	• C2C collaboration: Food digitalization (Connsensys, S3FOOD, SS4AF platform), Bioeconomy (Nutritional Ingredients platform, Vanguard Initiative), Internationalisation (Interreg projects) • EU collaborative RandD projects: APRIL (H2020, robotics), Glomicave (Omics Big Data), TOXDTECT (coordinators, FP7, intelligent packaging) • Food Pilot Plant: reproduces the areas, equipment, and flows in a food SME; the ideal place for validation and scaling up of food technology and digital solutions, and for vocational training, dissemination, awareness, and promotional activities Tech Transfer projects: yearly, around 50 active R&D projects

(continued)

Table 2 (continued)

Cluster	Intelligence systems initiatives
The Galician Food Cluster (Clusaga)	• S3FOOD project: enable the digital transition of SMEs food producers by equipping these companies to meet growing market demands for quality control, efficiency, higher standards of food safety and traceability through visits to technology centers, ideation, training and matchmaking sessions and the funding scheme • Innovation support: foster participation of cluster members in innovation and collaborative projects, primarily related to the development of innovative food products, new production processes and technologies, environmental sustainability and circular economy • HIS-FOOD: test the use of hyperspectral photonic technology for the determination of quality parameters in raw materials and final products in different food matrices • Smart Labels: utilize new printing technologies for the manufacture of intelligent tags CONNSENSYS: connect smart sensor systems for the food industry
Flanders' food	• S3FOOD: see the Galician Food Cluster description • Resilient and Sustainable Agrifood Systems: groups together all the research and innovation activities from the spearhead cluster aiming at the transformation process into an integrated and connected food system The technology used: Micro/nanoelectronics—Sensory systems—Photonics and imaging technologies—Robotics—IoT—AI—Interaction technologies—Big data, data analytics, data handling—Virtual, augmented and extended realit—Simulation, modelling and digital twins—Cloud computing—Additive manufacturing—Laser based manufacturing—New media technologies
Food & bio cluster Denmark	• BioSwitch: draws Brand Owners to Switch to Bio-Based in highly innovative ecosystems • ProEnrich: develop new functional proteins and bioactive ingredients from rape seed, olives, tomatoes, and side streams from the citrus industry for use in foods, cosmetics, feed, and as fixatives; • ValueWaste: unlock new value from urban biowaste • S3Food Smart Sensor Systems: see the Galician Food Cluster • Incubation: hosts start-ups and established companies with activities related to the agrifood and bioeconomy value chain The cluster coordinates two major national innovation networks: The Danish Innovation Network for Bioresources, and the Danish Food Innovation and it is involved in over 40 innovation projects

(continued)

Table 2 (continued)

Cluster	Intelligence systems initiatives
Foodservice cluster of catalonia	• SAFE SMART FOOD: create the conditions for a favorable ecosystem in the food and packaging sector to boost innovation, competitiveness and the recognition of intelligent regional specialisation strategies • PaaSiot: use disruptive technologies to change the food distribution to a more sustainable one • VISIONA: measure user's emotional experience in catering establishments through artificial vision with artificial intelligence • The Road to Reuse project promotes sustainability in restaurants with the implementation of returnable packaging through IoT LEONARDO: IoT design and development for remote supervision, management, and maintenance of professional coffee machines
Catalan cluster of meat and alternative protein	• MBOT-ITS: use of advanced technology and big data management to optimize dryers of cured sausages • SALADESP 4.0: implementation of technology for pork cutting room • LONCHA0DEFECT: an artificial concept for the control of defects in slices of sausages TH CARN 4.0: analyze and optimize meat industry processes' thermal and hygrometric conditions
INOVCLUSTER	• Unit + RD + Innovation: supports product development with new ingredients, processes, ways of packaging and innovative image • INOV2B: provide training for the generation of innovative ideas/products, for the implementation of new business initiatives, and for the creation of new companies • AddInov 4.0: implements new organizational methods and management practices, strengthens the presence in the digital market and the use of ICT, implements new management systems Farm to Fork: provide support services to innovative European SMEs–specialized in healthy food, techno-functional food and bio-based solutions, and sustainable bio-based solutions
NCE Heidner Biocluster	• BioTown: provides a self-developed and unique technology for increased and sustainable food production, covering the entire value chain from genetics to production of the raw materials at the farmer and to processing through the food manufacturers AgriFoodTech: aims to create a new Norwegian export industry for utilizing biological resources

(continued)

Table 2 (continued)

Cluster	Intelligence systems initiatives
Pôle Valorial	• REAMIT: improve the resource efficiency of agribusiness supply chains by minimizing waste using Big Data and IoT sensors • AHFES: improve competitiveness and growth of SMEs in the value chain of healthy eating and lifestyle through innovation, access to information and partners Other projects: AURORA F2F HEALTH MATTERS. FOOD SCALE UP, BIORURAL, B-RESILIENT, New Frontiers in Food
Portugal foods	• cLabel + : respond to the challenges that the food industry is facing through research and technological development and focuses on the concept of "clean label" • MobFood: respond to the challenges related to the promotion of a more competitive national food industry through research and technological development approach • NUTRIAGE: offers advanced solutions for healthy aging through nutrition within the Euroregion Galicia-North of Portugal framework Other projects: CHIC Project; SEAFOOD-Age; iWATERMAP; Agri Renaissance
Vegepolys	• RuralBioUp: strengthen the cooperation among regional key actors and knowledge holders, and empower them to establish an inclusive and long-lasting ecosystem to support the mainstreaming of bio-based business models in rural areas • SUAVE: boosts urban agriculture towards sustainable and resilient urban food systems • Agro BRIDGES: equip farmers with practical knowledge and tools to set up new commercial and marketing models based on short food supply chains • TRACK: boost the digitalization of agrifood value chains and stimulate new innovative solutions that meet the specific needs of the agrifood chain but also generate joint investments Others: PATH2DEA; HuMUS
Vitagora	• ToasterLAB: support innovative agrifood start-ups and scale-ups to allow great ideas to become real-world business success stories. ToasterLAB has accelerated more than 60 companies

Source Author's own research based on data extracted from the European Cluster Collaboration Platform (2023)

and participants in the supply chain are crucial. Creating an intelligent system to ensure a just-in-time flow of information between organizations and enterprises, parts of the agricultural cluster, and a permanent unification of regional infrastructure organizations in a cooperation network will guarantee collaboration between cluster members.

Also, a commitment towards innovation and permanent improvement is mandatory—each of the selected clusters has developed several tech projects covering all segments of the supply chain. Consequently, we recognize the need to commit to the diffusion of knowledge, innovation, technology, local externalities, and access to markets and information. The ultimate scope is to generate higher value-added operations along the value chain. The knowledge of the labor force—farmers most prominently—may vary from the benefits of using organic fertilizers (soil management) to the implementation of safer procedures (the production operations).

Subsidies, allocating targeted investments and grants on a competitive basis, tax benefits, the ease of administrative barriers, and the encouragement of developing quasi-public goods are also prone to influence the cluster's competitiveness.

5 Conclusion

In 2023, companies are more interconnected than ever due to the ease in the flow of goods and services and the impact of intelligent systems influencing every business sector activity. The main reason that motivates business cooperation is to improve profits, develop knowledge, and maximize efficiency—all of these and many other benefits can be achieved through the organization of clusters. The role of technology and intelligent systems for each step of the value chain is axiomatic for cluster competitiveness.

On the one hand, this paper provides a comprehensive overview of the Romanian cluster potential. We found that more than half of the Romanian counties have the highest potential to develop clusters in the primary industries. The counties that registered the most significant Cluster Index on the agriculture sector are Bihor, Covasna, Harghita, Botoșani, Neamț, and Vaslui.

On the other hand, going more into depth and looking over the factors that stimulate the cluster's competitiveness, we found that the top 14 European most competitive agricultural clusters identify as technology clusters. The actions taken by these clusters—a robust investment in knowledge and innovation, involvement in several technological projects, and implementation of intelligent systems—supported them in achieving efficiency and becoming world-class agricultural clusters.

Several research limitations narrow our paper's findings. For the first methodological approach—Cluster Index—we have used the latest data available to the research data issued by The National Institute of Statistics—2019. Also, we must admit that the best practices used by the selected clusters represent a personal view and a selection of information based on our comprehensive research. Interviews with each cluster representative would have offered our research a more documented perspective.

The lessons learned from the analyzed clusters show that there is no other door to efficiency and competitiveness than through technology. Consequently, we recommend that Romanian agricultural clusters concentrate on creating and implementing intelligent systems along the value chain, facilitating a multilateral collaboration at the national and international levels, and investing significantly in knowledge, technology, and innovation. Furthermore, the current paper may serve as a starting point for regional policymakers, cluster directors, and governments to define new strategies to support Romania's agricultural industry and clusters.

References

Abioye, E.A.: Precision irrigation management using machine learning and digital farming solutions. AgriEngineering **4**, 70–103 (2022)

Allal-Chérif, O., Simón-Moya, V., Cuenca Ballester, A.C.: Intelligent purchasing: How artificial intelligence can redefine the purchasing function. J. Bus. Res., **124**, 69–76 (2021)

Barenji, A.V., Wang, W., Li, Z., Guerra-Zubiaga, D.A.: Intelligent E-commerce logistics platform using hybrid agent based approach. Transp. Res. Part E: Logist. Trans-Portation Rev. **126**, 15–31 (2019)

Chen, M.-K., Wu, S.-W., Huang, Y.-P., Chang, F.-J.: The key success factors for the operation of SME cluster. Sustainability, **14**, (2022)

Chen, Q., Hu, X.: Design of intelligent control system for agricultural greenhouses based on adaptive improved genetic algorithm for multi-energy supply system. Energy Rep. **8**, 12126–12138 (2022)

Chowdhury, M. E., Tawsifur Rahman, A.K.-E.: Automatic and reliable leaf disease detection using deep learning techniques. AgriEngineering, **3**(2), 294–312 (2021)

Clipa, R.I., Pohoata, I., Filipeanu, D., Tudose, M.B.: Clusters and regional competitiveness in Romania. In: Annals of DAAAM for 2012 & proceedings of the 23rd international daaam symposium. 23. Vienna (Austria) (2012)

Delgado, M., Porter, M. E., Stern, S.: Defining clusters of related industries. J. Econ. Geogr., 1–38, (2015)

Duduian, M.: Innovative clusters: the case of Romania. Manag. Res. Pract. **3**(3), 1–11 (2011)

Essien, J.: Application of intelligent systems in rural agricultural practice. Int. J. Adv. Res. Publ., **3**(4) (2019)

European Cluster Collaboration Platform. https://reporting.clustercollaboration.eu/region/RO11. Last accessed 20 Jan 2023

Gajšek, B., Kovač, J.: Key factors for the successful operation of clusters: the case for Slovenia. Organizacija, **49**, (2016)

Goschin, Z., Constantin, D., Roman, M., Ileanu, B.: Regional specialization and geographic concentration of industries in Romania. South-East. Eur. J. Econ., 99–113 (2009)

Kolhe, S., Gupta, G.: Intelligence systems for agriculture domain. National conference on emerging trends in computer technology. Int. J. Comput. Appl., (2014)

Li, F., Su, Z., Wang, G.-M.: An effective integrated control with intelligent optimization for wastewater treatment process. J. Ind. Inf. Integr., (2021)

Lykhovyd, P.: Comparing reference evapotranspiration calculated in eto calculator (Ukraine) mobile app with the estimated by standard FAO-based approach. AgriEngineering **4**(3), 747–757 (2022)

Manukyan, I.: Formation and management of regional agri-food clusters in developing countries: case of "agrotransilvania" (Romania). Manag. Sustain. Dev. **13**(2), 34–40 (2021)

Marques, L.S.: Agricultural machinery telemetry: a bibliometric analysis. AgriEngineering **4**, 939–950 (2022)

Martínez-Marín, S., Puello-Pereira, N., Ovallos-Gazabon, D.: Cluster competitiveness modeling: An approach with systems dynamics. Soc. Sci., MDPI **9**(2), (2020)

Ministry of Economy. Tade and Business Environment. https://unece.org/fileadmin/DAM/ceci/ppt_presentations/2011/TOS_ICP4/Leucuta.pdf. Last accessed 09 Dec 2022

Nallari, R., Griffith, B.: Clusters of competitiveness. International Bank for Reconstruction and Development/ The World Bank, Washington DC, Washington DC (2013)

Păuna, C.B.: Cross-sectoral cooperation vs. cluster development at European level. In: 2nd international conference 'economic scientific research - theoretical, empirical and practical approaches. Elsevier B.V., Bucharest, Romania, pp. 175–183 (2015)

Porter, M.: The competitive advantage of nations. Harv. Bus. Rev. **68**, 73–93 (1990)

Porter, E.M.: Clusters and the new economics of competition, https://hbr.org/1998/11/clusters-and-the-new-economics-of-competition. Last accessed 11 Jan 2023

Sosnovskikh, S.: Industrial clusters in Russia: The development of special economic zones and industrial parks. Russ. J. Econ. **3**(2), 99–174 (2017)

Sternberg, R., Litzenberger, T.: Regional clusters in germany—their geography and their relevance for entrepreneurial activities. Eur. Plan. Stud., **12**(6), (2004)

Xiong, F., Qiao, K.: Intelligent systems and its application in agriculture. IFAC Proceedings Volumes **32**(2), 5597–5602 (1999)

Longitudinal Study of Research Interest on Sustainability Reporting: A Bibliometric Analysis

Miruna-Iuliana Cunea and Camelia Iuliana Lungu

Abstract This study presents a bibliometric analysis and a debate on the ongoing interest in European directives and international initiatives on corporate sustainability reporting. The objective of the study is to observe the interest of researchers in various topics related to sustainability reporting requirements over the last five years. The data used in this research was selected based on the PRISMA method to Web of Science (WOS) database. The database generated a total number of 572 articles. A plethora of features characterizing the research on sustainability reporting over the last five years reveal interesting results. The most researched topic in sustainability reporting research is Global Reporting Initiative, followed by non-financial reporting directives and integrated reporting framework. A VOS Viewer representation of the most used keywords related to sustainability reporting have been performed for the sampled data. The overall conclusion is that GRI is the favorite of companies that want to present their sustainability performance and the favorite topic approached by researchers. These results point out that authors affiliated to the Bucharest University of Economic Studies made a significant contribution especially to the GRI research. The main contribution of this study resides in the two-fold analysis of sustainability reporting, mirroring the continuous development of guidelines, initiatives, standards, and regulations at international and European levels in academics' research addressed from companies' perspectives and practices.

Keywords Sustainability reporting · Bibliometric analysis · Sustainability standards · Sustainability guidelines · Sustainability initiatives · Non-financial reporting directives

M.-I. Cunea (✉) · C. I. Lungu
Bucharest University of Economic Studies, Bucharest, Romania
e-mail: cuneamiruna16@stud.ase.ro

C. I. Lungu
e-mail: camelia.lungu@cig.ase.ro

M. Busu (ed.), *Rethinking Business for Sustainable Leadership in a VUCA World*, Springer Proceedings in Business and Economics,
https://doi.org/10.1007/978-3-031-50208-8_6

1 Introduction

The history of sustainability reporting began once the Brundtland Report was published in 1987. The classic approach sustained in the Report relating to "*meeting the needs of the present without compromising the ability of future generations to meet their own needs*" is seen to be increasingly relevant in corporate life (Moneva et al. 2006, p. 122). It did not take long until national and international institutions and organizations mobilized to create codes and standards about sustainability reporting and corporate governance. The first initiative to have a set of rules on extended corporate reporting was in the 1990s through the South African King I Code. Twelve years later, King II focused on introducing the concept of integrated reporting to disclose financial and non-financial information. In 2009, King III was redefined and represented a guide for companies that contains principles about sustainability. The code mentioned the obligation of companies to provide reports to stakeholders using committees to prepare integrated reports (Rahi et al. 2022; Adams 2017).

The regulations resulting from these codes comprised information adapted to the needs of the countries in which they operate. Adams (2017) demonstrated this need by addressing the differences in the annual reports published by companies listed on the stock exchanges from two countries on different continents, adopting opposite reporting guidelines. On the one hand, companies listed on the Johannesburg Stock Exchange used the King III Code as a reporting method, which contains the core sustainability principles; therefore, these companies were required to prepare integrated reports. On the other hand, companies listed on the Australian Stock Exchange were discouraged from providing information on sustainability. Debating topics such as value creation, integrated thinking, corporate social reporting, ESG risk, and the obligation to report information about their environmental impact, Adams (2017) highlights the significant role of sustainability reporting in assisting the complex contemporary decision-making process.

In recent years, sustainability reporting, called voluntary at the time, corporate social responsibility, and integrated reporting have also been the subject of interest at the institutional, business, and academic levels. The increasing consideration is enhanced by companies' benefits from choosing among existing sustainability reporting standards or developing their own frameworks (Christensen et al. 2019; Albu et al. 2013).

This research represents an investigation of the current seminal scientific literature that addresses various aspects of sustainability reporting over the last five years. It focuses on analyzing the academics' interest in international initiatives and standards, as well as in European directives related to sustainability reporting. The overall scope of the study is to observe the interest of researchers in issues related to various sustainability reporting requirements. Moreover, an in-depth analysis aims at identifying the most discussed sustainability related aspects. Two research questions are formulated: *What is the current interest in scientific research of the sustainability related guidelines, standards, and initiatives?* (RQ1) and *What is the most popular topic currently addressed in sustainability reporting research?* (RQ2).

The first part presents the evolution of the current sustainability reporting guidelines. The second part is a bibliometric analysis conducted on the most recent scientific research on sustainability reporting, published after the year 2019. It consists of a performance analysis based on publications and citation-related metrics (as in Donthu et al. 2021) complemented by a content analysis (as in Lungu et al. 2016) aimed at identifying the most frequent subject of sustainability reporting research.

The main contribution of this study resides in the two-fold analysis of sustainability reporting: the continuous development of guidelines, initiatives, standards, and regulations at the international and European levels, mirrored in academic research of company perspectives and practices. It also contributes to filling the literature gap on how to develop and apply sustainable reporting standards and guidelines by entities over the next few years.

The remainder of the paper is structured as follows. The literature review presents the evolution of the current state of sustainability reporting requirements. The research methodology describes the database and the research metrics. Next, the results of bibliometric analysis are presented and discussion on the implications of the most researched topic on sustainability reporting. In the end, conclusions, limitations of the study and proposals for future research are presented.

2 Literature Review

Corporate sustainability actions began to be more visible when economic, financial and health crises combined, reaching global dimensions. Over time, sustainability reporting management changed from initiatives (Global Reporting Initiative - GRI) to regulatory bodies (International Integrated Reporting Council - IIRC), international councils and boards (World Business Council for Sustainable Development – WBCSD and International Sustainability Standards Board - ISSB) and back to regulatory bodies (European Financial Reporting Advisory Group - EFRAG, Task Force on Climate-related Financial Disclosures - TCFD or US Securities and Exchange Commission - SEC).

All these actors had to be participants to sustainability actions, generating well-established guidelines, standards, and directives for corporate reporting as a solution to the need for comparability. However, these proposals are not aligned and could create challenges for entities that design consistent annual reports (KPMG 2022), raising concerns about framing diversity and promoting convergence in the field of corporate sustainability reporting.

2.1 GRI and IIRC, Promoters of Sustainability Initiatives

GRI was founded in 1997 and was considered the main guide for voluntary reporting for a significant period and the most important standard in reporting sustainability information (Halkos and Nomikos 2021). In 2009, on the initiative of the Prince of Great Britain, through the Prince Accounting Project, the GRI teamed up with the IIRC with the aim of creating a global set of accepted standards (Flower 2015; Halkos and Nomikos 2021). Although both the GRI and the IIRC are globally recognized as standards in sustainability, both have different directions in addressing sustainability issues, and each of them has been discussed as its applicability according to the needs of the business environment.

With more than 10,000 users from 100 countries (GRI 2022) GRI is still seen as a relevant institution for sustainability reporting standards. It ensures increased transparency of information provided by companies but also sets boundaries within which sustainability can unfold and restrict (Journeault et al. 2021).

Established in 2010 the IIRC was set on the idea that sustainability, governance, and remuneration reports, together with traditional financial statements and management vision, should be properly integrated to contribute to the creation of value in the short, medium, and long term. Flower (2015) notices that, during the last years, the IIRC has started to avoid using the term sustainability. Comparing GRI and IIRC principles, Flower (2015) reveals that words with similar meanings (such as conditions or capital) have been changed in IR framework when referring to the most important topics involving sustainability. Allegedly, the IIRC removed the term sustainability from its reporting framework because it wanted to avoid the burden of collecting data on sustainable activities that could reveal negative aspects.

2.2 European Directives Versus International Standards on Sustainability Reporting

In 2014, the European Commission implemented Directive 95/2014 (Non-Financial Reporting Directive) in all member states. The directive retracts the provisions of the European Directive 93/2013 regarding the presentation of non-financial and diversity information by certain large companies and groups. Currently, European companies with more than 500 employees are required to include non-financial information in their annual reports. However, the directive was seen as less helpful for many companies (Voss 2019); therefore, entities in extractive industries that produce carbon emissions will not be able to report information on sustainability issues under the directive. However, in 2020, the European Commission appointed EFRAG to undertake the preparatory work for a new directive on corporate sustainability reporting (CSRD-Corporate Sustainability Reporting Directive).

Established in 2001, EFRAG is a group that focuses on serving the public interest. At the end of 2021, its mission extended to offering technical advice to the European Union on sustainable development, by establishing a map of developing CSRD and preparing drafts of EU Sustainability Reporting Standards using transparency and stakeholder expertise (EFRAG 2021).

Entering into force in 2023, CSRD (EU Directive 2022/2464) is an updated directive to NFRD that cancels the previous provisions of reporting requirements by certain companies. The current directive applies to all large companies and listed companies (except for micro-entities). Those companies have the obligation to present beneficial information about their impact on people's activities. The sustainability reports prepared will be audited and presented digitally.

The new CSRD (EU Directive 2022/2464) regarding the reporting of companies in the field of sustainability highlights three main aspects: whether it is necessary to standardize non-financial reporting, the need to certify the reported non-financial information, and the scope of reporting and characteristics of the companies' subject to reporting requirements. This initiative aims to make reported information relevant, comparable, safe, and easily accessible to investors (European Commission 2021).

At international level, SASB (Sustainability Accounting Standards Board), an NGO established in 2011, aims to help businesses and investors with a common language about sustainability and financial impacts. Nine years later, IIRC merged with SASB and created the Value Reporting Foundation. At that time, IFRS announced the intention of consolidating the Value Reporting Foundation (VFR) with the Carbon Disclosure Standards Board (CDSB) to offer high-quality sustainability reporting standards. IFRS standards already require companies to provide information on their ESG risks and to make judgements on the long-term implications. The IFRS Foundation recognizes the investor's need to have financial and non-financial information interconnected.

Therefore, after the consolidation of VFR with CDSB in 2022, the IFRS Foundation established the ISSB (International Sustainability Standards Board) to create a global baseline on sustainability reporting, with an initial focus on climate disclosure requirements. Those standards are built after SASB standards which are industry-based, market-informed, and cost-effective and have been built with models from IASB (SASB 2022).

According to KPMG (2022), starting with 2022, the standards under ISSB would focus on corporate values and not only on climate-related regulations. Companies within the scope of these requirements would be companies based on the listing status. Countries such as Nigeria, Japan, and the UK have indicated that these proposals would be a key part of future requirements (KPMG 2022).

3 Methodology

The objective of this study is to analyze the academics' interest in international initiatives and standards as well as in European directives related to sustainability reporting. An in-depth analysis based on two research questions addresses the current interest of the researchers and the most popular topics on sustainability reporting. The first research question (RQ1) *What is the current interest in scientific research of the sustainability related guidelines, standards, and initiatives?* is conducted using a bibliometric analysis. A bibliometric analysis is a quantitative method of analyzing large volumes of data using a mixture of performance techniques, science mapping and enrichment techniques (Donthu et al. 2021). As in the archival study by Lungu et al. (2016) on CSR research, this study is conducted using bibliometric indicators on publications-related and citation-related metrics.

The data sampled used in the analysis of this study was selected using the PRISMA method for the Web of Science (WOS) database. The PRISMA method implies the identification, screening, eligibility, and inclusion of data (Echchakoui 2020). At the identification stage, articles published in the last five years (2019–2023), addressing the main international initiatives (GRI, IIRC and ISSB) and European Directives (NFRD, CSRD, and EFRAG) on sustainability reporting were selected. At the screening and eligibility stage, the results previously obtained were classified according to the Citation Meso Topics, limiting the database to the articles of economic or management interest, with a total of 687 articles on sustainability reporting initiatives and directives. Finally, in the inclusion step, the database was double-checked to include only articles that focused on sustainability reporting. The database was exported as Excel files, and 47 duplicates have been removed. Although according to WOS, the articles were placed in the categories of economics and management, it didn't generate the exact data needed for the investigation. Therefore, to determine the reliability of the database, the title, abstract, and keywords for each article were carefully read, and only the articles mentioning aspects related to sustainability reporting were kept. For example, of 63 articles that referred to CSRD, only four remained in the database because most articles referred to CSRD as corporate social responsibility disclosure, without indicating the Corporate Sustainability Reporting Directive. Similar content analysis was performed for the other groups of collected data and the final database was limited to a total number of 572 articles.

The second research question (RQ2) *What is the most popular topic currently addressed in sustainability reporting research?* is conducted with the same database used in RQ1. Additionally, the articles were exported as.csv files and analyzed with VOS Viewer to identify the most researched topic in sustainability reporting. The generated graphs illustrate the occurrence of keywords in the observed papers. To have a better visualization of the keywords, the minimum number of occurrences of a keyword was established at 25. From a total of 1,370 keywords, only ten met the threshold, being presented separately in a network and a density visualization analysis.

4 Results and Discussion

4.1 Results of the Bibliometric Analysis

To answer RQ1: *What is the current interest in scientific research of the sustainability related guidelines, standards, and initiatives?*, the publications included in the database were analyzed based on information about: publishing houses, authors' affiliations, type of publications, number of pages, and paper's citations.

The first approach was to identify the publishers that were more oriented towards research on sustainability reporting topics. In Table 1, the results were narrowed down, presenting the top five international publishers of articles related to sustainability reporting.

The highest number of papers have been published by Emerald Group—a global research group providing consultancy for companies. The Emerald Group published 155 articles on sustainability reporting over the last five years, 36 of them addressing topics on European directives. The publisher with the second most interest on sustainability reporting was MDPI, followed by the global leaders in publishing academic papers on any scientific-related subjects: Wiley, Elsevier SCI, and Routledge Journals Taylor & Francis.

The second approach refers to the details on the affiliation of the authors involved in publishing articles orientated to sustainability reporting. The results presented in Table 2 highlight the Bucharest University of Economic Studies with its significant contribution in GRI research. Moreover, NFRD and IIRC have also been subjects of interest for this university. The analysis based on the types of initiatives reveal a difference in the focus of researchers from various universities. Thus, whether the researchers from Bucharest University of Economic Studies were more interested

Table 1 Top five publishers interested in sustainability reporting

Publisher	International initiatives			European directives			Total articles
	ISSB	IIRC	GRI	NFRD	EFRAG	CSRD	
Emerald Publishing Ltd	1	33	85	33	1	2	155
MDPI	0	9	50	20	1	1	81
fWiley	0	7	52	18	0	0	77
Elsevier Sci Ltd	0	3	27	1	0	0	31
Routledge Journals, Taylor & Francis Ltd	0	5	15	8	1	0	29
Other Publishers	0	24	147	26	1	1	199
Total Published Articles	1	81	376	106	4	4	572

Source Authors' own research

in GRI and IIRC initiatives, the researchers from Babes Bolyai University focused more on non-financial reporting directive.

The third approach was based on the analysis of indicators related to the number of authors. In Table 3 it is represented the average, median, minimum, maximum number of authors that have published articles related to this research.

European directives (CSRD, NFRD and EFRAG) were studied by 316 authors in the last five years, and they published a total number of 116 articles. Authors interested in international initiatives on sustainability reporting published a total of 458 articles, most of which referred to GRI, with 1,114 authors of 376 papers. A number of 239 authors (in 81 papers) conducted research on IIRC, and only one on ISSB. The median values show that papers included each category are generally

Table 2 Affiliates—top five universities interested in sustainability reporting

Affiliation	International initiatives			European directives (EUD)			Total articles
	ISSB	IIRC	GRI	NFRD	EFRAG	CSRD	
Bucharest University of Economic Studies, Romania	0	2	5	2	0	0	9
Babes Bolyai University from Cluj-Napoca, Romania	0	1	2	5	0	0	8
University of South Australia	0	2	2	0	0	0	4
Universita degli Studi di Bari Aldo Moro, Italy	0	1	1	1	0	1	4
University of Indonesia	0	1	3	0	0	0	4
Affiliations with less than 4 papers each	1	74	363	98	4	3	543
Total	1	81	376	106	4	4	572

Source Authors' own research

Table 3 Characteristics of the number of authors

Descriptive statistics	EUD	GRI	IIRC	ISSB
Number of authors per article				
Mean	2.77	2.96	2.95	1
Median	3.00	3.00	3.00	1
Minimum	1.00	1.00	1.00	1
Maximum	5	8	6	1
Total number of authors	316	1,114	239	1
Total number of articles	114	376	81	1

Source Authors' own research

conducted by at least 3 authors per article, while small differences are noticed in term of means. The highest number of authors was reported for papers on GRI topics, with 8 authors, followed by topics on IIRC (6 authors), and topics on EUD (with 5 authors).

In terms of publication language, English is the most used language required by publishers, counting 521 articles in total, followed by Portuguese, Spanish, Russian, Malay, Chinese, German, and Turkish. Based on the types of publications, language diversity characterizes the papers published in journals, especially on GRI topics. Books and Conference proceedings include only papers in English. Detailed results are presented in Table 4.

The next analysis was conducted based on the number of articles published per year. According to Fig. 1 and Appendix 1, 2021 was the year when authors focused mostly on GRI topics of sustainability reporting, with 112 articles published, followed by European directives topic (with 24 papers on NDFR and 1 on CSRD), and IIRC (with 21 papers), as presented in Appendix 1. In terms of most addressed topics per year, the year 2022 is of reference for the research on IIRC and NRFD, at the expense of research on GRI topics.

The last analysis conducted is based on the most researched articles related to sustainability reporting. As shown in Table 5, the most cited article on European directives has 84 citations in all databases including Web of Science. This is a journal-type article published in 2021 that analyzes the impact of EU Directive 2014/95/EU on business reporting practices with reference to the determinants of Sustainable Development Goals and the related Scores. The article is an in-depth analysis of the effects of the Directive at companies' levels, using Italian public interest entities that apply non-financial reporting directive.

Table 4 Frequency of papers based on publication language

	EUD	GRI	IIRC	ISSB	Total
Conference proceedings	**7**	**17**	**3**	**0**	**27**
English	7	17	3	0	27
Journals	**107**	**358**	**77**	**1**	**543**
Chinese	0	1	0	0	1
English	104	345	71	1	521
German	1	0	0	0	1
Malay	0	1	1	0	2
Portuguese	0	5	4	0	9
Russian	1	2	0	0	3
Spanish	1	3	1	0	5
Turkish	0	1	0	0	1
Series/Books	**0**	**1**	**1**	**0**	**2**
English	0	1	1	0	2
Total	114	376	81	1	572

Source Authors' own research

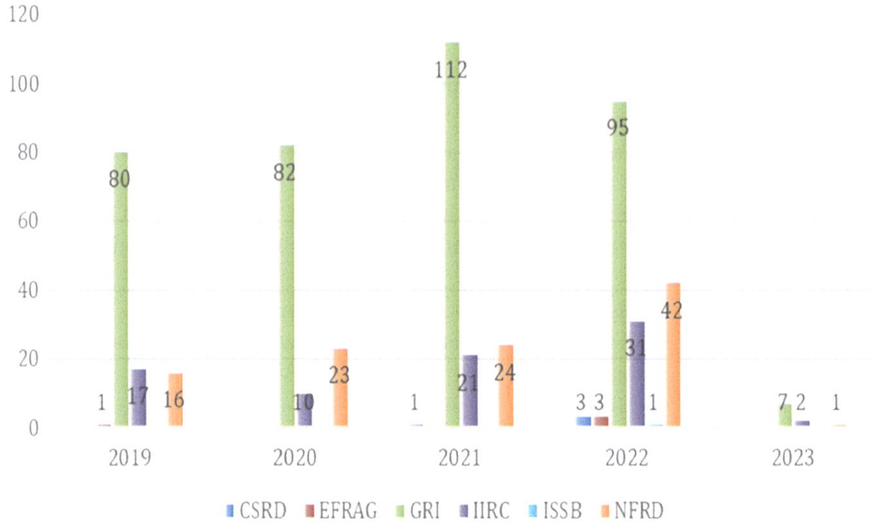

Fig. 1 Distribution of articles published by year and by topic. *Source* Authors' own research

Table 5 Most cited articles by topic

Topic	European directives	International initiatives
Article characteristics		
Article title	The determinants of business contribution to the 2030 Agenda: Introducing the SDG Reporting Score	New challenges for corporate sustainability reporting: United Nations' 2030 Agenda for sustainable development and the sustainable development goals
Author(s)	Pizzi, S; Rosati, F; Venturelli, A	Tsalis, TA; Malamateniou, KE; Koulouriotis, D; Nikolaou, IE
Pub. Year	2021	2020
Source (Journal) Title	Business strategy and the environment	Corporate social responsibility and environmental management
Publisher	WILEY	WILEY
Affiliations	University of Salento; Technical University of Denmark	Democritus University of Thrace; Democritus University of Thrace
Times cited in all databases	84	125
No. of pages	18	13
No. of references	99	126

Source Authors' own research

The most cited article on international initiatives of sustainability reporting has 125 times cited in all databases (Table 5). The research organized in only 13 pages aims at the development of a framework that evaluates the alignment of corporate sustainability practices oriented towards the SDG scope, extracting the topics from GRI requirements.

4.2 Results of Keywords Analysis of Sustainability Reporting Research

For the second research question (RQ2: *What is the most popular topic currently addressed in sustainability reporting research*?) the results are generated with VOS Viewer and show that GRI is the most researched topic, besides the *sustainability reporting* construct itself, based on keywords occurrences and the network overlay analysis presented in Figs. 2 and 3.

According to Donthu et al. (2021, p. 292), "*each node in a network represents an entity (…) where the size of the node indicates the occurrence of the keyword (…), the bigger the node, the greater the occurrence of the keyword, and the thicker the link between the nodes, the greater the occurrence of the cooccurrence (…)*". The results show that most of the papers have mentioned or included information related to GRI, non-financial reporting, integrated reporting, corporate social responsibility, and sustainability. A density visualization represented in Fig. 3 confirms the interest in sustainability reporting through those topics.

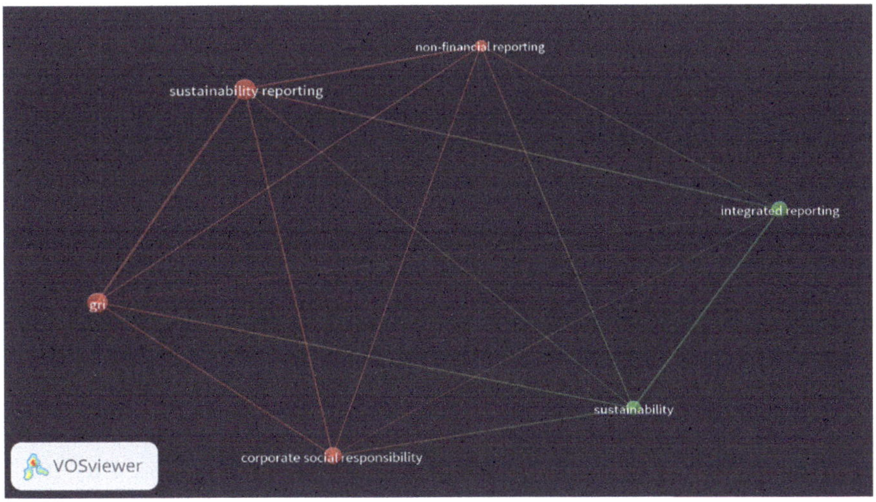

Fig. 2 Network visualization of keyword occurrence. *Source* Authors' own research based on VOS Viewer processing

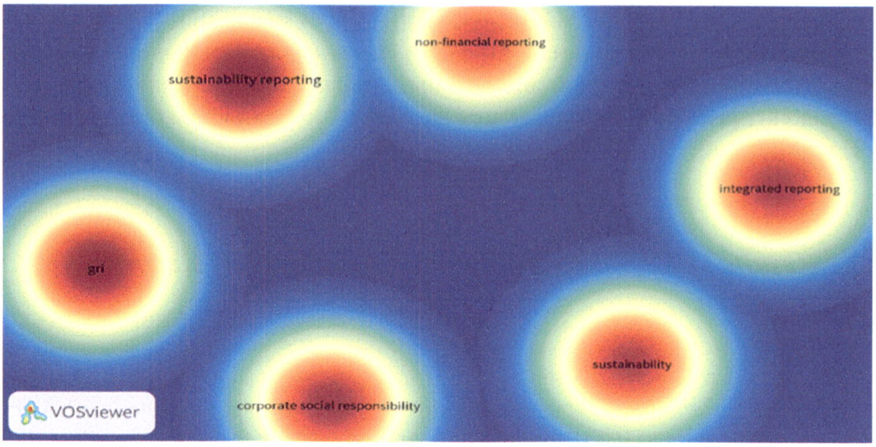

Fig. 3 Density visualization of keyword occurrence. *Source* Authors' own research based on VOS Viewer processing

Table 6 Keywords occurrences

Keywords	Total occurrences
Sustainability reporting	89
GRI	83
Sustainability	64
Non-financial reporting	51
Corporate social responsibility	72
Integrated reporting	67

Source Authors' own research

GRI has occurred 83 times in the articles included in this study, as presented in Table 6, and may be considered the most interesting topic for sustainability reporting research.

5 Conclusion

This study presents a bibliometric analysis and a debate on the ongoing interest in European directives and international initiatives related to corporate sustainability reporting. Research was conducted using database information extracted from Web of Science. The information collected was related to the following topics: GRI, IIRC, and ISSB, as international initiatives, and EFRAG, NFRD, and CSRD, as European directives perspective.

Bibliometric analysis has confirmed that there is a current interest in the European directives and international initiatives on sustainability reporting, more precisely, with an increased focus on GRI.

The first approach was to identify publishers specializing in articles on various sustainability reports. The results were condensed, presenting the top five international publishers of papers on sustainability reporting. The Emerald Group published 155 articles on sustainable reporting topics. However, in the past five years, this group has published most of its articles on European directives. In addition to the Emerald Group, MDPI published the second highest number of articles, focusing instead on IIRC and GRI. Other leading global publishers were Wiley, Routledge Journals Taylor & Francis, and Elsevier SCI. The second approach was to analyze the affiliation of authors interested in publishing research on sustainability reporting. Findings highlight the significant contribution of Romanian authors especially to the GRI research (The Bucharest University of Economic Studies), and to the European directives (Babes Bolyai University from Cluj-Napoca).

During the past five years, 316 authors have studied the sustainability reporting from the European directives perspectives and published 116 articles. Moreover, the authors interested in international initiatives published 458 articles, of which GRI papers registered 1,114 authors, IIRC research was conducted by 239 authors, and only one author focused exclusively on ISSB. Descriptive statistics reveal that, no matter the topic, the papers were usually written by at least three authors, with a maximum of 8 authors for the research on GRI topics.

English is the preferred language of publishers. Depending on the type of publications, the research on international initiatives is usually published in journal articles, making them easier to be accessed. However, a considerable number of papers researching the European directives have been published in conference proceedings.

A debate on the most cited articles on European directives presents the interest in non-financial reporting directives, analyzing the effect of NFRD on Italian companies. The most cited article on international initiatives is a smaller-sized article of 13 pages focusing on the development of a methodological framework starting from GRI perspectives.

VOS Viewer based analysis showed that most articles included information related to GRI, non-financial reporting, integrated reporting, corporate social responsibility, and sustainability. Density visualizations confirm the interest in sustainability reporting through these topics, and GRI as the most interesting topic for academics.

Even if most of the analyzed papers refer to GRI, there are new regulations, directives, and standards that need to be understood and implemented, through future research. Further research could include a longitudinal study of the evolution of the GRI guidelines from the sustainability performance and reporting perspective.

Appendix 1 Frequency Analysis of Articles Published by Years and by Topic

Year	International initiatives			European directives			Total Articles
	ISSB	IIRC	GRI	NFRD	EFRAG	CSRD	
2019	0	17	80	16	1	0	114
2020	0	10	82	23	0	0	115
2021	0	21	112	24	0	1	158
2022	1	31	95	42	3	3	175
2023	0	2	7	1	0	0	10
Total	1	81	376	106	4	4	572

Source Authors' own research

References

Adams, C.A.: Conceptualising the contemporary corporate value creation process. Account. Audit. Account. J. **30**(4), 906–931 (2017)

Albu, N., Albu, C.N., Dumitru, M., Dumitru, V.F.: Plurality or convergence in sustainability reporting standards?. Amfiteatru Econ. J. **15**(Special No. 7), 729–742 (2013)

Christensen, H.B., Hail, L., Leuz, C.: Adoption of CSR and Sustainability Reporting Standards: Economic Analysis and Review. NBER: National Bureau of Economic Research. United States of America. https://www.nber.org/system/files/working_papers/w26169/revisions/w26169.rev0.pdf. Last Accessed 01 Aug 2023 (2019)

Donthu, N., Kumar, S., Mukherjee, D., Pandey, N., Lim, W.M.: How to conduct a bibliometric analysis: an overview and guidelines. J. Bus. Res. **133**, 285–296 (2021)

Echchakoui, S.: Why and how to merge Scopus and Web of Science during bibliometric analysis: the case of sales force literature from 1912 to 2019. J. Market. Anal. **8**, 165–184 (2020)

EFRAG (2021) Sustainability Reporting Standards Roadmap. https://www.efrag.org/Activities/2010051123028442/Sustainability-reporting-standards-roadmap. Last Accessed 01 Aug 2023

European Commission (2021) Proposal for a Directive of the European Parliament and of the Council amending Directive 2013/34/EU, Directive 2004/109/EC, Directive 2006/43/EC, and Regulation (EU) No 537/2014. https://eur-lex.europa.eu/legal-content/EN/TXT/?uri=CELEX%3A52021PC0189. Last Accessed 01 Aug 2023

Flower, J.: The international integrated reporting council: a story of failure. Critical Perspect. Account. **27**, 1–17 (2015)

GRI (2022) The GRI Standards Enabling transparency on organizational impacts. https://www.globalreporting.org/media/wmxlklns/about-gri-brochure-2022.pdf. Last Accessed 01 Aug 2023

Halkos, G., Nomikos, S.: Corporate social responsibility: trends in global reporting initiative standards. Econ. Anal. Policy **69**, 106–117 (2021)

Journeault, M., Levant, Y., Picard, C.F.: Sustainability performance reporting: a technocratic shadowing and silencing. Crit. Perspect. Account. **74**, 1–23 (2021)

KPMG (2022) Comparing sustainability reporting proposals. https://assets.kpmg/content/dam/kpmg/be/pdf/2022/comparing-sustainability-reporting-proposals-talkbook.pdf. Last Accessed 01 Aug 2023

Lungu, C.I., Caraiani, C., Dascălu, C., Turcu, D., Turturea, M.: Archival analysis of corporate social responsibility research: the Romanian perspective. Account. Manag. Inf. Syst. **15**(2), 341–371 (2016)

Moneva, J.M., Archel, P., Correa, C.: GRI and the camouflaging of corporate unsustainability. Account. Forum **30**(2), 121–137 (2006)

Rahi, A.F., Johansson, J., Fagerström, A., Blomkvist, M.: Sustainability reporting and management control system: structured literature review. J. Risk Financ. Manag. **15**(12), 1–31 (2022)

SASB (2022) The Evolution of SASB Standards. https://www.sasb.org/about/. Last Accessed 01 Aug 2023

Voss, G.W.: The European Union's 2014 non-financial reporting: Mandatory ex post disclosure -but does it need improvement? In: Extractive Industries and Human Rights in an Era of Global Justice: New Ways of Resolving and Preventing Conflicts, pp. 359–381

Private, State-Owned or Hybrid? State-Owned Multinationals as Strategic Form of Corporate Ownership

Olga Cretu

Abstract According to recent research in International Business state-owned multinationals represent some of the world's most active cross-border investors. This phenomenon is specific to emerging markets where governments support the internationalization of state-owned enterprises. In this manuscript, we answer the research question: How does state ownership affect the strategies and performance of multinationals? To answer the research question, an analysis of the 20 largest Chinese state-owned multinationals listed in the Fortune Global 500 has been employed. We use quantitative research methods to conclude that state ownership plays a crucial role in the Chinese state-owned multinationals' cross-border strategies, especially under the project Belt and Road Initiative. We have identified a correlation between the active internationalization of Chinese state-owned enterprises and the implementation of the grandiose project Belt and Road Initiative. We argue that state-owned multinationals represent hybrid enterprises and can be considered a strategic form of corporate ownership. We contribute to internationalization theory by arguing that state-owned enterprises differ from private firms in their foreign entry modes and state-owned multinationals represent strategic entities to enter foreign markets. This research paper presents implications for practitioners and policymakers who are concerned about the corporate governance of state-owned enterprises from countries with a high concentration of state ownership, like China in particular.

Keywords State-owned multinationals · Internationalization theory · Foreign entry modes · State-owned enterprises governance · Strategic corporate governance · China

---

O. Cretu (✉)
SKOLKOVO School of Management, Moscow, Russia
e-mail: Olga_Cretu@skolkovo.ru

M. Busu (ed.), *Rethinking Business for Sustainable Leadership in a VUCA World*, Springer Proceedings in Business and Economics,
https://doi.org/10.1007/978-3-031-50208-8_7

1 Introduction

International Business (IB) research has recently started to revisit the theme of state ownership and state capitalism, as scholars are paying more attention to its growing role in the internationalization theory (Cuervo-Cazurra et al. 2014; Cuervo-Cazurra and Li 2020; Lazzarini and Musacchio 2018). The need to study the interactions between the government and the multinational business has significantly increased in the era of globalization and digitalization (Aguilera et al. 2020; Cuervo-Cazurra and Li 2020; He et al. 2016; Mariotti and Marzano 2019). State-owned enterprises (SOEs) started their active internationalization after the global economic crisis in 2008–2009; thus the formulation of 'the state as a cross-border investor' has become an important phenomenon, especially in emerging markets (He et al. 2016; Li et al. 2014; Meyer et al. 2014; Ren et al. 2019).

In 2011 UNCTAD, for the first time, dedicated to the state-owned multinationals (SOMNEs) a whole section in its respectable World Investment Report (WIR), presenting the actual trends of these companies' activity (WIR 2011). Nowadays, scholars argue about the advantages and disadvantages of state ownership for multinational enterprises. However, they all emphasize that a better understanding of the SOMNEs will contribute to the internationalization theory.

According to existing IB literature, the 'new MNEs' represent a trend in emerging markets, specifically China and Brazil (Cuervo-Cazurra and Li 2020). There has been much research devoted to Chinese multinational enterprises, which are, by and large, state-owned (Huang et al. 2016; Li et al. 2018; Meyer et al. 2014). But there is still a lack of research on the corporate governance of SOEs and the advantages and disadvantages of their ownership structure. From this point of view, the research question of this paper arises: *How does state ownership affect the strategies and performance of multinationals?*

To answer the research question, we conducted an analysis of the performance of the SOMNEs head-quartered in China in the context of Belt and Road Initiative (BRI) implementation. We have analyzed the internationalization peculiarities of the Chinese SOMNEs in the last ten years using the data on FDI, both outward (OFDI) and inward (IFDI), on Mergers & Acquisitions (M&As) and greenfield projects, the sectoral and geographical distribution of the SOMNEs' OFDI, as well the role of the state in promoting the SOEs' internationalization.

The Belt and Road Initiative (BRI), also known as the One Belt One Road (OBOR) initiative, is a global infrastructure development strategy launched by the Chinese government in 2013 (Huang et al. 2016). The initiative aims to promote economic connectivity and cooperation among countries in Asia, Europe, Africa, and beyond. The BRI involves building a network of transportation, communication, and energy infrastructure projects, such as ports, railways, highways, pipelines, and telecommunications networks. The initiative also includes a range of trade and investment policies to facilitate cross-border trade and investment, such as free trade agreements and investment promotion and protection agreements.

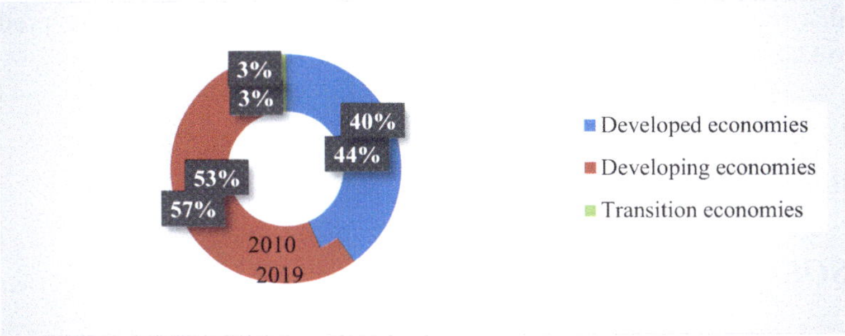

Fig. 1 Distribution of SOMNEs by region of origin in 2010 and 2019. *Source* Author's calculations based on data from World Investment Report (2011, 2019, 2021)

The Chinese SOMNEs are oriented toward acquiring as many foreign assets as possible. After the active absorption of foreign investment into the home economy, the Chinese companies started active internationalization thanks to the support of the state. As indicated above, the recent internationalization of SOEs is a process specific to developing countries. In 2010 there were 650 SOMNEs globally, while in the last decade, the total number practically tripled to 1600 entities in 2020 (WIR 2011, 2021). Among the top 100 largest non-financial MNEs in the world, 15% are SOMNEs (WIR 2021). More than half of SOMNEs are headquartered in emerging markets, and this proportion has increased in the last decade to a share of 57% (Fig. 1).

The main home countries of the SOMNEs are the emerging markets: China (18%), Malaysia (5%), India (4%), South Africa (4%), and Russia (3.4%) (WIR 2019). Only 30% of the SOMNEs are headquartered in developed countries. According to UNCTAD, there are 495 SOMNEs in Europe, mainly in Sweden (49 SOMNEs), followed by France (45 companies), Italy (44), Germany (43), Belgium (33), and Norway (32).

The impressive growth of OFDI from emerging markets in the last ten years illustrates an active expansion of MNEs, including SOMNEs. According to the eclectic paradigm, firms internationalize successfully with knowledge-based advantages like unique technologies and know-how, which are named "ownership advantages". Emerging market SOMNEs did not have these advantages when they started going abroad; thus, their successful internationalization was based on their location advantages, including the natural resources and labor at low costs (Huang et al. 2016; Li et al. 2018).

In the early 2010s, China started its policy of active internationalization of SOEs 'Go Global' predominantly to developed economies in search of new technologies and know-how, as R&D plays a crucial role in augmenting firms' performance (Cannizzaro and Weiner 2018). Several empirical studies demonstrated that the internationalization of firms from emerging markets contributes to their productivity and economic performance (Huang et al. 2016; Li et al. 2018; Kalotay 2018). In the case of Chinese SOMNEs internationalization policy supported by the government represented a successful strategy of acquiring technological and innovative assets that transformed China into one of the world's most innovative producers (Cuervo-Cazurra and Li 2020; Li et al. 2018).

2 Literature Review

2.1 SOEs' Internationalization

Internationalization theory constantly evolves as scholars continue to improve and update it following the actual trends of the global economy. Classically, there are several streams of the internationalization theory that explain the motives for the internationalization of MNEs and the SOEs too: the Buckley and Casson stream (the 1970s), the Rugman stream (the 1980s), and the Dunning stream (1970–2000s) (Santangelo and Meyer 2011).

The main motives of internationalization are traditionally considered: *the transaction costs* that can be reduced through the MNE hierarchy, *the resource-seeking* strategies of the companies, and the will to use the *country-specific advantages* to cut costs and raise profits. The famous OLI-paradigm created by Dunning (1998) generalizes and unifies all the previous theories into one: the MNEs can use 'owner-specific (O), location-specific (L) and the internalization (I) advantages' altogether and simultaneously (Narula et al. 2019). According to this paradigm, ownership advantages compensate for the liabilities of foreignness (capital, technology, and know-how), location advantages include the country-specific inputs and their costs (natural resources endowment, labor), and the internalization advantage arises when the firm intends to expand its ownership abroad rather than collaborate with local agents.

All the previous theories have been designed for and based on developed countries; they do not apply to EMNEs (Cuervo-Cazurra et al. 2014). According to some scholars, the internationalization of the EMNEs is possible thanks to the home country's comparative advantages, such as natural resources and cheap labor; thus, they do not possess firm-specific ownership advantages (Cuervo-Cazurra and Li 2020). Albeit others admit that the EMNEs internationalize because they have ownership advantages, these advantages are different from those of the developed countries' MNEs (Cuervo-Cazurra et al. 2014; Huang et al. 2016).

State ownership can be considered an essential factor for the EMNEs while internationalizing. For instance, one of the advantages of Chinese SOEs is their access to cheap capital, which has supported their internationalization (Cuervo-Cazurra and Li 2020). SOMNEs represent a more complex and exciting phenomenon as they can expand the existing internationalization theory (Cuervo-Cazurra et al. 2014). Many recent research papers demonstrated the renaissance of SOMNEs (He et al. 2016). This phenomenon is gaining momentum in the literature.

The research papers on state ownership and SOMNEs approach the agency theory, institutional theory, transaction costs economics, the resource-based view, and the resource dependence theory (Aguilera et al. 2020; Cuervo-Cazurra et al. 2014; Cuervo-Cazurra and Li 2020; Mariotti and Marzano 2019; Narula et al. 2019). SOMNEs pursue business and financial objectives and social and political ones (Aguilera et al. 2020; Mariotti and Marzano 2019); thus, they can invest in riskier projects with a lower financial value compared to private MNEs (Cuervo-Cazurra et al. 2014).

The government sees SOEs as a way to maintain social stability, without which the economy cannot function properly. For example, when social stability is low, SOEs are useful for hiring excess labor and bearing people's retirement benefits (Cuervo-Cazurra et al. 2014; Cuervo-Cazurra and Li 2020). The state uses the SOEs also to achieve social and ecological goals. SOMNEs are more likely to invest in greenfield projects than dealing with M&A projects and prefer larger projects than private MNEs (Cuervo-Cazurra et al. 2014; WIR 2019).

As SOMNEs present a mixed form of ownership, where the state has direct ownership stakes or indirect stakes through intermediary structures like state agencies, state-owned banks, or sovereign wealth funds (Cuervo-Cazurra and Li 2020; Kalotay 2018; WIR 2019), they tend to invest in countries where the state ownership is 'perceived as more legitimate' (Cuervo-Cazurra et al. 2014).

2.2 *State-Owned Versus Private Enterprises' Governance*

An actual and meaningful research direction is the analysis of the differences between state-owned and private enterprises (Lazzarini and Musacchio 2018). It is considered that SOEs are associated with lower economic performance than private enterprises (Cuervo-Cazurra et al. 2014; Dewenter and Malatesta 2001). It has been argued that SOEs pursue different goals when internationalizing in comparison with private companies. SOEs tend to invest in countries where the state ownership is 'perceived as more legitimate' (Cuervo-Cazurra et al. 2014). SOEs can benefit more from becoming multinational than private firms (Benito et al. 2016; Lazzarini et al. 2021).

Many researchers perceive state-owned multinationals as a hybrid form of ownership, where the state has direct ownership stakes or indirect stakes through intermediary structures like state agencies, state-owned banks, or sovereign wealth funds (Cuervo-Cazurra et al. 2022; Cuervo-Cazurra and Li 2021). It has been demonstrated that the SOEs internationalization depends on the level of state ownership. SOEs with a very low and very high level of state ownership have a limited level of internationalization than those with a medium level of state ownership (Kalasin et al. 2019). According to OECD, the governance practices of state-owned entities are usually considered deviations from the accepted standards for the governance of private firms, and the recommendation is to make the governance of state-owned entities resemble as much as possible the governance of private.

However, SOMNEs are hybrid organizations in their corporate governance, using elements of private corporate governance to enhance monitoring with elements of public administrative governance to enhance accountability. They use governance tools differently or employ additional instruments to develop a suitable corporate governance structure (Okhmatovskiy et al. 2021). This hybrid governance is reflected in governance guidelines, board structures, and nontraditional governance mechanisms adopted by state-owned entities (Grøgaard et al. 2019). Guidelines for SOEs differ in objectives from the codes of good governance for private firms, which are usually designed to address the separation of ownership and control (Aguilera et al. 2020).

Some corporate governance challenges may arise when we talk about SOEs. First, listed SOEs normally take the form of joint stock corporations and are thus subject to the classic agency issues present in privately owned listed companies. Second, the state as a controlling shareholder may hold particular ownership objectives linked to public policy, which can give rise to new forms of "private benefits" of control (Grøgaard et al. 2019).

SOEs can benefit significantly from internationalization, sometimes even more than the private firms (Benito et al. 2016). Being subject to monitoring by outside investors as well as the stricter governance and transparency requirements applied to listed corporations may improve SOEs' performance.

3 Methodology

Several scholars captured the driving forces capable of changing the Chinese outward FDI flow, such as using Chinese SOEs (Cuervo-Cazurra and Li 2020; Lazzarini et al. 2021; Huang et al. 2016). Regarding the impact of BRI, Buckley (2020) suggests that BRI increased the total outward investment from China. On the other hand, many researchers point out the many barriers the SOEs face, such as the lack of central coordination, corruption, the potential clash of different political regimes, and the financial viability of cross-border projects (Cuervo-Cazurra et al. 2014; Lazzarini and Musacchio 2018). Such difficulties tend to decrease the volume invested under the BRI umbrella.

The literature review also reflects the diverse ownership and entry mode options available for China, such as M&As, greenfields, or joint ventures (Cuervo-Cazurra and Li 2020). While some authors support the Chinese preference for M&As due to the high level of control that acquisitions present (Ren et al. 2019), a few other academics suggest an opposing point of view, regarding greenfield projects (Cuervo-Cazurra et al. 2014). From this point of view, we can formulate Hypothesis 1 and 2:

> Hypothesis 1: Chinese SOMNEs' active internationalization (OFDI) is corelated to the BRI implementation.

> Hypothesis 2: SOMNEs usually enter foreign markets through greenfield projects.

SOMNEs play an increasingly important role in today's global economy. China is the host country of an impressive quantity of SOMNEs, with the most state-owned enterprises in the world. There are three state-owned Chinese multinationals in the Top 10 MNEs in Fortune Global 500 in 2021. China is the leader with 135 MNEs on the list in 2021, 11 more than a year earlier. Table 1 presents the 20 largest Chinese SOMNEs. In 2021 China overpassed the US for the first time, which accounted for 126 companies.

In the Top 10 Fortune Global 500, there are three Chinese giants: *State Grid* (rank 2), *China National Petroleum* (rank 4), and *Sinopec Group* (rank 5). Chinese economy relies heavily on SOEs, which are considered to have lower production efficiency. Despite this, China has achieved outstanding economic growth in the last decades.

SOEs are different from private firms and have a market failure. According to UNCTAD data, 70% of SOMNEs are operating in the tertiary sector of the economy—in services activities, 23%—in manufacturing, and 8%—in the primary sector of the economy (WIR 2019). SOMNEs are predominantly concentrated in five industries: finance, utilities (especially electricity provision), transportation, mining, or represent integrated holdings (WIR 2019). This sectoral distribution reflects the tendency of SOMNEs to gain control in strategic sectors related to the management of natural resources and critical infrastructure projects. Governments tend to use SOMNEs to enter strategic sectors abroad (Cuervo-Cazurra et al. 2014). Here we can formulate Hypothesis 3:

> Hypothesis 3: SOMNEs' Investments significantly influence the Chinese OFDI, so they are strategic entities to expand abroad.

This research uses quantitative research methods. We use Statistical Modeling ANOVA and regression analysis in R to analyze the impact of the independent variables over the dependent variable of Chinese Outward FDI (OFDIChina). We use the dataset named the *China Global Investment Tracker (CGIT), published by the American Enterprise Institute. The CGIT has provided information about all known Chinese OFDI projects since 2005. CGIT contains over 4,000 transactions related to different sectors. The database provides a detailed list of those transactions, recording the Chinese parent company, the recipient country, and the affected industry. Additionally, CGIT separates data by country and region of every transaction, indicating if such agreement is associated with the BRI and whether or not it is a greenfield type of project.*

The dependent variable is the Chinese Outward FDI (*OFDIChina*) which represents the internationalization of the Chinese companies. Independent variables are the types of investments realized by the Chinese companies: greenfield projects (*Greenfield*), M&As projects (*MAs*), investments in the context of BRI projects (*BRI*), investments not in the BRI projects (*nonBRI*), investments realized by the SOMNEs (*SOMNEs*). The independent variable *SOMNEs* includes data of a dataset of 20 largest Chinese SOMNEs presented in Table 1. Taking all the variables we can formulate the regression equation as following:

$$\begin{aligned}\text{OFDIChina} &= \beta_0 + \beta_1\text{Greenfield} + \beta_2\text{MAs} \\ &\quad + \beta_3\text{BRI} + \beta_4\text{nonBRI} + \beta_5\text{SOMNEs} + \varepsilon_1 \qquad (1)\end{aligned}$$

Table 1 20 largest Chinese SOMNEs listed in «Fortune Global 500» in 2021

№	List Rank	Company name	Industry	Revenue (US$ Million)	Profit (US$ Million)	Assets (US$ Million)	Employees	M&As value in 2011–2021 (millions $)	Greenfield value in 2011–2021 (millions $)	Nr. of countries in which operates
1	2	State Grid Corporation of China	Electric power	386,617.7	5,580.4	666,088.5	89,636	38,990	500	12
2	4	China National Petroleum Corporation	Petroleum ind	283,957.6	4,575.2	626,616.7	1,242,245	54,620	11,200	26
3	5	China Petrochemical Corporation	Petroleum ind	283,727.6	6,205.2	343,289.0	553,833	33,190	2020	25
4	13	China State Construction Engineering	Construction	234,425.0	3,578.4	338,033.2	356,864	50,250	3120	6
5	20	Industrial and Commercial Bank of China	Commercial bank	234,425.0	3,578.4	338,033.2	356,864	3190	0	7
6	25	China Construction Bank	Commercial bank	172,000.2	39,282.5	4,311,456.6	373,814	2710	0	4
7	29	Agricultural Bank of China	Commercial bank	153,884.6	31,293.4	4,169,355.9	462,592	n/a	n/a	4
8	32	China Life Insurance Company	Insurance	144,589.1	4,648.1	776,308.8	183,417	n/a	n/a	2

(continued)

Table 1 (continued)

№	List Rank	Company name	Industry	Revenue (US$ Million)	Profit (US$ Million)	Assets (US$ Million)	Employees	M&As value in 2011–2021 (millions $)	Greenfield value in 2011–2021 (millions $)	Nr. of countries in which operates
9	35	China Railway Engineering Corporation	Construction	141,383.6	1,638.8	185,315.7	308,483	22,700	2800	5
10	39	Bank of China	Commercial bank	134,045.6	27,952.1	3,739,871.1	309,084	1120	100	5
11	42	China Railway Construction Corporation	Construction	131,992.3	1,485.5	190,916.1	364,632	50,980	2620	8
12	56	China Mobile	Telecommunications	111,825.5	12,920.1	304,527.8	455,721	1780	320	5
13	60	SAIC Motor	Automotive ind	107,555.2	2,961.0	140,906.5	143,261	430	0	5
14	61	China Communications Construction Company	Construction	106,867.7	1,165.0	306,555.0	213,438	89,210	8160	23
15	65	China Minmetals	Metalworking	102,014.8	491.1	150,651.9	205,015	13,240	1680	8
16	66	FAW Group	Automotive ind	101,075.8	2,866.5	74,933.4	121,002		0	N/A
17	69	China Resources	Pharmaceutical ind	99,437.6	4,330.2	275,691.9	370,955	1840	0	3
18	70	Shandong Energy	Coal mining	97,860.6	1,161.9	104,996.6	244,832	210	0	1
19	72	China Baowu Steel Group	Steelmaking	97,643.1	3,628.7	155,413.2	234,57	860	220	2
20	74	China Post	Mail, Package and Freight Delivery	96,304.1	4,698.4	1,518,542.8	827,231	0	0	1

Source Fortune Global 500, 2021 list and author's calculations based on China Global Investment Tracker Dataset

We followed a quantitative approach based on fundamental statistical analysis. In particular, the tools used were descriptive statistics and regression analysis in R. Our study included total annual investment in million dollars US as output variables. We took into consideration the series covering from 2013 to 2022. Additionally, we analyzed two separate periods, pre-BRI and post-BRI.

As previously stated, the main question in this study is how state ownership affects the MNEs' internationalization strategies and firms' performance. In the case of Chinese SOMNEs, these entities may represent strategic forms of governance because they may help understand if China is building or buying the BRI. The difference is more profound than it may appear because the investments should be of the greenfield type, thus creating new wealth rather than changing ownership on already existing assets.

We ran 3 regression models with all proposed independent variables during the ANOVA phase and the regression analysis in R. First, we checked the significance at the 5% level (if at least one of the proposed dependent variables was significant). Then, if at least one independent variable is significant ($p < 0.05$), each model variable is checked for significance. If at least one variable is not significant pooling procedure is executed until finding all significant independent variables. Finally, we reviewed the residual assumption consisting of all residuals coming from an identical independent normal distribution with a mean equal to zero and constant variance. New statistical modeling was tried if this assumption on the proposed debuted model was not met. In this research, linear investment was enough to achieve the results shown in the following section.

4 Results and Discussions

In this section, we present the findings of the research. We studied Chinese investments over the period following the official announcement of BRI in October 2013, all the way to December 2022. The purpose was to determine if greenfield projects were a priority investment policy, using a yearly analysis investments per project.

In Table 2 we present the ANOVA for linear regression for the BRI period (2013–2022) where the dependent variable is the Annual Outward FDI from China (*OFDIChina*) and the independent variables are the *Greenfrield, Mas, BRI, nonBRI* and *SOMNEs* from the regression equation.

Therefore, the regression equation for this period is the following:

$$\begin{aligned}\text{OFDIChina} &= 9.643 + \text{Greenfield} + \text{MAs} \\ &\quad - 2.312\text{BRI} - 2.364\text{SOMNEs} + 1.636\end{aligned} \tag{2}$$

ANOVA analysis allows us to *support Hypothesis 1* stating that the Chinese SOMNEs' active internationalization is corelated to the BRI implementation. Instead, the nonBRI investments are not significant to the regression.

To test Hypothesis 2, we employed an analysis of the greenfield and M&As realized by Chinese companies. To better illustrate the effect of the BRI implementation we split the data into two periods: pre-BRI (2005–2013) and BRI-period (2013–2022) as shown in Table 3. The total amount invested during 17 years for greenfield projects is 330,340 (25.57% of total) versus 961,720 US million dollars for M&A's (74.43% of total). Table 3 shows the information for the two-mean Hypothesis 2 test for the percentage of greenfield projects and the yearly average percentage for 2005–2013 and 2014–2022.

The two-mean test shows a p-value of 24.03%, higher than 5%. The annual average of greenfield projects in both periods is lower than that of the M&As. Consequentially, the lower volume of greenfield investments compared to the higher volume of M&As allows us to reject Hypothesis 2. M&As represent a priority when considering average project investments during the BRI period, thus contradicting the official narrative that BRI is a series of greenfield investments from Chinese enterprises. Therefore, the variable Greenfield might be deleted from the final regression. The final regression to test the Hypothesis 3 is the following:

$$\text{OFDIChina} = \beta_0 + \beta_1\text{MAs} + \beta_2\text{BRI} + \beta_3\text{SOMNEs} + \varepsilon_1 \quad (3)$$

Table 2 ANOVA for the initial regression equation

Regression Statistics					
Multiple R	1				
R Square	1				
Adjusted R Square	0.8				
Standard Error	1.636				
Observations	10				
ANOVA					
	df	*SS*	*MS*	*F*	*Significance F*
Regression	5	20,126,374,210	4,025,274,842	1.879	7.928
Residual	5	1.338	2.677		
Total	10	20,126,374,210			
	Coefficients	*Standard Error*	*t Stat*	*P-value*	
Intercept	9.643	1.909	0.050	0.961	
Greenfield	1	1.713	5.836	2.803	
MAs	1	3.352	2.982	8.037	
BRI	−2.312	1.197	−1.932	0.111	
nonBRI	0	0	65,535	0	
SOMNEs	−2.364	1.156	−2.045	0.002	

Source Author's calculations based on data from CGIT

We test the regression in R and obtain the following results, stating that the regression is a well-fitting regression model and all the coefficients are significant (Table 4).

Therefore, the regression equation for the BRI period is the following:

$$\text{OFDIChina} = 2996 + 0.8\ \text{MAs} + 0.6\ \text{BRI} + 0.5\ \text{SOMNEs} + 3899 \quad (4)$$

We need to test the regression for linearity. In R the linearity assumption can be checked by inspecting the **Residuals versus Fitted** plot (Fig. 2).

We obtained a horizontal line, without distinct patterns which is an indication for a linear relationship, what is good. It means that the relationship between X and the mean of Y is linear.

Next, we test for Homoscedasticity, so we employ the Breush-Pagan test. We obtained the BP = 3.3132, df = 3, p-value = 0.3458. The P-value is equal to 34.5%

Table 3 Annual percentage of greenfield versus M&A average investments, 2005–2021

	2005–2013	2014–2022
Percentage of annual average greenfield investments	39.44%	45.10%
Variance	0.64%	1.12%
Observations	9	9
Hypothesized mean difference	0	
df	13	
t Stat	−1.230527136	
P(T <= t) two-tail	0.240301233	
t Critical two-tail	2.160368656	

Source Author's calculations based on data from CGIT

Table 4 Regression 3 t-test of coefficients in R

(Intercept)	2996.700 (4383.214)
MAs	0.829*** (0.039)
BRI	0.654*** (0.101)
SOMNEs	0.543* (0.163)
R-squared	0.995
N	10

Significance *** = p < 0.001; ** = p < 0.01; * = p < 0.05
Source Author's calculations in R

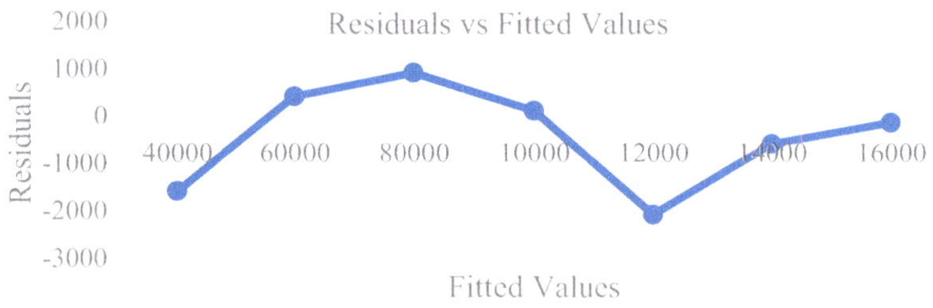

Fig. 2 Test for Regression 3 linearity in R using Residuals versus fitted plot. *Source* Author's calculations in R

which is higher than the 5% and even 10% significance level. Therefore, the errors in the regression are homoscedastic.

We need to test the regression for Independence. To test the Regression 3 for Independence in R we use the **Chi-square test**. It examines whether rows and columns of a contingency table are statistically significantly associated. We obtained the results X-squared = 218,428, df = 54, p-value < 2.2e−16. In Regression 3 the row and the column variables are statistically significantly associated (p-value = 0).

We tested the Regression 3 for Normality. In R we use the **Normal Q-Q** plot to examine whether the residuals are normally distributed. The residuals points follow the straight dashed line (Fig. 3), which is an indicator of normally distributed residuals. Therefore, for any fixed value of X, Y is normally distributed.

According to the analysis of the regression 3 all factors are significant, i.e. the M&As realized by the Chinese companies abroad, the Investments in BRI projects and the FDI realized by the SOMNEs. The results of presented by the regression 3 allow us to support Hypotheses 1 and 3 but to reject Hypothesis 2.

Chinese SOMNEs represent an important tool of the Chinese government in realization of the BRI project. The Government's strategies for supporting SOEs in their internationalization helped improve the SOMNEs' performance and active investments abroad (He et al. 2016). The reforms helped China transition from

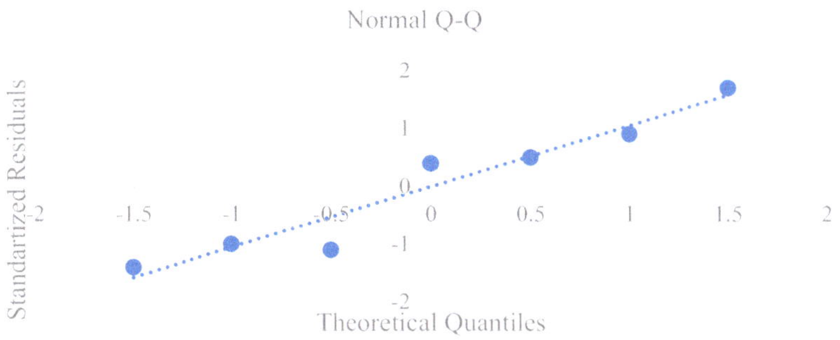

Fig. 3 Test for Regression 3 normality in R using Normal Q-Q plot. *Source* author's calculations in R

an investment-driven export economy to an innovation-driven economy reliant on domestic consumption (Li et al. 2018). This research paper proves the important role of SOEs in realizing the Chinese Government's major projects like BRI. The SOMNEs' investments significantly influence on the Chinese OFDI dynamics. According to Regression 3 every million US dollars of SOMNEs' FDI augment the Chinese OFDI with 5 thousand US dollars, while the other regressors remain unchanged. The government's active support represents a successful case of collaboration between the state and the business, a win–win model when state ownership is a strong advantage.

5 Conclusion

In the last decade, scholars have argued about the advantages and disadvantages of SOMNEs for international business. However, all of them affirm the increasing role of these entities in the world economy and FDI flows. SOEs are considered inefficient, instead, SOMNEs are big profitable companies because they represent a hybrid form of governance.

The regression analysis of the Chinese OFDI and the top-20 Chinese SOMNEs demonstrated that Chinese SOMNEs' active internationalization started with governmental support and during the implementation of BRI. The test of the Hypothesis 1 confirms the positive correlation between the Chinese investment in the BRI project and the overall Chinese OFDI. By testing the Hypothesis 2 we argue that the Chinese SOMNEs prefer M&As rather than greenfield projects as a foreign entry mode. This fact suggests that China is practically buying not building the BRI, thus this is more of a geopolitical project. Is state ownership an advantage or a disadvantage for SOMNEs? After the analysis we accepted the Hypothesis 3 stating that Chinese SOMNEs significantly influence the Chinese OFDI. Therefore, the state ownership represents a strong advantage for the SOMNEs since they became hybrid organizations: listed, profit-oriented but with the state as a major shareholder. Government can provide SOMNEs exclusive rights to operate in specific industries or geographical areas (predominantly in natural resources and utilities), can grant them financial support like subsidized credits or tax exemption, can support them with diplomatic networks with foreign governments while investing abroad (Aguilera et al. 2020; Cuervo-Cazurra et al. 2014; Lazzarini and Musacchio 2018). The SOEs reforms are updated continuously and adapted to the new realities.

However, this research paper has a series of limitations. First, the analysis includes only the 20 largest Chinese SOMNEs. These companies' investments are impressive but a more detailed analysis of the total amount of Chinese SOMNEs would improve the analysis and reduce the bias. Second, the regression would improve if we could add more coefficients. Future research might add to the regression more independent variables like exports, GDP per capita and control variables. In this case the research analysis might be more complex and less biased.

Most research papers consider the SOMNEs almost like MNEs with just an additional, extremely powerful stakeholder whose interests need to be protected and who can interfere with the MNE's policy. However, this study can draw a different picture: SOMNEs represent hybrid organizations, with strong advantages in today's world of disruptions. SOMNEs corporate governance represents a promising future research avenue. Thus, It would be interesting to continue this research in the context of Romania. According to OECD, Romania is a country with high ownership concentration, with high economic and social disparities (Cretu 2022; OECD 2022). Romanian biggest SOEs are profitable and are concentrated in the energy sector and utilities: *Hidroelectrica, Romgaz, Nuclearelectrica, SAPE*, etc. But SOE's partial privatization and internationalization contribute to a firm's financial and social performance (Cannizzaro and Weiner 2018; Cuervo-Cazurra et al. 2014). To reduce the government's influence and become more independent, it is highly recommended that the SOEs internationalize. Managers and policymakers should draw their forces towards the diversification of activities and ownership. Undoubtedly, the SOMNEs have significantly expanded the internationalization theory recently.

This research paper represents an initial attempt to make a complex comparison of SOMNEs from different countries, so it has substantial future research potential. Thus, future research can be more valuable by using quantitative methods to compare the EMNEs from two or more emerging countries. A quantitative analysis will help determine the role of state ownership in the evolution and performance of the SOMNEs. Instead, qualitative research will allow a better understand of the SOMNEs corporate governance and formulate recommendations for practitioners. The challenge for this research can represent the lack of data, as many SOMNEs do not reveal their financial data, and the emerging markets' statistical entities do not have evidence of the MNEs, especially SOMNEs.

Another research avenue of huge interest can be the comparison of the SOMNEs from developed and developing countries. We know that many SOMNEs are based in Europe. Thus, a research manuscript would be very timely to figure out the similarities and differences between SOMNEs from developed and developing countries. Valuable insights could be obtained by analyzing their internationalization strategies, what are their models of growth, and their most valuable assets.

References

Aguilera, R., Duran, P., Heugens, P.P.M.A.R., Sauerwald, S., Roxana, T., VanEssen, M.: State ownership, political ideology, and firm performance around the world. J. World Bus. **56**(1), 101113 (2020)

Benito, G.R., Rygh, A., Lunnan, R.: The benefits of internationalization for state-owned enterprises. Glob. Strateg. J. **6**(4), 269–288 (2016)

Buckley, P.J.: Book review-China's belt and road initiative: Changing the rules of globalization [Zhang, W., Alon, I., & Lattemann, C. (Editors)]. J. Int. Bus. Stud. **51**(2), 279–281 (2020)

Cannizzaro, A.P., Weiner, R.J.: State ownership and transparency in foreign direct investment. J. Int. Bus. Stud. **49**(2), 172–195 (2018)

Cretu, O.: Romania: on the way to improving inherited business models. In: Casas-Klett, T., Cozzi, G. (Eds.) Elite Quality Report 2022: Country Scores and Global Rankings. Seismo, Zurich (2022). https://doi.org/10.33058/seismo.30769.6221

Cuervo-Cazurra, A., Li, C.: State ownership and internationalization: the advantage and disadvantage of stateness. J. World Bus. 101112 (2020)

Cuervo-Cazurra, A., Inkpen, A., Musacchio, A., Ramaswamy, K.: Governments as owners: state-owned multinational companies. J. Int. Bus. Stud. **45**(8), 919–942 (2014)

Dewenter, K.L., Malatesta, P.H.: State-owned and privately owned firms: an empirical analysis of profitability, leverage, and labor intensity. Am. Econ. Rev. **91**(1), 320–334 (2001)

Grøgaard, B., Rygh, A., Benito, G.R.G.: Bringing corporate governance into internalization theory: state ownership and foreign entry strategies. J. Int. Bus. Stud. **50**, 1310–1337 (2019)

He, X., Eden, L., Hitt, M.A.: The renaissance of state-owned multinationals. Thunderbird Int. Bus. Rev. **58**(2), 117–129 (2016)

Huang, Y., Xie, E., Li, Y., Reddy, K.S.: Does state ownership facilitate outward FDI of Chinese SOEs? Institutional development, market competition, and the logic of interdependence between governments and SOEs. Int. Bus. Rev. **26**(1), 176–188 (2016)

Kalasin, K., Cuervo-Cazurra, A., Ramamurti, R.: State ownership and international expansion: the S-curve relationship. Global Strategy J. (2019)

Kalotay, K.: State-owned Multinationals: an emerging market phenomenon? J. Comp. Econ. Stud. **13**, 13–38 (2018)

Lazzarini, S.G., Musacchio, A.: State ownership reinvented? Explaining performance differences between state-owned and private firms. Corp. Gov. Int. Rev. **26**(4), 255–272 (2018)

Lazzarini, S.G., Mesquita, L.F., Monteiro, F., Musacchio, A.: Leviathan as an inventor: an extended agency model of state-owned versus private firm invention in emerging and developed economies. J. Int. Bus. Stud. **52**, 560–594 (2021)

Li, M.H., Cui, L., Lu, J.: Varieties in state capitalism: outward FDI strategies of central and local state-owned enterprises from emerging economy countries. J. Int. Bus. Stud. **45**, 980–1004 (2014)

Li, J., Xia, J., Shapiro, D., Lin, Z.: Institutional compatibility and the internationalization of Chinese SOEs: the moderating role of home subnational institutions. J. World Bus. **53**(5), 641–652 (2018)

Mariotti, S., Marzano, R.: Varieties of capitalism and the internationalization of state-owned enterprises. J. Int. Bus. Stud. **50**(5), 669–691 (2019)

Meyer, K.E., Ding, Y., Li, J., Zhang, H.: Overcoming distrust: how state-owned enterprises adapt their foreign entries to institutional pressures abroad. J. Int. Bus. Stud. **45**(8), 1005–1028 (2014)

Narula, R., Asmussen, C.G., Chi, T., Kundu, S.K.: Applying and advancing internalization theory: the multinational enterprise in the twenty-first century. J. Int. Bus. Stud. **50**(8), 1231–1252 (2019)

OECD Corporate Governance Working Papers №.27. Corporate ownership and concentration. A. Medina, A. de la Cruz, Y.Tang (2022). https://doi.org/10.1787/bc3adca3-en

Okhmatovskiy, I., Grosman, A., Sun P.: Hybrid governance of state-owned enterprises. In: Wright, M., Wood, G.T., Cuervo-Cazurra, A., Sun, P., Okhmatovskiy, I., Grosman, A.(Eds.) Oxford Handbook of State Capitalism and the Firm (2021)

Ren, M., Manning, S., Vavilov, S.: Does state ownership really matter? The dynamic alignment of China's resource environment and firm internationalization strategies. J. Int. Manag. **25**(3), 100667 (2019)

Santangelo, G.D., Meyer, K.E.: Extending the internationalization process model: increases and decreases of MNE commitment in emerging economies. J. Int. Bus. Stud. **42**(7), 894–909 (2011)

World Investment Report. Non-Equity Modes of International Production and Development. UNCTAD, UN, New York, Geneva (2011)

World Investment Report. Investment and the Digital Economy. UNCTAD, UN, New York, Geneva (2019)

World Investment Report. Investing in Sustainable Recovery. UNCTAD, UN, New York, Geneva (2021)

The Role of Community Characteristics in Entrepreneurship in Vietnam: An Importance-Performance Analysis

Khanh Hung Doan

Abstract Currently, communities are considered an essential factor in the development of entrepreneurship. More specifically, the characteristics and features of the community not only contribute to maintaining, but also promote and support the development of entrepreneurship in the locality. Therefore, investing in community features and characteristics is meaningful in many countries. However, the impact of each community characteristic is different in entrepreneurship. This study aimed to explore the evaluations of community characteristics in supporting and developing entrepreneurship. From this, we can have a specific view of the relationship between community and entrepreneurship. The study was conducted based on an online questionnaire survey of 201 community members in Vietnam. Importance-Performance Analysis method was used to assess the influence of community characteristics. The results of the IPA analysis show that "Finance and community funds" is a feature that is considered important and needs to be developed. Other community features about Community policies, Networks and relationships, Training and education activities, Security guaranteed, and Rules of the community should be maintained and steadily developed. Other characteristics of the community regarding the Organizational structure of the community, Culture and social norms, Social capital from the community, and Assessment and testing activities should limit development. Finally, communities should reduce their investment in the Space and working environment of the community in Vietnam. These results contribute to a better understanding of the role of community characteristics in entrepreneurship. Through this research, stakeholders such as communities, local managers, and policymakers will have solutions to improve the role and effectiveness of community characteristics. Furthermore, this study promotes potential research on community and entrepreneurship in the future. The paper ends with a discussion of the results and implications of the study.

K. H. Doan (✉)
Bucharest University of Economic Studies, 010374 Bucharest, Romania
e-mail: dkhung@hueuni.edu.vn

Hue University, 530000 Hue City, Vietnam

M. Busu (ed.), *Rethinking Business for Sustainable Leadership in a VUCA World*, Springer Proceedings in Business and Economics,
https://doi.org/10.1007/978-3-031-50208-8_8

Keywords Community · Community characteristics · Entrepreneurship · Importance–Performance analysis · Vietnam

1 Introduction

Many studies show that individuals have the desire to become entrepreneurs, thereby enriching themselves and society. The emergence of more and more new business models, such as innovative business models (Bellini et al. 2019), business through digital platforms (Ruggieri et al. 2018), etc., have also contributed to the promotion of entrepreneurship. However, nowadays, entrepreneurs and individuals always face many risks and challenges in the formation and development of entrepreneurship (Priede-Bergamini et al. 2019; Doan 2022). Regarding the factors affecting entrepreneurship, many scholars have long recognized that entrepreneurship is influenced by the surrounding context (Welter 2011). In different environments, circumstances and times, entrepreneurial behaviour and intentions will also be different (Ambad and Damit 2016; Maresch et al. 2016). Any change in nature, accidental or intentional, can affect businesses and organizations (Păunescu et al. 2018a). However, in addition to the advantages of development, many risks and challenges to the economy, institutions, society, and business environment affect businesses (Păunescu et al. 2018b; Priede-Bergamini et al. 2019; Păunescu and Molnar 2020; Păunescu and Mátyus 2020). Therefore, entrepreneurs need to change and negotiate flexibly in these different social contexts, times, and places (Watson 2009).

There are many reasons for the formation of entrepreneurship, in which the community plays a significant role in the establishment and development of entrepreneurship (Collins et al. 2016). Communities can influence entrepreneurial behaviour through specific contextual characteristics of the community (Lyons et al. 2012). Additionally, communities are not only resource-rich but also effective platforms for doing business (Martin et al. 2013). Therefore, the community is a solid foundation for entrepreneurship development (Lyons et al. 2012). Thus, community characteristics and entrepreneurship have an essential relationship with each other (Darwish and Van Dyk 2018).

More specifically, Hindle and Moroz (2010) determined that many aspects and characteristics of the community influence any entrepreneurial process and the entrepreneurship of entrepreneurs. On the other hand, Korsching and Allen (2004) considered community supports as a catalyst for business ventures. Communities provide significant services with the aim of supporting entrepreneurship (Darwish and Van Dyk 2018) through various measures and subsidies for entrepreneurship (Prijon 2012). Lyons et al. (2012) determined that communities also engage in activities that facilitate investment and are effective for entrepreneurship. However, it is significant for community developers and leaders to keep in mind that the support of community attributes also makes a difference. The more supportive the community, the greater the number of potential entrepreneurs who will start new businesses (Gruidl and Markley 2014). Thus, building the right environment and community characteristics to promote entrepreneurship and promote the process of starting a business is an essential task for the each government. From there, communities can create the best environment for entrepreneurship, helping to support, maintain, and promote entrepreneurship development.

Vietnam is a developing country with a rapidly transforming economy, with many types of businesses operating (Doan 2021). Therefore, promoting and improving entrepreneurship is essential in promoting local development and raising people's incomes. Thus, along with increasingly fierce competition and globalization, the role of the community is increasingly important in enhancing competitiveness and developing entrepreneurship (Darwish and Van Dyk 2018; Doan 2023). In addition, specific assessments of the impact of community characteristics on entrepreneurship help business leaders and policymakers gain a more detailed view of the relationship between community and entrepreneurship in Vietnam. This paper attempts to supplement a study on understanding the influence of the community on entrepreneurship by assessing the role of the characteristics of the community in promoting entrepreneurship in Vietnam. The proposed research question in this study are: what characteristics of the community influence entrepreneurship? and what role do those characteristics play in entrepreneurship in Vietnam today? This study uses the Importance-Performance Analysis (IPA) method for 201 samples collected from community members in Vietnam. From there, the paper aims to present the actual status of the influence of community characteristics on entrepreneurship. Implications are also provided for community managers and local managers to improve efficiency, quality, and satisfaction and to attract more people to participate in local entrepreneurship.

The paper is organized into the following sections: the first part provides a brief literature review on the role of the community in entrepreneurship. Herein, there is an emphasis on the influence of community characteristics on entrepreneurship. In the second part, the research methods are provided. The third part covers the research results. The last part is the conclusion and discussion of the research results of the paper, establishing future research directions as well as the limitations of the study.

2 Literature Review

2.1 The Role of Community Characteristics in Entrepreneurship

Many studies have shown that we cannot assume that the impact of community on entrepreneurship is the same everywhere. It is because entrepreneurship is not just an individual quality, but must include its response to opportunities that appear to be exploited for profit (Lichtenstein and Lyons 2010). As aforementioned, community characteristics, such as favourable natural resources, population size, and geographical features, can determine how well a business operates. In this situation, the community and entrepreneurship are mutually beneficial, and both reap benefits from the interaction (Lyons et al. 2012).

However, in addition to the positive effects, communities can have a negative impact on the success of enterprises and entrepreneurship (Lyons et al. 2012). In particular, the lack of motivation to create new entrepreneurs in the community will harm the sustainability of entrepreneurial activities and the development of entrepreneurship, which will also have a negative impact on the community (because entrepreneurship correlates with economic growth) (Lyons et al. 2012). In summary, the nature and role of any relevant community as a factor influencing the entrepreneurship process is an issue that needs to be carefully considered in the research and evaluation of entrepreneurship (Hindle and Moroz 2010). Apart from being equipped with beneficial raw materials, many communities do not fully understand how to support entrepreneurs to start up and promote entrepreneurial activities (Lyons et al. 2012).

In fact, many aspects and characteristics of the community have had an influence on any entrepreneurial process and the entrepreneurship of entrepreneurs (Hindle and Moroz 2010). Collins et al. (2016) also assessed that community factors and characteristics are the main factors shaping the operation and success of enterprises. Furthermore, Wilkinson (1991) believed that issues related to building and strengthening the field of the community have important implications for future entrepreneurship research.

Peredo and Chrisman (2006) gave three factors that influence interaction models among different community members and stakeholders, including the entrepreneurship of entrepreneurs. These factors are social structure, social capital, and cultural values. Quality of life factors (for example, education, healthcare, climate, entertainment, etc.) are also significant in many decisions about the business location of the enterprise (Pittman 2006). In addition, local norms, beliefs, laws, rules, etc., not only can shape the ability of entrepreneurs, but can also significantly influence who becomes an entrepreneur, including why they become entrepreneurs and what goals they want to achieve through their entrepreneurial ventures (Lyons et al. 2012; Păunescu and Molnar 2020). However, when a community is not ready to grow, it becomes more difficult to attract new entrepreneurs, as well as maintain and expand existing businesses and entrepreneurship (Phillips and Pittman 2009). Therefore, communities need to have policies in place to be ready for the development of entrepreneurship.

In addition, entrepreneurship is also deeply rooted in local culture (Dana and Dana 2005). Culture is one of the important factors affecting the entrepreneurial activities of entrepreneurs (Warren et al. 2017; Porfírio et al. 2018). However, due to its characteristics, the community culture can both positively and negatively affect entrepreneurship (Lyons et al. 2012).

Currently, a number of researchers who have studied the role of the community in connecting entrepreneurs and businesses with financial resources and funds. This occurred not only for other individuals but also for the community and society as a whole. Research by Hui et al. (2014) and Inbar and Barzilay (2014) identified community efforts in crowdfunding. Research by Giudici et al. (2012) identified that community funds and finance play an important role in the entrepreneurship finance of enterprises.

In addition, many studies have investigated the role of the community in creating spaces that promote the creativity, innovation, and entrepreneurship of individuals (Rieken et al. 2019). Local communities have seen the development of spaces that serve as opportunities for innovation, technology, and entrepreneurship (Ensign and Leupold 2018). Since then, the maker space has become a hotbed for technological innovation and entrepreneurship, allowing inventors to bring their ideas from the sketch to the market (Mortara and Parisot 2018). Community infrastructure and facilitators are the main elements of entrepreneur support spaces (Rieken et al. 2019).

Communities also provide support between entrepreneurs through networks (Ketilson and MacPherson 2002; Manyara and Jones 2007). Building and strengthening an entrepreneur's network is essential to their success (Lyons et al. 2012). Fortunato and Alter (2011) proposed that the vast network of weak relationships will give these people access to the best information regarding economic and environmental opportunities. Therefore, having a large network will lead to a higher level of entrepreneurship development.

Research by Reynolds et al. (2000) also identified the factors that determine the interaction between entrepreneurship and the community: entrepreneurship opportunities, entrepreneurial capacity, social legitimacy, finance, and information technology. These conditions have a direct impact on the level of entrepreneurial activity. Furthermore, these conditions depend on the individual entrepreneur as well as the community (Reynolds et al. 2000).

Many community managers are looking to expand the number of new startups in the community and create a culture of entrepreneurship among community residents. Thus, the community's support and characteristics must be focused on the needs and desires of entrepreneurs if they are to be successful (Gruidl and Markley 2014). From there, it can enhance the role of the community in entrepreneurship. Communities can be actively involved in supporting training and skill development, community development, cultural development, and empowerment (Memmott 2010; Torri 2010). This partnership between business individuals and the community will yield positive synergies to foster creativity, innovation, growth, and development (Fortunato and Alter 2015). Phillips and Pittman (2009) determined that as a company grows, it will need community support. Therefore, the characteristics of the community that are consistent with entrepreneurship are significant. Furthermore, communities must be flexible and responsive to meet the different needs of entrepreneurs (Gruidl and Markley 2014).

3 Methodology

3.1 Research Purpose

The purpose of the research was to find out community members' evaluations about community characteristics in supporting entrepreneurship in Vietnam.

3.2 Data Collection

The survey subjects here are community members in Vietnam. To ensure the accuracy of the data we collect, we think it is important that community members know their communities well. In addition, they also have time to learn and understand the characteristics of the community, from which they can get their evaluations and comments related to the community. Therefore, the subjects of the survey conducted are people in the Vietnamese community who have participated in community activities for some time.

Data were collected through online questionnaires. The survey questionnaire on the Google Docs platform was selected for use. The survey was conducted in November and December 2022 for members of the community in Vietnam. The sampling method is convenient, not probabilistic. Finally, 201 complete and valid samples were used to conduct the analysis.

3.3 Research Scale

To capture quantitative information on stakeholders' perceptions of how the community's impact, role, and influence on their decision-making, a 6-point Likert questionnaire (1 = very disagree to 6 = strongly agree, or 1 = least affect to 6 = most influential) was used in most of the questionnaire. Currently, in practice, when using the Likert scale, researchers often will use 5-point or 7-point Likert questions to collect quantitative information about research problems. Scholars assumed that when using a 5-point or 7-point Likert question, participants had the opportunity to answer difficult questions about which they were uncertain by choosing a central score (Weaver et al. 2007; Bui 2009). Therefore, the 5-point or 7-point Likert questions can lead to misleading and poor findings (Dawes 2008). To avoid this bias, the 6-point Likert question is suggested as an alternative that will eliminate respondents choosing a central score (Gwinner 2006). The 6-point Likert question also allows researchers to address ethical issues because respondents have two opportunities to answer any questions they are undecided about. Score-3 represents a negative response (for example, partly disagree), and score-4 represents a positive response (for example, partly agree).

3.4 Data Analysis

In this study, the data analysis methods used are mainly descriptive statistics and Importance-Performance Analysis (IPA model). The IPA model was proposed by Martilla and James (1977). It involves evaluating different aspects of an organization's features in terms of customer perceptions of performance and the importance of that performance (I-P gaps). The IPA model helps managers determine the essentials of the criteria, as well as the strengths and weaknesses of the characteristics of products and services provided. In this paper, the IPA model helps community managers make specific assessments and realize the importance of community characteristics and features to develop and support community members' entrepreneurship.

More specifically, the development of the IPA model is carried out by comparing two criteria that shape the selection decision, including (1) The relative importance of the features and (2) The degree of implementation of the features. However, the IPA model does not yet have a specific definition for "importance" and "performance" (Oh 2001). Therefore, for this study to be consistent with the purpose of the paper, "importance" for community characteristics is the opinion of community members about how important community characteristics are to entrepreneurship. In addition, "performance" for community characteristics is the extent to which the community performs on those characteristics as assessed by community members. According to Barsky (1995), the lowest importance level of the feature indicates the ability to have little influence on the general perception of customers. On the contrary, the high importance will enormously influence their perception. This analysis categorises the measured characteristics and provides managers with useful information about the strengths and weaknesses of the services and features they provided. From there, the manager will have the right strategic decisions to improve the ability and the quality of the service and features provided.

Typically, such functions in the IPA model are represented in a 2×2 grid, where each quadrant can be summarized into a specific implication for management (see Fig. 1).

Quadrant 1—High importance, low implementation (focus on development). The attributes in this quadrant are considered very important to the customer, but the supplier's performance is poor. This result helps supply managers pay attention to these attributes and focus on developing them to satisfy customers.

Quadrant 2—High importance, high performance (continue to maintain). Attributes in this quadrant are considered essential to customers, and suppliers have also performed very well. Administrators should continue to maintain and promote this strength.

Quadrant 3—Low importance, low performance (development constraints). Attributes located in this quadrant are considered low-performance and unimportant. Managers should limit resources to develop these attributes.

Quadrant 4—Low importance, high performance (reducing investment). Attributes in this quadrant are considered unimportant to the customer, but the

Fig. 1 The IPA scheme. *Source* Martilla and James (1977)

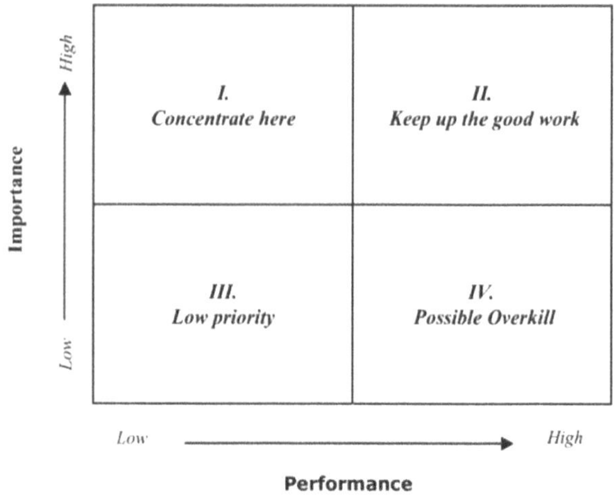

supplier's performance is good. The current overinvestment can be considered futile. Managers should use this resource to focus on developing other attributes.

The IPA has become a suitable tool for understanding and assessing the satisfaction and effectiveness of organizational attributes (Matzler et al. 2003). By examining the points in each quadrant, the manager can infer the attributes that the customer feels should be the highest priority and the lowest priority for improving his characteristics. From there, it is often used to provide solutions to strategic-level decisions.

4 Results and Discussions

4.1 Description of the Samples

Statistical results on the demographic characteristics of respondents that are members of the community are shown in Table 1. The observations include 23.9% male and 76.1% female. In general, the gender distribution among community members in this study is female. The fact that female members of communities make up a large proportion is consistent with the results of several studies on community-based enterprises and practices in Vietnam. It is also evident in the characteristics of the community in Vietnam, where the number of female members accounts for a higher proportion. Furthermore, due to cultural features in Vietnam, men will dominate the economic activities in the family, so they have less time to participate in community activities than women.

The age of the community members surveyed is between 20 and 60 years old in the community. The proportion of community members aged from 20 to 30 years old accounted for the highest proportion of the sample with 81.1%, followed by 30–50 years old with 17.4%, and Older than 50 years old with 1.5%. This disparity is

Table 1 Socio-demographic data in the samples

		N	Percent (%)			N	Percent (%)
Age	< 30 years old	163	81.1	Marital status	Married	38	18.9
	30–50 years old	35	17.4		Not married	163	81.1
	> 50 years old	3	1.5	Work experience	< 5 years	159	79.1
Location	Rural area	74	36.8		≥ 5 years	42	20.9
	Urban area	127	63.2	Gender	Male	48	23.9
Total	**201**	**100**		Female	153	76.1	

Source Author's own research result

because the questionnaire was conducted through the Google Forms tool, so young people are more tech-savvy than the rest. In addition, the study also focused on young community members to assess the current status of the community' influence on stimulating entrepreneurship today.

Therefore, in terms of work experience, respondents with less than five years of work experience also accounted for the highest proportion (79.1%), and those with more than five years of work experience (20.9%). Furthermore, unmarried people account for a large proportion (81.1%) compared to married people (18.9%). Regarding residence, 63.2% of the respondents are currently in urban areas and 36.8% are in rural areas.

4.2 Community Support Activities on Entrepreneurship

The study measured the frequency of organizing activities to support entrepreneurship. The results in Fig. 2 show that although these support activities are organized, the frequency of community support activities is not high (mean value is 5.15 on a scale of 10). However, the percentage of communities that regularly organize support activities (rated from 8 or more) is not good (10.5%). It is a positive sign that many communities do not regularly organize activities to support entrepreneurship. More specifically, analyzing the difference in regularly organizing activities to support entrepreneurship, the statistical results show that rural areas (mean = 5.47) often organize these activities more than urban areas (mean = 4.97). However, these differences were not statistically significant (see Table 2).

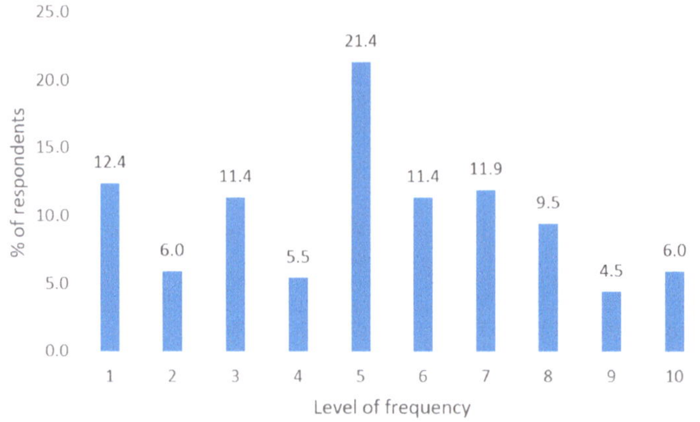

Fig. 2 Frequency of organizing community support activities. *Source* Author's own research result

Table 2 Frequency of organizing community support activities by location

		N	Average value of evaluation*	P value
Total		201	5.15	
Location	Urban area	127	4.97	0.183**
	Rural area	74	5.47	

*1: very rarely, 10: very frequently, **Result from Independent Sample T-test
Source Author's own research result

4.3 Importance – Performance Analysis for Community Characteristics

Importance - Performance Analysis was used in this study to find gaps between the importance and performance of every single attribute and provide a visual overview of these gaps through the IPA grid. It is the basis for implications for improving the status and roles of community characteristics in Vietnam.

The Importance of Community Characteristics

Reliability analysis was applied to test the goodness of all variables using Cronbach's alpha. Cronbach's Alpha is taken to analyze the degree of consistency between items in a structure. The higher the Cronbach Alpha coefficient, the better. In this study, the value of each variable is above 0.9, so they are valid for analysis (Nunnally and Bernstein 1994; Konting et al. 2009; Taber 2018). In addition, the ranking results of the community characteristics used in this study and their weighted importance are shown in Table 3.

"Security guaranteed", "Rules of the community", and "Community policies" are the three characteristics that receive the highest relevance for members of the community. In addition, "Finance and community funds" needs attention because it has the

Table 3 Importance ranking of community's characteristics

	% very important	% important	% slightly important	Total %
Security guaranteed	20.4	27.9	25.4	73.7
Rules of the community	22.9	26.9	22.9	72.7
Community policies	24.4	25.4	22.4	72.2
Training and education activities	23.4	26.4	20.9	70.7
Space and working environment	17.9	29.4	23.4	70.7
Finance and community funds	26.4	21.9	22.4	70.7
Networks and relationships	22.9	27.4	19.4	69.7
Culture and social norms	18.9	27.9	21.9	68.7
Social capital from the community	24.4	19.9	23.9	68.2
Assessment and testing activities	18.4	26.9	22.9	68.2
Organizational structure of the community	21.9	22.9	22.9	67.7

Source Author's own research result

most "very important" rating (26.4%). The characteristics considered less important by community members are "Social capital from the community", "Assessment and testing activities", and "Organizational structure of the community". It makes sense because these specific attributes are less likely to appear in the minds of community members. Community members can only perceive these characteristics of the community when there is a history of participation and activity in the community or when they have heard them before.

Considering the six levels of responsiveness as an interval scale, some more detailed results are more noticeable. The interval ranking by mean (see Table 4) provides a slightly different order from the previous ratings shown in Table 3. In Table 3, attributes with "very important" are ranked higher on the interval scale.

"Rules of the community", "Community policies", "Finance and community funds" and "Training and education activities" are features that community members consider important in the community. Among them, "Rules of the community" and "Finance and community funds" along with "Security guaranteed" and "Training and education activities" are equally appreciated by members of the community. Besides, out of the total of eleven community characteristics, "Assessment and testing activities" was rated the lowest by community members. Assessment and testing activities play an important role in the functioning of the community. This activity contributes to stabilising other activities. In addition, these activities contribute to finding the

Table 4 Importance of community characteristics for entrepreneurship and local development

	Rank	Mean*	Std. Deviation
Organizational structure of the community	10	4.15	1.453
Culture and social norms	9	4.16	1.413
Social capital from the community	7	4.19	1.448
Networks and relationships	6	4.25	1.442
Finance and community funds	3	4.27	1.479
Space and working environment	8	4.17	1.402
Training and education activities	5	4.26	1.443
Assessment and testing activities	11	4.12	1.425
Security guaranteed	4	4.26	1.383
Community policies	1	4.29	1.441
Rules of the community	2	4.27	1.432

*1 very unimportant, 6: very important
Source Author's own research result

wrong, from which we can take measures to overcome the limitations in the community. However, in this study, community members did not appear to be fully aware of the importance of this trait. It may be because these activities are often lacking or the communities have not yet paid attention to them.

The Performance of Community Characteristics

Community member assessments of the performance of community features will be presented in this section. In addition, the differences in the evaluation of community characteristics were also provided between groups of community members (by gender, place of residence, age, and time of participating in community activities).

Table 5 displays the rating of community characteristics based on the proportion of respondents who strongly agree, agree, and slightly agree with the community characteristics surveyed in Vietnam. Of the community characteristics, "Social capital from the community" is the characteristic with the least feedback from community members, and "Security guaranteed" is the characteristic with the most positive feedback.

Considering the six levels of responsiveness as an interval scale (see Table 6), there are some notable, and more detailed results. The interval ranking according to the mean of scale in Table 6 provides a slightly different ordering from the previous ratings shown in Table 5. Attributes with "very important" weight are ranked higher on the interval scale. Table 6 presents the mean values of performance and statistical changes.

"Security guaranteed", "Training and education activities", "Community policies", "Rules of the community", and "Networks and relationships" are the five characteristics that best represent community members. In addition, "Space and working environment" and "Networks and relationships" received the same high

Table 5 Performance ranking of community characteristics

	% strongly agree	% agree	% slightly agree	Total %
Security guaranteed	20.4	26.9	23.4	70.7
Community policies	21.4	25.9	22.4	69.7
Rules of the community	16.9	28.4	23.9	69.2
Training and education activities	22.4	23.9	22.4	68.7
Culture and social norms	16.4	25.9	25.4	67.7
Networks and relationships	18.9	27.4	20.9	67.2
Space and working environment	18.4	26.4	21.4	66.2
Finance and community funds	19.9	21.9	23.9	65.7
Assessment and testing activities	16.4	26.9	22.4	65.7
Organizational structure of the community	15.9	22.9	25.4	64.2
Social capital from the community	14.4	24.4	24.4	63.2

Source Author's own research result

Table 6 Performance of community characteristics for entrepreneurship and local development

	Rank	Mean*	Std. Deviation
Organizational structure of the community	11	3.97	1.426
Culture and social norms	8	4.04	1.435
Social capital from the community	10	3.97	1.382
Networks and relationships	5	4.10	1.468
Finance and community funds	7	4.08	1.422
Space and working environment	5	4.10	1.432
Training and education activities	2	4.19	1.444
Assessment and testing activities	8	4.04	1.435
Security guaranteed	1	4.21	1.409
Community policies	3	4.16	1.479
Rules of the community	4	4.11	1.408

* 1 totally disagree, 6: totally agree
Source Author's own research result

rating. It shows that the communities in Vietnam have a friendly and peaceful environment and space, and the networks and relationships are also good. These factors help create an atmosphere, a working space for members of communities in Vietnam.

Out of eleven characteristics of community in Vietnam, "Organizational structure of the community" and "Social capital from the community" received the lowest ratings from community members. Local communities in Vietnam are known to regularly organize activities for people so that communities can have supportive policies for community members. Each community has different structures, but these structures are the basis for operating their own activities. Furthermore, social capital from the community is also an important basis. However, in this study, community members do not appear to be fully aware of this characteristic of communities. It may be due to specific activities for community members and promotional activities that represent how the community works.

Important-Performance Matrix

Examining the relationship between the importance and performance of community characteristics from the perspective of community members is necessary. It is to measure the gap between importance and performance and identify solutions for each community feature that enhance the effect of community features on entrepreneurship and local development. The reason is that the comparison results between the importance and the performance evaluation generate valuable information.

The result in Table 7 illustrates the assessment and gaps of community members in the relationship between the importance and performance of community characteristics in Vietnam. Data show that the performance of eleven characteristics of the community is not commensurate with their importance. The average ratings for performance are all less than their importance. Herein, there are five gaps out of 11 that are statistically significant, which are "Organizational structure of the community", "Social capital from the community", "Networks and relationships", "Finance and community funds", and "Rules of the community".

More details, the study analyzed the difference in importance and performance assessment between the group from different places of residence in urban and rural

Table 7 Importance-Performance gaps

	Importance *	Performance **	Gap	P value ***
Organizational structure of the community	4.15	3.97	**−0.18**	**0.004**
Culture and social norms	4.16	4.04	−0.12	0.064
Social capital from the community	4.19	3.97	**−0.22**	**0.002**
Networks and relationships	4.25	4.10	**−0.15**	**0.028**
Finance and community funds	4.27	4.08	**−0.19**	**0.018**
Space and working environment	4.17	4.10	−0.07	0.303
Training and education activities	4.26	4.19	−0.07	0.323
Assessment and testing activities	4.12	4.04	−0.08	0.266
Security guaranteed	4.26	4.21	−0.05	0.502
Community policies	4.29	4.16	−0.13	0.093
Rules of the community	4.27	4.11	**−0.16**	**0.023**

* 1 very unimportant, 6 very important, ** 1 strongly disagree, 6 strongly agree, *** Results from Pair Sample T-Test

Source Author's own research result

areas. Research results show that community members from both urban and rural areas are disappointed in the characteristics of the community in Vietnam. They rate the performance of each attribute below their importance rating for that attribute (see Figs. 3 and 4).

Specifically, for urban areas, there are two characteristics that are evaluated as more important than performance, "Space and working environment" and "Community policies". However, this gap is not high. Overall, community members in rural areas seem more disappointed. The gap between importance and performance is quite large, and the total negative gap between importance is 2.19 for rural areas and 0.89 for urban areas. This result is contrary to the study by Chrisman et al. (2002). In their research, Chrisman et al. (2002) evaluated that the urban–rural distinction

Fig. 3 Importance–Performance gap in urban area. *Source* Author's own research result

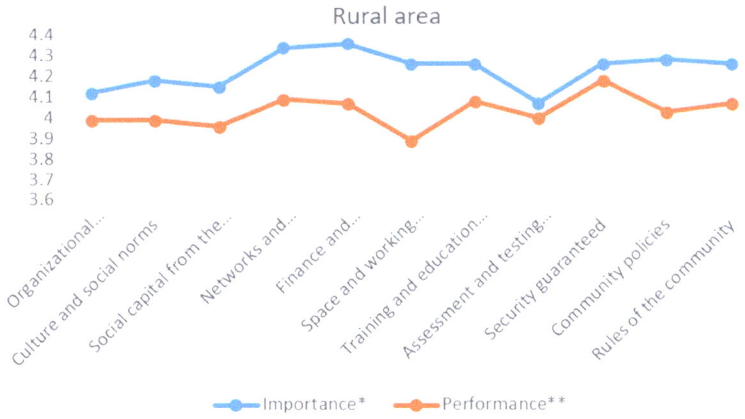

Fig. 4 Importance–Performance gap in rural area. *Source* Author's own research result

was not related to any environmental problems nor the perception and effectiveness of entrepreneurship.

IPA analysis is a simple but useful method that helps managers determine which characteristics need to be improved to increase the satisfaction of the respondents. From a research perspective, this study also supports the application of IPA as an alternative framework to assess community members' perceptions of community characteristics in promoting entrepreneurship. Figure 5 illustrates the results of the importance-performance grid for community characteristics. The average importance levels of the eleven community characteristics and the average level of performance of these attributes are calculated. The position of each attribute on the importance-performance grid is determined using the mean score of importance and performance as the coordinates (Martilla and James 1977; Lynch et al. 1996). After the calculations, they are plotted on a two-dimensional grid. Thus, the mean of importance and performance are used to determine the position of the axes on the grid (importance mean score = 4.22 and performance mean score = 4.09). Each property in the grid is then analyzed by locating the appropriate quadrant in which it is placed (see Fig. 5).

The Importance-Performance matrix shows that the attributes and characteristics of the community that are considered the most important by community members are in quadrants 1 and 4.

- Quadrant 1 (focus on development) should be immediately reviewed by community managers (very high importance - low performance). It is easy to visually see that "5. Finance and community funds" appears to be the most significant feature for the community in promoting entrepreneurship and local development. It can be said that the accessibility and connections of finance sources can be a decisive factor in the success or failure of an enterprise (Parwez 2017). Community involvement and support in crowdfunding have contributed to better funding results than other factors

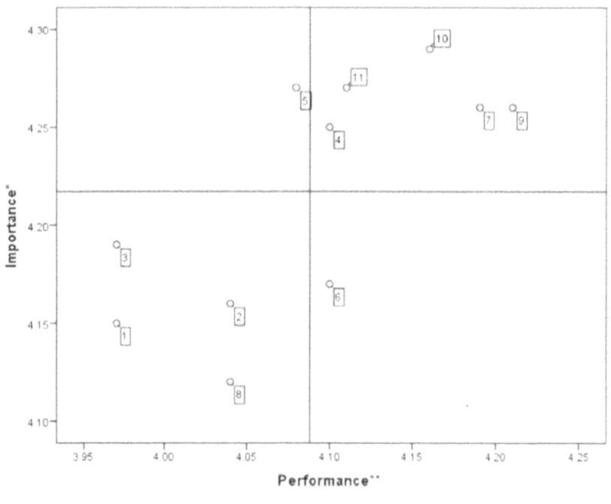

1. Organizational structure of the community
2. Culture and social norms
3. Social capital from the community
4. Networks and relationships
5. Finance and community funds
6. Space and working environment
7. Training and education activities
8. Assessment and testing activities
9. Security guaranteed
10. Community policies
11. Rules of the community

Fig. 5 Importance—Performance matrix. *1 very unimportant, 6 very important, **1 strongly disagree, 6 strongly agree. Source: Author's own research result

(Josefy et al. 2017), contributing to better project success (Hui et al. 2014). However, in Vietnam, capital from the community is still limited, and the ability to connect to funds is still weak. In addition, the number of projects connecting and attracting capital from the community is still few (Thu et al. 2020).

- Quadrant 2 (continue to maintain) is paramount to the community, and they also perform well, especially the "10. Community policies" in addition to other features such as "4. Networks and relationships", "7. Training and education activities", "9. Security guaranteed", "11. Rules of the community".

Community characteristics belonging to quadrants 3 and 4 have a lower priority, including attributes of lower importance than other groups.

- Quadrant 3 (development constraints) with attributes below average importance and above average performance indicates that these attributes can be used in community activities. The characteristics of the communities in this group are "1. Organizational structure of the community", "2. Culture and social norms", "3. Social capital from the community", and "8. Assessment and testing activities".

- Quadrant 4 (reducing investment) demonstrates that the performance of attributes is below average but is not considered important, including "6. Space and working environment". Therefore, urgent efforts do not require to improve these properties. These community facilities and spaces provide inspiration and new ideas for future projects, learning new concepts, and m2014eeting like-minded people (Kuznetsov and Paulos 2010). In Vietnam, the "New Rural" program has been invested and implemented by the state (Diep 2014). The goal is to build infrastructure and boost the rural economy. However, the construction is still spread out, and the planning is still spontaneous, spreading, etc. Therefore, it is necessary to reduce this investment and focus on the urgent needs of local people and communities (Diep 2014).

Importance-Performance Analysis has implications for both managers and academics. From a management perspective, the results can provide clear guidance for improving community characteristics by identifying key areas of intervention based on the perceptions of community members. In particular, community policymakers should put more effort into improving "5. Finance and community funds". It is an essential feature for communities in Vietnam in promoting entrepreneurship and local development. In addition, Vietnamese communities need to limit investment for "6. Space and working environment", to focus on investing in improving other characteristics of the community.

5 Conclusions

Communities play an important role in supporting and developing entrepreneurship. These community roles are essential to promote not only entrepreneurship, but also local development. To support the development of entrepreneurship, the characteristics of the community are important factors that form the basis for the development of entrepreneurship. In fact, the community supports all stages of business development, from the start of the business through to the growth of the enterprises. This paper has provided an assessment of the characteristics of the community for entrepreneurship development in Vietnam. From the research results analyzed above, we can see that community characteristics have a significant contribution to the development of entrepreneurship. Communities in Vietnam organize activities that support entrepreneurship. However, the frequency of these activities is not high.

In a general assessment of community characteristics for entrepreneurship, the performance of the eleven characteristics of the community is not commensurate with their importance. The average levels for performance are all less than their importance. The community characteristics with a high gap in importance and performance are the "Organizational structure of the community", "Social capital from the community", "Finance and community funds". In addition, community members from rural areas seem more frustrated because the gap between importance and performance is quite high. In addition, the total negative gap between importance and performance is higher in rural areas than in urban areas. It is a significant implication for local and community managers.

The results of the IPA model show that "Finance and community funds" is the most essential feature of the community for entrepreneurship and needs to be developed. In fact, financial resources and funds have always played an important role in promoting entrepreneurship. Currently, in Vietnam, there have been many policies and solutions to support the government in funding start-ups and promoting entrepreneurship. However, access to these capital sources is still limited. At the local community level, this problem is weak because there are not many mechanisms, policies, and supports. Community characteristics on "Community policies", "Networks and relationships", "Training and education activities", "Security guaranteed", and "Rules of the community" should be maintained and steadily developed when appreciated for importance and performance. Communities should continue to improve their policies for entrepreneurship; focus on entrepreneurship training and education activities for community members; continue to maintain and expand the networks and relationships that support community entrepreneurship; and strengthen activities to ensure safety and security in the community.

The characteristics of the community on "Organizational structure of the community", "Culture and social norms", "Social capital from the community", "Assessment and testing activities" need limited development because of low level of importance and performance assessment. Finally, communities should reduce investment in "Space and working environment". Currently, along with the development of information technology and the Internet, the working space has expanded, so investing in "Space and working environment" is no longer effective in Vietnam. Therefore, community managers should focus on improving the existing supply capacity and quality, reducing the investment in the "Space and working environment" of their community. These findings contribute to the current knowledge and practice of research on the influence of communities on entrepreneurship in Vietnam in particular and research on the role and characteristics of the community in general.

Although some efforts have been made to conduct the study, there are some limitations to this study. The first limitation of the study is the application of the convenience sampling method. Convenience sampling is often criticized for its representativeness of the collected data because it constitutes a limitation for attempts to generalize and make inferences. Therefore, the results of the study may be biased. In addition, the number of collected research samples is with 201 samples collected. The third limitation is related to the subjects surveyed in different community groups. Further research directions are proposed to require comparisons between different community groups. From there, it is possible to examine differences in the influence of community characteristics on entrepreneurship, especially in localities where there are developed and undeveloped communities. In addition, further studies on the behaviour of community support for entrepreneurship can use specific scales or can be combined with the characteristics of local people and entrepreneurs. From there, we can make timely and accurate assessments of the community characteristics and roles to regulate and support the development of entrepreneurship.

Acknowledgements This paper was co-financed by The Bucharest University of Economic Studies during the PhD program.

References

Ambad, S.N.A., Damit, D.H.D.A.: Determinants of entrepreneurial intention among undergraduate students in Malaysia. Procedia Econ. Finance **37**, 108–114 (2016)

Barsky, J.D.: World-Class Customer Satisfaction. IL, Irwin Professional Pub, Chicago (1995)

Bellini, F., Dulskaia, I., Savastano, M., D'Ascenzo, F.: Business models innovation for sustainable urban mobility in small and medium-sized european cities. Management & Marketing. Challenges for the Knowledge Society **14**(3), 266–277 (2019)

Bui, D.T.: Tourism industry responses to the rise of sustainable tourism and related environmental policy initiatives: The case of Hue city, Vietnam (Doctoral dissertation, Auckland University of Technology) (2009)

Chrisman, J.J., Chua, J.H., Steier, L.P.: The influence of national culture and family involvement on entrepreneurial perceptions and performance at the state level. Entrep. Theory Pract. **26**(4), 113–130 (2002)

Collins, J., Morrison, M., Krivokapic-Skoko, B., Butler, R., Basu, P.K.: Indigenous small businesses in the Australian Indigenous economy (2016)

Dana, L.P., Dana, T.E.: Expanding the scope of methodologies used in entrepreneurship research. Int. J. Entrep. Small Bus. **2**, 79–88 (2005)

Darwish, H., Van Dyk, L.: Bottom of Pyramid 4.0: modularising and assimilating industrial revolution cognition into a 4-tiered social entrepreneurship upliftment model for previously disconnected communities. J. Ind. Integr. Manag. **3**(2), 1850010 (2018)

Dawes, J.: Do data characteristics change according to the number of scale points used? An experiment using 5-point, 7-point and 10-point scales. Int. J. Mark. Res. **50**(1), 61–104 (2008)

Diep, D.T.B.: New rural program in Vietnam: Situation and Solutions. Vietnam J. Soc. Sci. **8**(81), 61–69 (2014)

Doan, K.H.: The National Policy Environment for Starting a Business: An Assessment in Vietnam. In Resilience and Economic Intelligence Through Digitalization and Big Data Analytics, Sciendo, 232–243 (2021)

Doan, K.H.: The differences in the impact of entrepreneurship education on entrepreneurial knowledge: A cross-country analysis. Management & Marketing. Challenges for the Knowledge Society, 17(1), 73–97 (2022)

Doan, K.H.: A Bibliometric and Visualization Analysis of Community and Entrepreneurship Research. In: Busu, M. (eds) Digital Economy and the Green Revolution. Springer Proceedings in Business and Economics. Springer, Cham, 45–64 (2023)
Ensign, P.C., Leupold, P.: Grassroots opportunities for innovation, technology, and entrepreneurship: Makerspaces in non-urban communities. In 2018 Portland International Conference on Management of Engineering and Technology (PICMET), IEEE, 1–7 (2018)
Fortunato, M.W.P., Alter, T.R.: The individual-institutional opportunity nexus: An integrated framework for analyzing entrepreneurship development. Entrepreneurship Research Journal, 1(1), Article 6 (2011)
Fortunato, M.W.P., Alter, T.R.: Community entrepreneurship development: an introduction. Community Dev. **46**(5), 444–455 (2015)
Giudici, G., Nava, R., Rossi Lamastra, C., Verecondo, C.: Crowdfunding: The new frontier for financing entrepreneurship? SSRN Electronic Journal (2012)
Gruidl, J., Markley, D.M.: Entrepreneurship as a community development strategy. An introduction to community development, 220–235 (2014).
Gwinner, C.: Infosurv white paper 5-point vs. 6-point Likert scales, Infosurv Online Research Service (2006)
Hindle, K., Moroz, P.: Indigenous entrepreneurship as a research field: developing a definitional framework from the emerging canon. International Entrepreneurship and Management Journal **6**(4), 357–385 (2010)
Hui, J.S., Greenberg, M.D., Gerber, E.M.: Understanding the role of community in crowdfunding work. In Proceedings of the 17th ACM conference on Computer supported cooperative work & social computing, 62–74 (2014)
Inbar, Y., Barzilay, O.: Community impact on crowdfunding performance. Available at SSRN 2524910 (2014)
Josefy, M., Dean, T.J., Albert, L.S., Fitza, M.A.: The role of community in crowdfunding success: Evidence on cultural attributes in funding campaigns to "save the local theater." Entrep. Theory Pract. **41**(2), 161–182 (2017)
Ketilson, L.H., MacPherson, I.: A report on Aboriginal co-operatives in Canada: current situation and potential for growth, BC Institute for Co-operative Studies, Vancouver (2002)
Konting, M.M., Kamaruddin, N., Man, N.A.: Quality Assurance in Higher Education Institutions: Exit Survey among Universiti Putra Malaysia Graduating Students. Int. Educ. Stud. **2**(1), 25–31 (2009)
Korsching, P.F., Allen, J.C.: Locality based entrepreneurship: a strategy for community economic vitality. Community Development Journal **39**(4), 385–400 (2004)
Kuznetsov, S., Paulos, E.: Rise of the expert amateur: DIY projects, communities, and cultures, 6th Nordic Conference on Human-Computer Interaction: Extending Boundaries, Reykjavik, 295–304 (2010)
Lichtenstein, G.A., Lyons, T.S.: Investing in entrepreneurs: A strategic approach for strengthen-ingyour regional and community economy. Santa Barbara, CA: Praeger/ABC-CLIO (2010)
Lynch, J., Carver, R., Virgo, J.: Quadrant Analysis as a Strategic Planning Technique in Curriculum Development and Program Marketing. J. Mark. High. Educ. **7**(2), 17–32 (1996)
Lyons, T.S., Alter, T.R., Audretsch, D., Augustine, D.: Entrepreneurship and community: The next frontier of entrepreneurship inquiry. Entrepreneurship Research Journal, 2(1) (2012)
Manyara, G., Jones, E.: Community-based tourism enterprise development in Kenya: an exploration of their potential as avenues of poverty reduction. J. Sustain. Tour. **15**(6), 628–644 (2007)
Maresch, D., Harms, R., Kailer, N., Wimmer-Wurm, B.: The impact of entrepreneurship education on the entrepreneurial intention of students in science and engineering versus business studies university programs. Technol. Forecast. Soc. Chang. **104**, 172–179 (2016)
Martilla, J.A., James, J.C.: Importance-Performance Analysis. J. Mark. **41**(1), 77–79 (1977)
Martin, B.C., McNally, J.J., Kay, M.J.: Examining the formation of human capital in entrepreneurship: A meta-analysis of entrepreneurship education outcomes. J. Bus. Ventur. **28**(2), 211–224 (2013)

Matzler, K., Sauerwein, E., Heischmidt, K.: Importance-performance analysis revisited: the role of the factor structure of customer satisfaction. Serv. Ind. J. **23**(2), 112–129 (2003)

Memmott, P.: Demand-responsive services and culturally sustainable enterprise in remote Aboriginal settings: a case study of the Myuma Group, Desert Knowledge CRC Report No. 63, Desert Knowledge CRC, Alice Springs (2010)

Mortara, L., Parisot, N.: How do fab-spaces enable entrepreneurship? Case studies of'makers'–entrepreneurs. Int. J. Manuf. Technol. Manag. **32**(1), 16–42 (2018)

Nunnally, J.C., Bernstein, I.H.: Psychometric Theory, 3rd edn. McGraw- Hill, New York (1994)

Oh, H.: Revisiting importance–performance analysis. Tour. Manag. **22**(6), 617–627 (2001)

Parwez, S.: Community-based entrepreneurship: evidences from a retail case study. J. Innov. Entrepreneurship **6**, 1–16 (2017)

Păunescu, C., Mátyus, E.: Resilience measures to dealing with the COVID-19 pandemic Evidence from Romanian micro and small enterprises. Management & Marketing. Challenges for the Knowledge Society, 15(s1), 439–457 (2020)

Păunescu, C., Molnar, E.: Country's entrepreneurial environment predictors for starting a new venture - evidence for Romania. Sustainability **12**(18), 7794 (2020)

Păunescu, C., Popescu, M.C., Blid, L.: Business impact analysis for business continuity: Evidence from Romanian enterprises on critical functions. Management & Marketing. Challenges for the Knowledge Society, 13(3), 1035–1050 (2018a)

Păunescu, C., Popescu, M.C., Duennweber, M.: Factors determining desirability of entrepreneurship in Romania. Sustainability **10**(11), 3893 (2018b)

Peredo, A.M., Chrisman, J.J.: Toward a theory of community-based enterprise. Acad. Manag. Rev. **31**(2), 309–328 (2006)

Phillips, R., Pittman, R.H.: A framework for community and economic development. An introduction to community development, 3–19 (2009)

Pittman, R.H.: Location, location, location: winning site selection proposals. Manag. q. **47**(1), 12 (2006)

Porfírio, J.A., Mendes, T.C., Felício, J.A.: From entrepreneurship potential in culture and creative industries to economic development: the situation of UK and southern European countries. International Entrepreneurship and Management Journal **14**(2), 329–343 (2018)

Priede-Bergamini, T., López-Cózar-Navarro, C., Benito-Hernández, S., Rodríguez-Duarte, A., Platero, M.: The dual effect of the age of the entrepreneur on the innovation performance of the micro-enterprises. Int. J. Entrep. Ventur. **11**(1), 81–102 (2019)

Prijon, L.: Role of private sector in local and regional development. Selected aspects of local and regional development, 86–103 (2012)

Reynolds, P.D., Hay, M., Camp, S.M., Bygarve, W.D., Autio, E.: Global Entrepreneurship Monitor: 2000 Executive Report. Kansas, MO: Ewing Marion Kauffman Foundation (2000)

Rieken, F., Boehm, T., Heinzen, M., Meboldt, M.: Corporate makerspaces as innovation driver in companies: A literature review-based framework. Journal of Manufacturing Technology Management (2019)

Ruggieri, R., Savastano, M., Scalingi, A., Bala, D., D'Ascenzo, F.: The impact of Digital Platforms on Business Models: an empirical investigation on innovative start-ups. Management & Marketing. Challenges for the Knowledge Society, 13(4), 1210–1225 (2018)

Taber, K.S:. The use of Cronbach's alpha when developing and reporting research instruments in science education. Research in Science Education, 48(6), 127 (2018).

Thu, N.Q., Hoang, T.T., Tan, H.K., Son, N.H.: Crowdfunding activities for startups: Approach from the theory of relationship marketing. Journal of Asian Business and Economic Studies **29**(5), 05–22 (2020)

Torri, M.: Community-based enterprises: a promising basis towards an alternative entrepreneurial model for sustainability enhancing livelihoods and promoting socio-economic development in rural India. J. Small Bus. Entrep. **23**(2), 237–248 (2010)

Warren, L., Mika, J., Palmer, F.: How does enterprise assistance support Māori entrepreneurs? An identity approach. J. Manag. Organ. **23**(6), 873–885 (2017)

Watson, T.J.: Entrepreneurial action, identity work and the use of multiple discursive resources: the case of a rapidly changing family business. Int. Small Bus. J. **27**(3), 251–274 (2009)
Weaver, P.A., Weber, K., McCleary, K.W.: Destination evaluation: the role of previous travel experience and trip characteristics. J. Travel Res. **45**(3), 333–344 (2007)
Welter, F.: Contextualizing entrepreneurship—conceptual challenges and ways forward. Entrep. Theory Pract. **35**(1), 165–184 (2011)
Wilkinson, K.P.: The community in rural America. Social Ecology Press, Appleton, WI (1991)

Intervention Economic Policy in Question: The Impact of Privatization of State-Owned Enterprises on Cost-Efficiency in Ethiopia. Insights from the Case of Ethio-Telecom

Alula Nerea Geberemeskel, Dominique Lepore, Mattia Tassinari, and Maruf Mohammad Sirajum Monir

Abstract In general, the paper questions interventionist economic policy approach by emphasizing the privatization policy of State-owned enterprises in the telecommunication sector of Ethiopia. Specifically, it investigates the impact of privatization policy on the cost-efficiency of Ethio-Telecom by comparing it with other telecom operators in liberal markets. Moreover, it also identifies sources of inefficiency and indicates paths for improvement. The study used secondary panel data of three decision-making units (DMU) or Organizations employing cluster sampling techniques for the years between 2007 and 2015. The study employs a Cost efficiency Data Envelopment Analysis to compare the impact of State-Owned Enterprises (SOE) to that of privately-owned enterprises in a competitive market. The results show that Ethio Telecom is inefficient as compared to the benchmark operators selected. Moreover, the main source of inefficiency comes from issues related to resource allocation. Furthermore, slack movements recommended to reduce capital expenditure and employee number while increasing total assets to attain 94% of the cost efficiency realized by a benchmark DMU.

Keywords DEA · Cost efficiency · Privatization · Intervention policy · Monopoly · SOE

A. N. Geberemeskel (✉) · M. M. S. Monir
Bucharest University of Economic Studies, 010374 Bucharest, Romania
e-mail: gebremeskelalula20@stud.ase.ro

D. Lepore · M. Tassinari
University of Macerata, 62100 Macerata MC 30, Italy

M. Busu (ed.), *Rethinking Business for Sustainable Leadership in a VUCA World*, Springer Proceedings in Business and Economics,
https://doi.org/10.1007/978-3-031-50208-8_9

1 Introduction

The role of government in the economy has been disputed over time starting from socialist and capitalist ideological tenets. Free market or neoliberal principles have become the leading world economic order after the end of the cold war. The dominant economic order has redefined the role of government in the economy as a "night watchman" role (Thorsen and Lie 2006). Though debatable, the new order brought significant economic growth, mainly, prominent neoliberal policy measures, such as liberalization, deregulation, and privatization, which was introduced to help business to insure higher profit rates aiming to drive production and way out from stagflation. As a result, business profit rates were improved despite it doesn't match the growth rate which had been exhibited during the golden age of capitalism which was influenced by Keynesian economic policy measures from 1950 to 1970 (Konicz 2011).

However, in Africa, the neoliberal economic paradigm had mixed results and was revealed to be not as effective during the implementation of structural adjustment programs (SAP). Since the 1980's many African countries like Ghana, Zambia, Angola, and Nigeria failed. Some scholars argued that perhaps SAPs never intended to assist African development (Fraser 2015). As a result, Africa was considered a confused continent when choosing its path for economic growth and development. Nevertheless, at the corners of Southeast Asia, the rise of economies like Malaysia, Taiwan, Singapore, South Korea, and later China provided effective alternatives to the neoliberal paradigm; resulting in visible and significant progress in economic and social development.

Elements of developmental growth path were seen since many post-colonial states as African states were considered as "developmentalist" almost by definition (Gibbon 1997). In this line of studies, Tian (2007) shows that the computer industry in China, which operates in a largely laissez-faire market environment is more efficient than the automobile industry, which enjoyed government protection and uses entry barriers and high tariffs.

Likewise, development was certainly a central preoccupation, and aware of the need for some nationalist-cum-developmentalist ideology for both nation-building and development (Mkandawire 1999) countries like Rwanda, Uganda, and Ethiopia are employing this path and have registered significant and continuous economic growth; as a result, this model has again emerged as point of debate concerning its effectiveness in terms of viable economic development policy in Africa (Gebremariam and Bayu 2017).

Literature suggests that the interventionist economic development approach has resulted in corruption, bad governance, and monopoly of power. A study made by Ioan (2009) across 135 countries has found that corruption is high when there is government intervention in the economy. Thus, interventionism is still debatable in many countries that have different political and social contexts. Even if the active involvement of the state in the economy has often demonstrated severe political failures in sector management and operation (Nepal and Jamasb 2015), also raises questions about the motives of privatization (Nepal and Foster 2015).

Even if there is no consensus among government, opposition parties, elite citizens and multilateral development discuss the success of interventionist developmental state economic policy in Ethiopia. In the last three decades, since Ethiopian People's Revolutionary Democratic Front (EPRDF) took power in 1991, the policy is under enforcement. The points of debate over this policy have been changing over time. The first point of argument used to emanate workability concerns related to whether this approach can ensure economic growth in general or if it fits Ethiopia's context. Current answers to such questions seem to be obsolete since Ethiopia's interventionist economic policy has exhibited economic growth over the past patrimonial two and a half decades.

The second argument is whether corruption, the hegemony of political power, bad governance, and economic inefficacies are related to this model. Groups in favor of the first point of the argument are supporters of the incumbent administration whereas the second strand is opposition political parties and intellectuals affiliated with the opposition camp. Considering concerns of the opposition camp, they argue against the interventionist approach since they assume it will affect the efficiency of the economy moreover sectors that could yield a higher benefit from competition, such as the telecom sector. Moreover, according to Human Rights Watch (2014) report the government has been using internet monopoly to restrain lawful political demonstrations by interrupting the internet, and phone networks and surveilling high-profile diasporas. Similarly, according to European Parliament (2021), a motion for a resolution in the currently ongoing civil war in the Tigray region of Ethiopia revealed government shut down all telecom services to hide war crimes.

The incumbent government of Ethiopia argues that selective interventionist economic policy on sectors that have been identified as strategic is necessary to fasten economic growth. Moreover, they can be sources of development if financial and leadership assistance is provided to expand and generate ransom incomes in their sectors.

According to Investment Proclamation No. 769/2012, the government is entitled to monopoly power in the telecom sector (Federal Democratic Republic of Ethiopia 2012). This sector is protected both from local and international competitors. Telecommunications service was introduced in 1894. Ethiopian Telecommunication Corporation (ETC), which was established by the Council of Minister Regulation No 47/1999 as the only incumbent state-owned telecom operator enterprise (Estifanos 2010), characterized by weak service quality and expensive charge which has resulted in frustrations of clients (Adam 2010). Recently, in a regular meeting of EPRDF the Executive Committee, which was conducted on June 5, 2018, announced decisions to privatize greater shares of Telecom to assure better development and continuous growth which would benefit the society at large (Council of Secretariat of the Office of EPRDF 2018).

Notwithstanding the argument on the necessity of partial privatization decision to fasten economic growth, which is contentious in different economic and political ideologies, the scope of the research covers investigations on one of the rationales for SOE's partial privatization policy shift through measuring the impact of interventionist economic policy on SOE's cost efficiency in Ethio Telecom. Primarily, the role of the government is reviewed, focusing on the impact on cost efficiency. Second, the case of Ethio Telecom is presented and discussed. Lastly, conclusions and policy implications are drawn.

2 Literature Review

2.1 The Evolving Role of a Government in Economy

The intervention of governments has been part of many civilizations since the role of rulers was to keep peace in society. However, the level and type of intervention could be different depending on the socio-political and economic context of a country. Liberal thinkers like John Locke perceive government intervention as interference or intrusion in citizen's life with the exception of protecting property rights; on the contrary, socialists perceive government intervention as a necessity to establish peace and order, economic stability, and justice in a society. According to Van der Waldt (2015), the notion of a government's intervention lies in three arenas—political, social, and economic. In this vein, there are two main ideological perspectives. At the one extreme of the ideological spectrum; neo-liberals seek to reduce the State's role to the level of 'protector' of private property rights (Huffman 2013). This view was advocated in the seventeenth century by capitalist protagonists, such as Adam Smith. On the contrary, socialite or neo-Marxist thinkers advocated for governments to assume full ownership and control over the economy. In this vane, Kropotkin (1995, p 97) argues that state intervention is a prominent feature of capitalism. According to him, "Nowhere has the system of 'non-intervention of the State' ever existed". Similarly, Polanyi (1944) notes that even though many ideological proponents of capitalism thunder loudly against state intervention, the fact is that capitalism itself generates causes that lead to the need for such economic intervention.

Even if the necessity of intervention is inevitable, the crucial question shall be how to determine the degree of intervention concerning each political economy ideology. Keeping this perspective, there are basic roles of the government in developing market economies, which are much more different from the roles of government in developed countries' market economies in terms of intent, level, and mechanism of intervention. Going beyond ideological divisions, Benhassine and Raballand (2009), claim that the degree of selectivity—in terms of sectors or targeted categories of firms—and the extent of price subsidies embedded in such interventions determine how much the intervention departs from neutral policies. These two aspects of interventions respond to different types of justifications, and they do not automatically need to go together, even if it frequently happens. Depending on these two factors, the Authors suggest that interventions are shown to vary in their distortions, their benefits, and

Table 1 Types of government Intervention

	Traditional instruments	Market-based approaches
Providing public services	Direct provisions	Competitive tendering & user choice
Influencing private markets	Regulation tax and subsidy	Trading schemes, self-regulation

Note Types of government in a market, from Office of Fair Trade (OFT) United Kingdom. Copyright 2009 by Crown

their opportunity costs. Furthermore, according to the Food and Agriculture Organization of the United Nations (FAO), the intent of government interventions should be to facilitate market competition and help the market achieve national policy objectives. In this sense, policies should aim for effective workability with the existing capitalist system, not to replace it. Since the government attempts to replace free market systems have often raised the costs due to the anti-competitive nature of the market as a result distorted market prices and resource allocations could lead to an adverse impact on the economy. According to Sekkat (2009), the existence of State-Owned Enterprises is an obstacle to competition, based on his findings for Jordan, Morocco, and Egypt.

According to Van der Waldt (2015), the level of intervention may depend on various factors: such as the growth trajectory of the country; demographics—culture groups; historic realities—colonization; the maturity of its democracy (political system); geographical circumstances—size, location, growth hubs, urban–rural settings; and availability of natural resources. Thus, a country, which registered low performance regarding these factors, may require strong economic protection that leads to government intervention. According to the OFT (2009), the government intervenes in a market in a versatile manner depending on the objective, level, and type of a problem, the type and maturity of the economic system, and effectiveness and efficiency of intervention mechanisms. In the economy, the types of government intervention are divided into traditional and market-based approaches corresponding with intervention tools as shown in Table 1.

2.2 *Direct Intervention Through SOEs and Implication on Cost Efficacy*

The impact of interventionist economic policy to measure SOEs' performance from a cost-efficiency point of view becomes relevant since it is one of the strongest rationales why the interventionist economic policy is criticized as a manifestation of government failure, which is a potential cause for the recommendation of privatization.

In this aspect, there is diverging research literature. Research that collected results from 71 state-owned companies demonstrated comparatively lower efficiency in 56 cases (Mueller 2003). Similarly, Shirley and Walsh (2000), who surveyed performance differences between SOEs and private corporations in 52 studies in the monopolistic utility sectors, shows that only five state-owned enterprises outperformed private corporations.

A study based on a share of firms in China from 2001 to 2006 that aimed to test the relationship between government intervention and investment efficiency shows that SOE's investments were less efficient than non-SOE. Furthermore, intervention related to political connections with SOEs has resulted in a significantly negative impact on efficacies because of weak investment sensitivity in SOEs (Chen et al. 2011).

In addition, Nepal and Jamasb (2015) in reference to the electricity sector acknowledge how the reliance on a market-driven model reflects a political belief in the efficacy of the markets and that the active involvement of the state in the sector across developing countries has often demonstrated severe political failures in sector management and operation as evident among some transition and most South Asian countries.

Instead, by considering data on private, state-invested (i.e., partially state-owned), and state-owned enterprises in the telecommunications industry, Castelnovo et al. (2019) suggest that public ownership has a negative impact on firm-level total factor productivity. This effect is however mitigated by high external institutional quality and even reversed in some countries with a favorable institutional environment. Whereas, studying China, Xie and Zhang (2020) reveal that the positive relationship between government intervention and corporate cash holdings of firms is alleviated as the anti-corruption intensity increases and that this alleviation effect is more relevant for SOEs, firms in high intervention areas and firms without political connections.

Furthermore, according to Widodo (2017), the existence of conflicting and multiple objectives (socio-political) is identified as the major reason for SOE inefficiency. Moreover, the combination of rent-seeking behavior and low productivity will cause a higher cost for the SOEs—which increases the potential for cost inefficiency (Li et al. 2014).

There are also arguments supporting that ownership does not necessarily affect efficiency. According to Sueyoshi (1998), efficiency results are not necessarily positive in the property rights theory, and this assertion depends upon the type of industry studied. Therefore, whether a firm is state-owned or privately owned; efficiency can only be achieved as long as the firm operates in a competitive market, gives full autonomy to the management to make crucial business decisions based on market signals, and where performance-based compensation is provided.

To the same extent, Nellis (1994) argued that efficiency does not depend on ownership. In this context, Gao et al. (2019) found that the impact of fiscal decentralization on over-investment, under-investment, and investment efficiency is not different between SOE and non-SOE. Moreover, Nepal and Foster (2015) by comparing the economic performance between the privately and state-owned electricity networks in Australia across three dimensions: prices, quality, and investment, find that privately owned networks are not worse off than the state-owned networks in terms of performance. The long-term concerns on future investments, security of supply, climate change, and economic regulation of networks will continue to prevail once the short-term efficiency gains from privatization are exhausted. These concerns imply that the role of the state will still be significant, although transformed, raising questions on the motives of privatization. Instead, looking into the telecommunication sector, Wallsten (2003) tests whether establishing a regulatory authority before privatizing incumbent telecommunications firms is significant. The Author finds that established regulatory authorities prior to privatization recorded increased telecom investment and telephone penetration with respect to countries that did not.

2.3 *Telecommunication Sector in Ethiopia*

Telecommunications service in Ethiopia used to be administered under the Ministry of Posts Telephone and Telegraph until 1952. However, in 1952 the Ethiopian Telecommunication Board was established autonomously with the role of regulating and operating telecommunication services, changed to ETA.

In 2002 according to Proclamation No. 281/2002, ETA was divided into Ethiopian Telecommunication Agency (ETA)—a separate federal telecommunication regulatory entity and Ethiopian Telecommunication Corporation (ETC). The government-owned the monopoly operator, which was later called Ethio Telecom.

Ethiopian Telecom provides (PSTN) Public Switched Telephone Network used for fixed line and dial-up internet, Mobile telephone, Internet, and data communication services. In 2018–2019 total number of subscribers reached 43.6 million. Mobile voice subscribers are 41.92 million (89.95%), Data and Internet users are 22.3 million (53.44%), and fixed Services 1.2 million (2.75%), showing a teledensity—number of telecom connections for every hundred inhabitants in a specific geographic area—of 44.5%. (Ethio Telecom 2019). Over the last two decades, apart from government direct budget finance from national resources, the role of international debt finance was instrumental. Extensive expansion projects were executed partly by China's Technical and financial loans funded through export credit facilities. According to China development bank (CDB) and China Exim Bank (export–import bank) offered 1.5 billion USD in 2006 Ethio telecom which the project is commissioned by ZTE corporation.

However, the Ethiopian telecom market monopolized by Ethio Telecom is characterized by excessive pricing (especially in the broadband and international direct dialing segment), poor service quality, inefficiency, vendor lock-in, and the absence of choice (Adam 2010). In the debate on Ethiopia's stated controlled telecommunications sector, public–private partnership has been addressed as a sustainable model for telecommunication operation (Workneh 2018).

3 Methodology

This research may contribute insight for data-driven decisions from the vantage point of an economic criterion of privatization in general and cost-efficiency dimensions in specific. Thus, this research shall investigate and measure interventionist policy as to cost efficiency criterion, answering three main research questions:

- What is the impact—magnitude, and direction—of the governmental intervention on the cost efficiency of Ethio Telecom?
- What are the sources of cost efficiency in the case of Ethio Telecom in comparison to privately owned telecom operators in a competitive market?
- What are alternative paths for cost efficiency improvements and their policy implications?

To investigate the research questions highlighted, the research employed an ontological research paradigm since it is "the study of being" Crotty (1998, p. 10). In other words, it answers "What is" on the matter or phenomena under investigation. This paradigm is aligned with positivist research philosophy, linked to reality or realism.

Realism is the view that objects have an existence independent of the knower Cohen et al. (2007, p. 7). According to Levin (1988), positivists believe that reality can be observed and described objectively. Thus, the researcher's view and role in comprehending the phenomena is positivist as the researcher claims both the objective and single reality besides the objective of the research, which in this case is to know "what is" the level and direction of intervention policy impact impartially. As a result, the role and influence of the researcher's perspective on the research are restricted in the research process to keep objectivity in accordance with the ontology research paradigm.

The positivist research paradigm underpins on quantitative methodology to maintain objectivity since it prescribes a fixed design, which favors to more restrictive option (Tuli 2010). Moreover, as it is represented in Fig. 1, this research is designed following a deductive research approach—general to specific—since the research aims to investigate research questions that emanate from the theoretical discourses in the context of Ethio-telecom at an organizational level.

3.1 Data and Model

Data Envelopment Analysis (DEA): DEA is a linear mathematical programming model used to evaluate the relative efficiency (in terms of magnitude and direction) of decision-making units (DMU's) or organizations. This model is based on the mathematical concept of Lagrange function. The following nomenclature is adopted from Sueyoshi (1997).

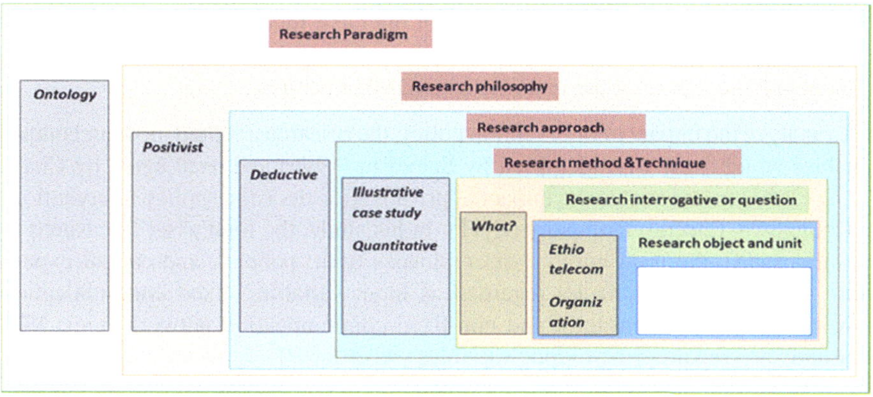

Fig. 1 Methodological framework. *Source* Adapted from (Saunders et al. 2009, p. 108)

$$Min\ C_k = \sum_{i=1}^{m} p_i x_i$$

$$s.t - \sum \lambda_j / X_{ij} + x_i \geq 0,\ i = 1, \ldots, m$$

$$\sum \lambda_i Y_{rj} \geq Y_{rk},\ \ r = 1, \ldots, s$$

$$L \leq \sum \lambda_j \leq U$$

$$\lambda_j, x_i \geq 0,\ j = 1, \ldots, n;\ i = 1, \ldots, m$$

where: the sum of λ_j is restricted by its upper ($U \geq 1$) and lower ($0 \leq L \leq 1$) bounds; and p_i is the input price for the ith input variable. Since pi and x_i are incorporated, an optimal x*i, minimizes the total cost of C*k. Consequently, the level of cost efficiency is determined by C k*/C k

DEA Cost Efficiency model: This model indicates to what extent the cost target DMU can be reduced while producing the same level of output relative to other DMU's or CERS is a relatively cost-efficient set of units) is a set of relatively efficient DMU and respective weights are used as a benchmark to measure the efficiency levels of other DMU (decision-making unites). CERS, based on:

- Source and slack the product or mixture of the technical and allocation efficiencies
- Efficiency Score (θ*) 0–1, 1 relative highest efficiency, 0 Relatively Inefficient

Model Orientation- The Input oriented DEA was considered to analyze the efficiency performance of Ethion Telecom in terms of how the minimum cost of inputs can be reduced to produce a current level of output, based on:

- Number of Inputs: Three

- Variable Retunes to Scale: VRS is one of the DEA measurements which is used to test non-proportional change of input and output. It's selected due to the monopolistic nature of the market where DMU operates.

To analyze the impact of intervention policy, the research adopted input and output variables which were first identified by Sueyoshi (1996) and used again by Chao-Chung (2007) to investigate the impact of privatization decision (policy intervention) on Chunghwa Telecom Company (CHT). In his study, the total asset (measured in NT$ thousand), the total number of employees (unit: people), and capital expenditure (unit: NT$ million) are regarded as input variables; fixed communication network (unit: NT$ thousand), and non-fixed communication network (unit: NT$ thousand) are regarded as output variables. Also let P1 = owner's equities/total assets, P2 = total labor cost/employee, and P3 = depreciation cost/capital expenditure be the input price of total assets, employee cost, and capital expenditure be the input variables, respectively. The variables highlighted in (Table 2) are adopted from (Chao-Chung 2007) .

This study is based on data from three telecom operators in Africa: namely, Ethio Telecom, safari, and Vodacom for the period 2007–2015. The sources of data are comprehensive annual reports of Ethio Telecom, Safaricom Kenya, and Vodacom South Africa under sections of the statement of financial position and profit and loss. According to Shahooth and Al-Delaimi (2006) to ensure sufficient degrees of freedom for making a meaningful analysis, this study used three DMUs as per recommendations made by Shahooth.

In addition, the number of subscribers is considered as a selection criterion to minimize the impacts of scale efficiency (efficiency that arises from economies of scale) on cost efficiency. However, for the purpose of generalization of results and to control the researcher's involvement in data selection, randomized cluster sampling is employed as a probability sampling technique as advised in the positivist pregame (Chirkov 2015) Finally, the cost efficiency impact of intervention policy is measured using a linear programming software called DEA max 7.11.

Table 2 Definition of variables

Variables	Measurement	Input–Output	Price
Operating revenue	USD $	Output	Operating income
Total asset	USD $	Input	Owner's equities/total assets
Total number of employees	Head count	Input	Total labor cost/employee
Capital expenditure	USD $	Input	Depreciation cost/capital expenditure

Source Authors' compilation

4 Results and Discussion

As for the first research question (RQ1), the average cost efficiency score resulted to be negative, since Safaricom and Vodacom registered scores of 0.7303 and 0.9413 which are higher by 0.1109 and 0.3218 from Ethio telecom's efficiency score of 0. 6194 respectively. Therefore, the impact—magnitude, and direction—of intervention economic policy on the cost efficiency of Ethio telecom results to be negative. More specifically, Ethio telecoms performance in terms of cost is inefficient by 11% and 32% as compared to Safari and Vodacom Ethio Telecom is operating at 68% efficiency to generate average annual operating revenue of 704.9 million USD, as presented in Fig. 2.

As for the second RQ2, a decomposition analysis of cost efficiency was performed to understand the sources of cost efficiency of Ethio Telecom in comparison to a privately-owned telecom operator in a competitive market. Precisely, the decomposition cost efficiency scores are examined in terms of technical and allocative efficiency scores to know from where efficiency or inefficiency comes. The results show that:

- Allocative efficiency contributed 0.68 whereas technical efficiency contributed 0.89 to achieve a total cost efficiency score of 0.62 in a year b/n 2007 and 2015.
- Cost inefficiency which arises from technical parts is only 11% whereas cost inefficiency which arises from allocative parts amounts to 32% thus the main source of inefficiency is from allocative.
- Compared to Safaricom and Vodacom the reverse is true, are indented.

As indicated in Fig. 3, technical efficiency has contributed more than allocative efficiency since its efficiency score is near one relative to the allocative efficiency score. Also, as we look at the trends, technical efficiency has less perturbation than allocative efficiency.

Thus, from the above figure, we can generalize that technical efficiency has contributed to a higher value in a constant manner to the cost efficiency of Ethio

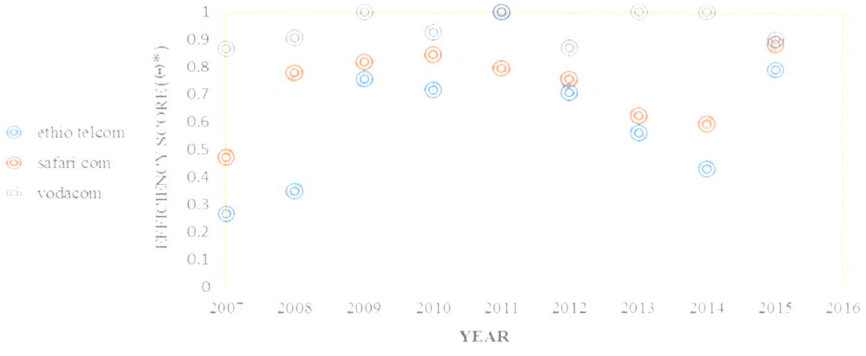

Fig. 2 Cost efficiency score trend from 2007 to 2015. *Source* Own computation

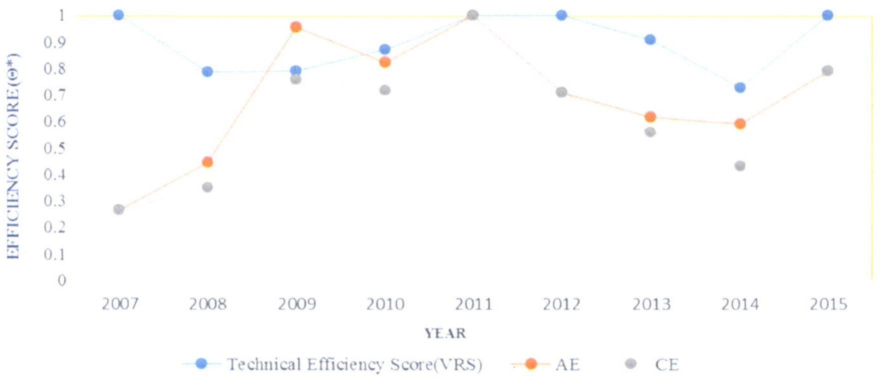

Fig. 3 Decomposed cost efficiency trend. *Source* Own computation

Telecom than allocative efficiency. In other words, we can conclude that allocative efficiency is unstable and a major source of cost inefficiency as compared to technical efficacy. Lastly, in reference to cost efficiency improvements and policy implications (RQ3) the analysis proves evidence on which type of cost arrangements should be adjusted to improve inefficient DMU (Ethio Telecom) to reach the level of efficient DMU (Safaricom) Or, it shows a direction and magnitude that has to be altered to make inefficient DMU as efficient as relatively efficient DMU. The results show that:

Ethio Telecom can be 94% cost efficient as Vodacom does if it can reduce the cost of inputs by 32% through reducing capital expenditure by 241.9 million USD, employee number by 4,297 and increase total assets by 7.9 million USD.

Figure 4 highlights how much Ethio Telecom must reduce or increase inputs to operate cost efficiently as a benchmark DMU (Vodacom). In line with this interpretation, on average Ethio Telecom has to reduce capital expenditure by 241.9 million USD and employee number by 4,297 whereas the total asset has to be increased by 7.9 million USD.

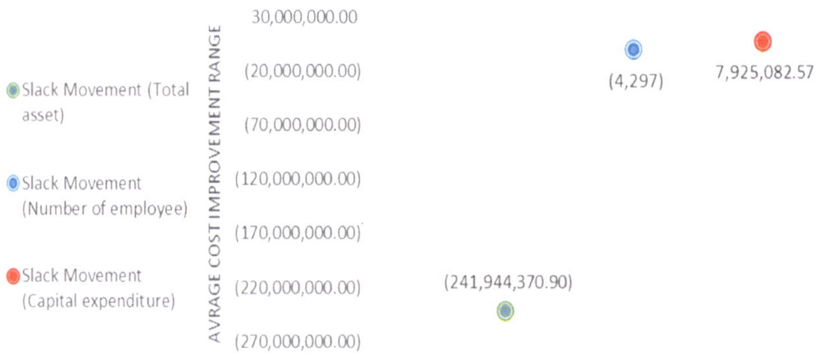

Fig. 4 Normalized average slack value in USD. *Source* Own computation

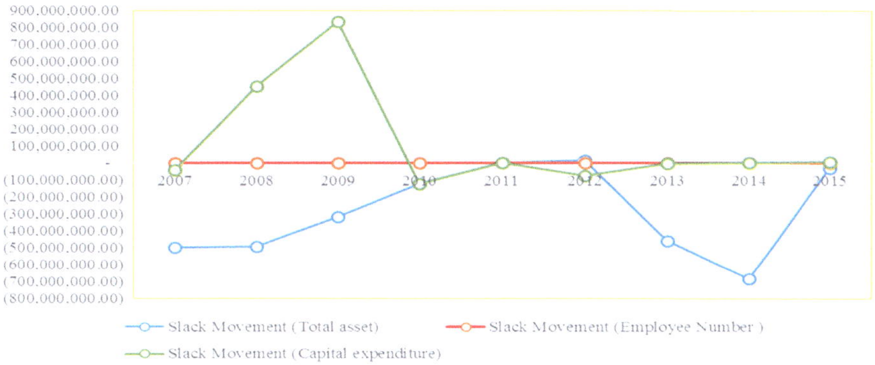

Fig. 5 Slack movement trend of Ethio Telecom in USD from 2007 to 2015. *Source* Own computation

Figure 5 shows a trend of cost remedial action that makes Ethio telecom input costs as efficient as Vodacom. Slack movement in total assets has recommended spending or costs to increase until 2011. Whereas, capital expenditure and employment numbers are recommended to reduce until the same year. However, in the year 2011 slack value movement indicators did not recommend any value to increase or decrease from the actual performance of 2011 due to a relative efficiency score of one ($\theta^* = 1$).

5 Conclusion

In general, governments introduce various interventions aiming at reaching sustainable economic development. In this research, we question FDRE's interventionalist approach. Special emphasis is given to privatization policy to investigate its impact on the cost-efficiency of SOE enterprises by comparing Ethio Telecom with other telecom operators in liberal markets. A reason that strikes this research is the massive privatization program announced by the Executive Committee of EPRDF on June 5, 2018, and currently executed by the prosperity party.

The privatization program has resulted in good and bad outcomes in different countries. For example, SAP is ranked among the worst outcomes in Africa. Thus, this research intended to investigate objectively from a cost-efficiency perspective the impact of the policy intervention for Ethio Telecom. Therefore, it also identifies sources of cost efficiency and indicates alternative paths of cost efficiency improvement for Ethio Telecom. For this aim, data is collected from two telecom operators (Vodacom and Safaricom) using cluster random sampling. Then, type two cost DEA which constitutes varieties of characteristics such as input-oriented and VRS is employed using DEA Max 7.11. Moreover, to assert the objectivity of the research ontological research paradigm and positivist research philosophy is used.

Outputs of the research indicate that Ethio Telecom- which operates in a monopoly regime resulted in an average efficiency score of 0.61, which indicates 32% cost inefficiency as compared to Vodacom's benchmark cost efficiency score of 0.94. Scores of cost efficiency are identified by decomposing cost inefficiency into technical and locative inefficiency. In this regard, the source of inefficiency of Ethio Telecom is aroused from allocating inefficiency as compared to technical inefficiency since cost inefficiency which arises from technical efficiency is only 11% whereas allocative inefficiency amounts to as high as 32%.

Finally, alternative paths for improvement are recommended using slack values. In this regard, slack movements have recommended input adjustment that reduces capital expenditure by 241.9 million USD and employee number by 4,297 whereas total asset is recommended to increase by 7.9 million USD on average to attain 94% cost efficiency realized by Vodacom which is a benchmark operator.

Thus, according to the results, to promote Ethio Telecom cost efficiency, it is necessary to consider shifting from a protected monopoly to a competitive market structure. The study suggests that allocative inefficiency could be improved by minimizing government intervention in the telecom sector by promoting both privatization and competitiveness in a market.

References

Adam, L.: Ethiopia ICT sector performance review. In: Towards Evidence-Based ICT Policy and Regulation, pp. 32–33 (2010)

Benhassine, N., Raballand, G.: Beyond ideological cleavages: a unifying framework for industrial policies and other public interventions. Econ. Syst. **33**(4), 293–309 (2009)

Castelnovo P., Del Bo, C.F., Florio, M.: Quality of institutions and productivity of state-invested enterprises: international evidence from major telecom companie. Eur. J. Polit. Econ. **58**, 102–117 (2019)

Chao-Chung, K. (ed.): Measuring the production and cost efficiency in telecommunications industry: The Taiwan case. Proceedings of the Eastern Asia Society for Transportation Studies; Eastern Asia Society for Transportation Studies Vol 6, The 7th International Conference of Eastern Asia Society for Transportation Studies, (2007)

Chen, S., Sun, Z., Tang, S., Wu, D.: Government intervention and investment efficiency: evidence from China. J. Corp. Finan. **17**(2), 259–271 (2011)

Chirkov, V.: Fundamentals of Research on Culture and Psychology: Theory and Methods. Taylor & Francis (2015)

Cohen, L., Manion, L., Morrison, K.: Research Methods in Education, 6th edn. Routledge, London (2007)

Council of Secretariat of the Office of EPRDF.: Eprdf.org.et. http://www.eprdf.org.et/web/en/-/statement-of-eprdf-s-executive-committee (2018). Accessed 21 Dec 2018

Crotty, M.: The Foundations of Social Research. Sage, London (1998)

Estifanos, T.: Unctad.org/sections. unctad.org/sections/wcmu/docs (2010). Accessed 20 Dec 2018

Ethio telecom Report. https://www.ethiotelecom.et/2018-19-efy-p-reporte/ (2019). Accessed 01 Nov 2019

European Parliament.: https://www.europarl.europa.eu/doceo/document/B-9-2021-0484_EN.html (2021)

Federal Democratic Republic of Ethiopia.: Federal NegaritGazeta Investment Proclamation No. 769/2012 (2012)
Fraser L.: Did structural adjustment programs assist African development? http://www.eir.info/2015/01/13/did-structural-adjustment-programmes-assist-africandevelopment/ (2015)
Gao, S., Wang, L., Liu, N., Zhang, M.: Fiscal decentralization and corporate investment: empirical evidence from China. J. Econ. Policy Reform **22**(1), 51–68 (2019)
Gebremariam, M., Bayu, M.: Ethiopia: a democratic developmental state? ILIRIA Int. Rev. **7**(2), 153–170 (2017)
Gibbon, P.: Civil society, politics and agricultural development. In Lindberg, S., Sverisson, A. (Eds.) Social Movements in Development. Macmillan, London (1997)
Huffman, J.: Private Property and State Power: Philosophical Justification. Palgrave Macmillan, New York, NY (2013)
Human Rights Watch.: They know everything we do: Telecom and internet surveillance in Ethiopia. Human Rights Watch (2014)
Ioan, M.: The effect of the government intervention in economy on corruption. Eurasian J. Bus. Econ. **2**(3), 55–71 (2009)
Kropotkin, PA.: Evolution and environment. Monereal: Black Roses Books (1995)
Konicz, T.: The end of the golden age of capitalism and the rise of neoliberalism, dio. https://www.indybay.org/newsitems/2011/01/20/18669647.php (2011)
Levin, D.M.: The Opening of Vision: Nihilism and the Postmodern Situation. Routledge, London (1988)
Li, S., Lin, Y.C., & Selover, D.D.: Chinese state-owned enterprises: are they inefficient? Chin. Econ. **47**(5–6), 81–115 (2014)
Mkandawire, T.: Shifting commitments and national cohesion in African countries. In: Wohlgemuth, L., Gibson, S., Klasen, S., Rothchild, E. (Eds.) Common Security and Civil Society in Africa. Nordiska Afrikainstitutet, Uppsala, pp. 14–41 (1999)
Mueller, D.: Public Choice III. Cambridge University Press (2003)
Nellis, J.: Is privatization necessary? Public Policy for the Private Sector. Public policy for the private sector, pp. 1–4 (1994)
Nepal, R., Jamasb, T.: Caught between theory and practice: government, market, and regulatory failure in electricity sector reforms. Econ. Anal. Policy **46**, 16–24 (2015). https://doi.org/10.1016/j.eap.2015.03.001
Nepal, R., Foster, J.: Electricity networks privatization in Australia: an overview of the debate. Econ. Anal. Policy (2015). https://doi.org/10.1016/j.eap.2015.10.001
Office of Fair Trade.: Types of government in a market why competition matters—A guide for policy-makers (2009)
Polanyi, K.: The Great Transformation: The Political and Economic Origins of Our Time. Beacon Press, Boston (1944)
Saunders, M., Lewis, P., Thornhill, A.: Research Methods for Business Students. Pitman Publishing, Edinburgh Gate (2009)
Sekkat, K.: Does competition improve productivity in developing countries? J. Econ. Policy Reform **12**(2), 145–162 (2009)
Shahooth, K., Al-Delaimi, K.: Using data envelopment analysis to measure cost efficiency with an application on Islamic Banks. Sci. J. Admin. Dev. **4**, 135–156 (2006)
Shirley, M.M., Walsh, P.: Public versus private ownership: the current state of the debate. Policy Research Working Papers. Washington DC: World Bank (2000)
Sueyoshi T.: Divestiture of nippon telegraph and telephone. Manag. Sci. **42**(9), 1326–1351 (1996)
Sueyoshi, T.: Measuring efficiencies and return to scale of Nippon Telegraph & Telephonein production and cost analyses. Manag. Sci. **43**(6), 779–796 (1997)
Sueyoshi, T.: Privatization of Nippon Telegraph and Telephone: was it a good policy decision? Eur. J. Oper. Res. **107**, 45–61 (1998)
Thorsen DE, Lie A. What is neoliberalism. Oslo, University of Oslo, Department of Political Science, Manuscript. 1–21 (2006)

Tian, L.: Does government intervention help the Chinese automobile industry? A comparison with the Chinese computer industry. Econ. Syst. **31**(4), 364–374 (2007)

Tuli, F.: The basis of distinction between qualitative and quantitative research in social science: reflection on ontological, epistemological and methodological perspectives. Ethiop. J. Educ. Sci. **6**(1), 103–106 (2010)

Van der Waldt, G.: Government interventionism and sustainable development: the case of South Africa. Afr. J. Public Affairs **8**, 35–51 (2015)

Wallsten, S.: Of carts and horses: regulation and privatization in telecommunications reforms. J. Policy Reform **6**(4), 217–223 (2003)

Widodo, A.P.P.: Board member duality, agency costs, performance, and accountability: the evidence in parent and subsidiary of state-owned enterprises in Indonesia. Doctoral dissertation, Victoria University (2017)

Workneh, T.W.: State monopoly of telecommunications in Ethiopia: origins, debates, and the way forward. Rev. Afr. Polit. Econ. **45**(158), 592–608 (2018)

Xie, J., Zhang Y.: Economic systems, anti-corruption, government intervention, and corporate cash holdings: evidence from China. Economic Systems, 100745. In Press, 13 February 2020, Corrected Proof (2020)

AI for Car Damage Detection and Repair Price Estimation in Insurance: Market Research and Novel Solution

Vladimir Ghita, Denis Iorga, Laurentiu-Marian Neagu, Mihai Dascalu, and Gheorghe Militaru

Abstract The current study focuses on Artificial Intelligence (AI) products for car damage detection and repair price estimation in the insurance industry. The opportunity to introduce such products in the Romanian local market is highlighted following a scoping review of potential benefits and challenges, existing commercial solutions, and existing market research. The lack of market research data concerning the challenges faced by Romanian customers and employees of insurance companies during the insurance claim process is identified as a gap. To address it, the current work presents the results of a pilot survey-based market research that sought to understand the challenges faced by two main Romanian stakeholders of the insurance claim process, namely drivers involved in car accidents (N = 20) and car damage inspectors (N = 15). The result are used to define and advance a novel architecture for an AI system for car damage detection and repair price estimation (InsureAI) that aims to: (a) streamline the communication process between customers and representatives of insurance companies, (b) minimize appointment and traveling challenges related to car damage inspection, (c) reduce the complexity of the repair decision and price estimation related to the insurance claim procedure, (d) address the challenges of customers of car insurance companies in organizing and filing in the necessary

V. Ghita (✉) · L.-M. Neagu · M. Dascalu · G. Militaru
University Politehnica of Bucharest, Splaiul Independentei 313, 060042 Bucharest, Romania
e-mail: vladimir.ghita@grx.ro

L.-M. Neagu
e-mail: laurentiu.neagu@upb.ro

M. Dascalu
e-mail: mihai.dascalu@upb.ro

G. Militaru
e-mail: gheorghe.militaru@upb.ro

V. Ghita · L.-M. Neagu · M. Dascalu
Global Resolution Experts S.A., Nutu Ion Street 44, 4th Floor, 050762 Bucharest, Romania

D. Iorga
University of Bucharest ISDS, Regina Elisabeta Boulevard 4-12, 030018 Bucharest, Romania
e-mail: denis.iorga@drd.unibuc.ro

M. Busu (ed.), *Rethinking Business for Sustainable Leadership in a VUCA World*,
Springer Proceedings in Business and Economics,
https://doi.org/10.1007/978-3-031-50208-8_10

information for starting a new claims file, as well as (e) enhance the overall process transparency. Additionally, we detail the envisioned user flow and the user interface prototype.

Keywords Artificial intelligence · Car damage · Car price estimation · Car insurance · Market research

1 Introduction

Recent advancements in Artificial Intelligence (AI) development have led to a proliferation of AI-based solutions in various industries. One such industry is the car insurance industry, which has recently seen the emergence of numerous commercial solutions that leverage AI to detect car damages and estimate repair costs. Such solutions can streamline the insurance claim process, thus leading to benefits for car insurance companies. At the same time, such solutions can also lead to new challenges for stakeholders of the car insurance claim process. Nonetheless, the current state of existing commercial solutions is yet to be documented.

To this end, the current work first expands on the potential benefits and challenges of using AI for car damage detection and repair price estimation in the insurance sector. Next, a scoping market review is conducted to identify existing commercial solutions for the insurance sector. Next, a scoping review of existing market research on this topic is conducted. Given the identified gaps, the current work then presents the method and results of a pilot local Romanian market analysis concerning the challenges and problems encountered by both customers and employees in the car insurance industry. Finally, the challenges identified through a quantitative approach based on surveys are used to define the high-level functional architecture of a novel AI system for car damage detection and repair price estimation.

2 Literature Review

2.1 Benefits and Challenges

Using AI for car damage detection and repair price estimation greatly impacts the car insurance sector, and the implications are twofold. On one hand, implementing such AI-based solutions enhances the efficiency of the insurance claim process by reducing the required time and human resources. On the other hand, it may raise unforeseen complications and challenges. The current section reviews existing literature to expand on AI's potential benefits and challenges for car damage detection and repair price estimation.

Starting from the work of Corea (2019), we identified three business areas with potential benefits and positive impacts of using AI for car damage detection and repair price prediction: costs, customers, and risk awareness. First, new AI systems for car damage detection and repair price estimation can lower costs for both companies and customers (i.e., staff and travel costs). Second, such systems can engage and retain customers, especially tech-savvy ones. Third, new quantifications of risk can be obtained to benefit insurance companies and customers.

Nevertheless, AI solutions for car damage detection and repair price estimation can also pose challenges. At least five potential typologies of challenges posed by such solutions can be derived using the work of Bhatt et al. (2021), Lutz (2019), and Fabris et al. (2021), namely: damage detection errors, a large volume of data required for developing performant systems, implementation challenges (e.g., hyperparameter tunning), explainability problems, and the amplification of inequalities. Given the potential of AI solutions for car damage detection and repair price estimation to yield both advantageous outcomes and challenges, it is essential to guide their development by capitalizing on the benefits and mitigating the associated risks.

2.2 *Existing Commercial Solutions*

This section presents a scoping review of existing AI commercial solutions for automatic car damage detection and repair price estimation. The analysis was carried out by referring to an existing solution in the Romanian market, namely Qapter. Following a series of scoping searches, five international alternatives that address the same type of market segment as Qapter were identified, each briefly presented below.

Qapter, an AI solution for automatic car damage detection and repair price prediction, is offered by Solera (2023). The key functionalities are grouped into two modules: Qapter Mobile Estimating and Qapter Intelligent Damage Detection. As the name implies, the latter allows for automated car damage detection from images using AI, while the former estimates the repair price. Audatex Romania, as part of the Solera group, introduced the Qapter solution to the Romanian market in 2019 (Conceatu 2022).

Tractable provides an AI-based solution that leverages artificial vision to evaluate images of cars and properties, streamlining the inspection and repair process (2023). Tractable's customer base primarily consists of B2B (business-to-business) entities operating in the auto or property insurance industry, as well as auxiliary segments such as the automotive recycling industry. With regards to the latter, Tractable differentiates itself from Qapter through its functionality of identifying salvageable parts from a damaged car. The company achieved unicorn status in 2021, securing significant investments for its services in the auto and property insurance segments (Nonninger 2021). According to a Deloitte Private (2021) report, Tractable is one of the fastest-growing companies in the UK insurtech industry.

CCC Estimate STP is designed to automate the claim process for damaged cars (2021). The solution is developed by CCC Intelligent Solutions, the leading US provider of vehicle collision software and workflow tools to insurance companies, automakers, accident repairers, and parts suppliers (Advent International: CCC Intelligent Solutions 2017). Similar to Qapter, it provides car damage detection and repair price estimation from images. Unique to this solution is that it is also addressed to the B2C (business-to-customer) market through a downloadable application that allows automatic car damage detection and price estimation. However, to the authors' best knowledge, the application is not downloadable in Romania.

GeniusINSPECT is another AI solution designed for car damage detection and repair price estimation addressed to the insurance industry by Claim Genius (2022). Like previously presented solutions, GeniusINSPECT allows the detection of car damages from images and the estimation of the repair price. The solution adheres to the "human in the loop" principle (Zanzotto 2019) by allowing the insurer to modify the classifications and predictions made by the AI system. Distinct from the other car damage detection and repair price estimation solutions, GeniusINSPECT comes as a white-label solution.

Mitchell Intelligent Estimating is a solution for vehicle damage identification and repair cost estimation from images provided by Mitchell International under the umbrella of Enlyte (Mitchell 2020). The Mitchell Intelligent Estimating solution is aimed at the auto insurance segment and uses AI algorithms to automate the insurance claims process. While this solution 'offers similar functionalities to Qapter, the Mitchell Intelligent Estimating product distinguishes itself through an open ecosystem that allows integration with other solutions for detecting damage from images. For example, public information illustrates integration with the Tractable (Huetter 2021a) or the Claim Genius solutions (Huetter 2021b).

WeProov Claim is a solution that uses AI technologies for claims management and auto insurance claims (2022). The proposed solution involves a mobile application for vehicle inspection, the generation of inspection reports, and a management interface for monitoring the condition of vehicles. The WeProov solution distinguishes itself by developing the option of using real-time video instead of images for the car damage detection (2020).

The findings of the scoping market review reveal that, while international alternatives to Qapter Solera are available, each with unique differentiators, no information was found regarding their utilization in the Romanian territory. This indicates the opportunity to develop an alternative solution to Qapter Solera to avoid monopolizing the car damage detection and evaluation processes in the Romanian market. However, such a solution needs to be built based on the needs of the stakeholders in the insurance industry. These needs are often identifiable through market research analysis.

2.3 Existing Studies Concerning the Feasibility of the Application

Existing market research concerning the use of AI for car damage detection and repair price prediction is mainly focused on the adoption of such tools by organizational players in the insurance industry. As such, this section describes the main findings of three identified market research studies.

A Business Insider Intelligence study of 9 companies in the insurance industry (Nonninger 2019) suggested that, among others, AI-based technologies can bring $390 billion in cost reduction for insurance companies through streamlining front-office, middle-office, and back-office processes. Moreover, the same study illustrated that the AI-based technologies having the greatest impact on insurance companies are related to conversational agents, personalized insurance, and claims reporting on the processes.

The analysis carried out by Kumar et al. (2019) on the use of AI in insurance highlighted, among other things, that about half of the insurance companies surveyed in an Everest Global study are focusing on implementing AI to optimize processes and improve customer experience. According to this study, 11 global insurance companies were considered early adopters: Ageas and Zurich, AIG, Lemonade, Liberty Mutual, USAA and Allstate, Aviva, AXA, and Swiss Re.

A study conducted by Solera (2022)—the company owning Qapter—on 2000 entities in the global automotive industry (parts manufacturers/distributors, repair shops, insurers) suggested, among others, that 34% of the respondents identify the Covid-19 pandemic as a catalyst in the adoption of automated solutions in the auto insurance claims industry due to their capability of allowing employees to work remotely, without the need to travel; 33% of respondents identify customer demand for digital experiences as an important catalyst for the adoption of automated solutions in the auto insurance claims industry; 2/3 customers would choose an insurer that offers a fast digital insurance claim experience; and 73% of customers would choose an insurer that uses AI for faster insurance processing.

As it can be observed from existing market research concerning the use of AI for car damage detection and repair price prediction, the primary focus is on insurance companies and cost reductions, while less attention is placed on the employees of such organizations or the actual perspective of customers. Moreover, to the best of our knowledge, open data on AI solutions targeting these two segments, namely car damage inspectors and customers of insurance companies, is also lacking.

3 Methodology

Our literature review revealed two main gaps about the utilization of AI for the detection of car damages and the subsequent estimation of repair costs: (1) first, no alternatives to Qapter Solera were identified within the Romanian market, (2) there is a lack of available market research data concerning the potential problems and challenges that could be addressed by an AI solution for car damage detection and repair price estimation in Romania. In light of these findings, this study aims to bridge these two aforementioned gaps via a pilot market research focused on identifying the potential issues faced by customers and employees of insurance companies in Romania. Based on the resulting insights, the study then seeks to advance a novel architecture for an AI-powered solution that addresses the detection of vehicular damages and the estimation of repair costs.

The market research consisted of two parts. The first part focused on customers of insurance companies. It involved conducting a survey with former and current customers of insurance companies. The survey aimed to identify customers' challenges when interacting with insurance companies when filing a new claim. The development of the survey was conducted in a multistep fashion. First, three unstructured interviews were conducted with drivers involved in car accidents who had undergone the insurance claim process. Next, an analysis of the insurance claim procedure was conducted for the R.C.A ("Răspundere Civilă Auto") insurance (with or without direct settlement), which is compulsory for all Romanian drivers, and the optional CASCO ("Casualty and Collision") insurance. Both the interviews and the document analysis were used to identify key indicators of issues within the insurance claim process, which were then used to design the survey.

The process described above resulted in the discovery of eight potential indicators or problems that respondents could encounter during the insurance claim process, namely: (a) communication challenges with the insurance company representative, (b) challenges in organizing and filing the necessary documents, (c) difficulties with opening a claims file, (d) issues with scheduling damage inspections, (e) communication challenges with the car damage inspector, (f) lack of trust in the estimates provided by the car damage inspector, (g) ambiguity regarding the overall procedure, and g) extended duration of the procedure. The indicators were transformed into statements (e.g., "Communication with the insurance company representative was difficult") for which respondents (drivers involved in car accidents and insurance claims) were asked to express their answer on a 5-point Likert scale (i.e., 1- total disagreement, 2- disagreement, 3- neither agreement nor disagreement, 4- agreement, and 5- total agreement). Respondents could also choose the "Do not know/ Prefer not to answer" option. An open-ended question was also included to identify other issues not included in the previous list of statements.

The survey also included a series of filter and validation questions. The respondents were required to indicate whether they had been involved in a vehicle accident within the last ten years. This question was designed to direct our study toward recent procedures for insurance claims. In order to ensure the validity of responses, an inter-item validation strategy was employed; more specifically, respondents were asked to indicate the number of car accidents they were involved in at the end of the survey. Moreover, participants were asked to disclose whether they had directly contacted an insurance company about their claim. The underlying objective of this question was to exclude respondents who had experienced a car accident but were not engaged in the insurance claim process. Demographic information was also collected.

The second part addressed car damage inspectors employed in insurance companies. Unlike existing market research presented in this work (Nonninger 2019; Kumar et al. 2019; Solera 2022), we focused our research specifically on the issues of car damage inspectors within insurance companies rather than representatives of the companies. For this purpose, a second survey was developed based on two unstructured interviews. One of the interviewed respondents was a car damage inspector, while the second was an insurance company associate. The development of the survey sought to identify key issues in = the insurance claim management process.

The procedure presented above resulted in a list of sentences for which respondents (i.e., car damage inspectors) selected their answers on a similar 5-point Likert scale. Again, respondents could choose the "Do not know/Prefer not to answer" option. The identified indicators of issues within the insurance claim management process were: (a) customer communication issues, (b) issues with obtaining and organizing the necessary documents, (c) difficulties in traveling for car damage inspection, (d) the complexity of identifying car damages, (e) making a repair decision, (f) providing an initial price estimation, and (g) the time required to fill in the claims file. Additionally, an open-ended question at the end sought to identify other issues not included in the previous statements.

The survey addressed to the car damage inspectors also included filter and validation questions. The respondents were required to indicate whether they were active as car damage inspection specialists. This question sought to exclude respondents that work in an insurance company but are not responsible for the car damage inspection. In order to ensure the validity of responses, an inter-item validation strategy was employed. Specifically, respondents were asked to indicate their tenure as a car damage inspector and the number of managed insurance claim files at the end of the survey. Demographic information was also collected.

Both surveys were disseminated to a convenience sample based on the researchers' personal networks as a pilot study. A total of 43 potential customers answered; out of these, only 20 were deemed as customers involved in a car accident as drivers and were personally involved in the insurance claim process. The mean age of the 20 valid respondents was approximately 35 years old, with 55% male and 45% female; 75% were involved in one car accident in the past 10 years, whereas 25% were involved in between 2 and 5 car accidents. As regards car damage inspectors, a total of 15 valid answers were collected via the dedicated survey. Out of these, all the respondents except one were male. The mean age of the respondents was 38.5 years old. Moreover, 72% had a tenure of over 5 years, 18% had between 2 and 5 years of tenure, and 10% had between 1 and 2 years of tenure. Approximately 80% of the respondents managed over 500 claim damage files, while 20% managed less than 500 claim files.

4 Results and Discussion

4.1 Results Analysis

This section introduces the results of the pilot market study described in the previous section and the architecture of a novel AI solution for car damage detection and repair price estimation—InsureAI. Only valid responses were considered in this pilot market analysis. The mean and standard deviation were computed for each indicator of challenges faced by customers and employees of insurance companies. The results highlight the main challenges that the proposed solution should address.

The results from Table 1 suggest that customers' most pressing issues are communication problems with the insurance company representative, challenges in organizing and filing the necessary documents, and issues with scheduling damage inspections. On top of these, one respondent that owned a CASCO insurance policy also highlighted the lack of towing services included in the offer from the insurance company, and the customer was unaware of this before an incident.

In terms of the issues reported by the employees of car insurance companies, the results in Table 2 illustrate that the most pressing issue faced by car damage inspectors refers to difficulties in traveling for a car damage inspection. This is closely followed by customer communication issues and the complexity of making a repair decision. No other specific issues were identified through the open-ended question.

Taken together, these results suggest that developing a market problem-oriented AI system for car damage detection and repair price estimation in the Romanian ecosystem should consider improving the communication issues between representatives and customers of insurance companies. It should also consider issues related to appointments and travels for car damage inspections. Furthermore, it should also consider the transparency and complexity of the repair decision and price estimation, as well as the challenges of customers in organizing and filing the necessary documents.

Table 1 Mean and standard deviation for challenges faced by customers of insurance companies (N = 20)

Indicator	Mean	SD
Communication issues with the insurance company representative	3.11	1.36
Challenges in organizing and filing the necessary documents	3.05	1.30
Difficulties with opening a claims file	2.61	1.28
Issues with scheduling damage inspections	3.00	1.41
Communication challenges with the car damage inspector	2.77	1.35
Lack of trust in the estimates provided by the car damage inspector	3.00	1.37
Ambiguity regarding the overall insurance claim procedure	2.77	1.26
Extended duration of the procedure	2.83	1.46

Table 2 Mean and standard deviation for challenges faced by car damage inspectors currently employed in insurance companies (N = 10)

Indicator	Mean	SD
Customer communication issues	2.72	0.90
Issues with obtaining and organizing the necessary documents	2.27	0.90
Difficulties in traveling for car damage inspection	3.27	1.10
Complexity of identifying car damages	2.45	1.12
Complexity of taking a repair decision	2.90	1.22
Complexity of making the initial price estimation	2.27	1.00
Time required to fill in the claim file	2.45	0.93

4.2 Envisioned Application Architecture

Considering the limitations spurred by the convenience sample, these insights were then used to develop the architecture of a novel AI solution for car damage detection and repair price estimation. The solution proposed is called InsureAI, and its high-level architecture is presented in Fig. 1. We described the envisioned workflows where the user exchanges information with the frontend through Requests and Responses while the backend processes the requests.

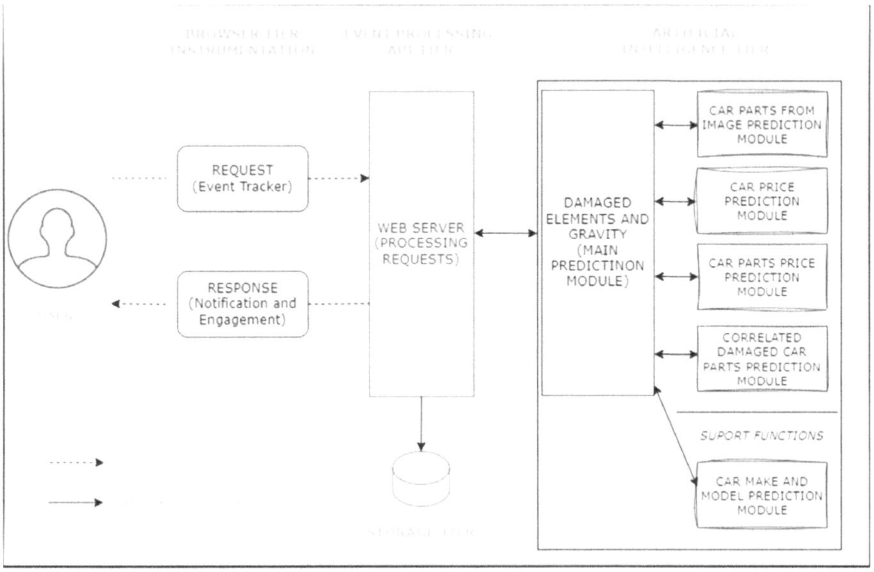

Fig. 1 InsureAI high-level architecture of the InsureAI solution

The InsureAI platform is currently under development using modern web technologies. The frontend tier is built using React framework,[1] a Javascript library for building user interfaces. The backend tier is built using FastAPI,[2] a Web framework for developing RESTful API in Python. FastAPI is a high-performance, fast-to-code, intuitive, easy, and robust backend framework.

The storage tier uses MySQL,[3] an open-source relational database management system. The Artificial Intelligence tier exposes a main prediction module—Damaged elements and gravity—and a list of sub-modules used as workers for the main task. The main prediction module receives an image of a car as input, and returns a list of damaged car parts, their gravity, and the repair cost per damage; this central component considers 5 sub-modules. First, the Car Parts from Image Prediction Sub-Module receives an image with a car as input, and returns a list of car parts identified, with their corresponding bounding boxes from the original image. Second, the Car Price Prediction Sub-Module receives as input the main characteristics of a car (i.e., manufacturer, model, car manufacturing year, and the number of kilometers) and returns a price range. Third, the Car Parts Price Prediction Sub-Module receives as input the main characteristics of a car (i.e., maker, model, manufacturing year) and the name of a corresponding part and returns a minimum and maximum predicted price for the car part. Fourth, the Correlated Damaged Car Parts Prediction Sub-Module receives the name of a car part as input and returns a list of potentially damaged additional parts with corresponding probabilities. Fifth, the Car Make and Model Prediction Support Module receives an image of a car as input and returns the predicted car make and model for validation purposes.

The Artificial Intelligence tier will be the backbone of the InsureAI product. The platform will allow three types of users: (a) driver, who reports a car accident and send the accident details to an insurance company, (b) damage inspector, who views a list of reported car accidents with their statuses, reviews a car accident dossier, and changes the status within the platform, and (c) admin, who authorizes the access of specific users in the system and adds new damage inspectors.

The driver flow within the InsureAI platform is described further. A driver has to set up an account in the system, which is free of charge and easy to set up. The registration in the system can be performed using 3rd party applications (Google and Facebook), or custom, by using email and password. After account creation, the user can enter personal information like first and last name, gender, birthday, phone number, current city, a short bio, and official documents (i.e., national identification document and driving license). The user can either manually input the data from the identification documents, or the documents can be photographed, and the data will be automatically extracted using Optical Character Recognition techniques. After automatic extraction, the driver can edit the extracted data if any error occurs.

The driver can add all owned cars to the InsureAI platform. The user will enter basic details for all cars: make, model, and number of kilometers. Moreover, the user

[1] https://reactjs.org.

[2] https://fastapi.tiangolo.com.

[3] https://www.mysql.com.

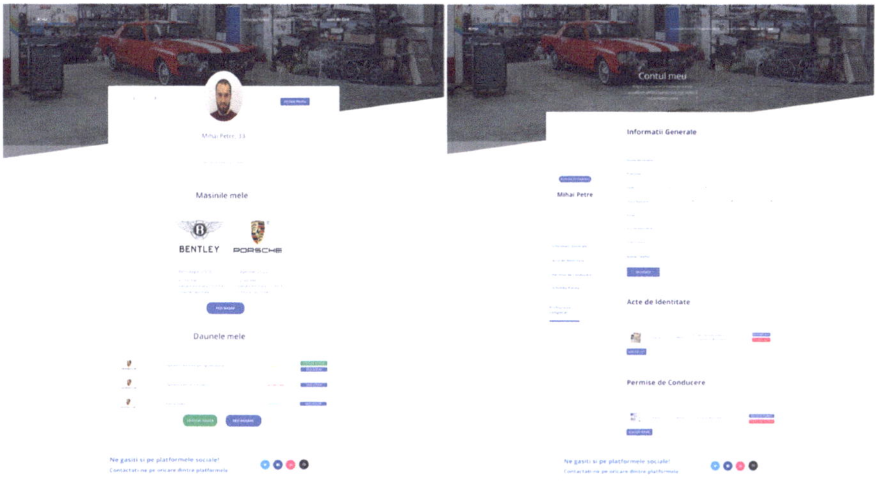

(a) Home Screen (b) User Account Screen

Fig. 2 InsureAI platform sample screens

must also add information from the official car document, which can be photographed and automatically detected or manually written by the user. The main profile includes personal information, the list of cars, and the list of damages, as seen in Fig. 2 introduces the page where users can edit their data.

When new car damage occurs, the user opens a new damage file. Photos are uploaded with the damaged car, and the backend server automatically predicts the damaged areas and the repair cost. At this stage, we adopt the human-in-the-loop approach where the user can edit the predictions made by the Artificial Intelligence module: the list of damaged elements, the bounding boxes of recognized damaged areas, and the gravity of a damaged car part; these updates performed by the driver will be logged and shown to the damage inspector. The driver can review the summary of a car damage dossier and submit it for review within the InsureAI platform.

Next, the damage inspector reviews a specific claims file. The damage inspectors are special accounts created by the administrator, who have full rights over the opened files. A damage inspector can edit the information, ask for clarifications, and approve or reject the claims file. As such, our InsureAI platform will streamline the process of reporting car damage to an insurance company, track all the changes, make it transparent, and improve the work of damage inspectors through Artificial Intelligence state-of-the-art models.

5 Conclusions

The current work illustrated the potential for developing an AI solution for car damage detection and evaluation. According to our literature review, several business areas will benefit from developing and implementing the proposed software solution. Also, it was determined that similar endeavors exist by analyzing comparable commercial systems. However, none of the potential competitors were identified as being active in the Romanian market, except for one, which means that InsureAI could be successfully proposed as an alternative solution for various local industries that would require automation of car damage assessments.

Furthermore, the possible adoption of such a solution was argued using two surveys targeting two types of users, each with unique interests and operational issues. Although the results were based on a rather limited number of respondents, the surveys highlighted interest in using the application and that certain groups of people would benefit from its development.

By using state-of-the-art technologies and implementing various Machine Learning models, InsureAI will address the most important issues mentioned by the survey responders when determining the application's feasibility. Once the application development is finalized, we intend to provide free access to the machine learning models, aiming to boost the capabilities of the academic community and the local industry in terms of image object detection.

Acknowledgements This work was funded by the "Automated car damage detection and cost prediction – InsureAI"/"Detectia automata a daunelor si predictia contravalorii aferente - InsureAI" project, Contract Number 30/221_ap3/22.07.2022, MySMIS code: 142909.

References

Advent International: CCC Intelligent Solutions. https://www.adventinternational.com/advent_case_studies/ccc-information-services-inc/ (2017). Accessed 18 Feb 2023

Bhatt, D., Patel, C., Talsania, H., Patel, J., Vaghela, R., Pandya, S., Modi, K., Ghayvat, H.: CNN variants for computer vision: history, architecture, application, challenges and future scope. Electronics **10**(20), 2470 (2021)

CCC: CCC Estimate—STP. Automate claims from photo to estimate. https://cccis.com/insurance-carriers/claims-solutions/apd/repair-management/estimate-stp/ (2021). Accessed 19 Feb 2023

Claim Genius: GeniusINSPECT For Claims. https://claimgenius.com/geniusinspect-claims/ (2022). Accessed 17 Feb 2023

Conceatu, C.: Audatex Romania lanseaza o noua aplicatie de evaluare a daunelor auto. https://www.1asig.ro/Audatex-Romania-lanseaza-o-noua-aplicatie-de-evaluare-a-daunelor-auto-articol-3,100-61894.htm (2019). Accessed 12 Sept 2022

Corea, F.: How AI is changing the insurance landscape. Applied artificial intelligence: where AI can be used in business, pp. 5–10. SpringerBriefs in Complexity. Springer Cham (2019)

Deloitte Private: The UK Fast 50. https://www.deloitte.co.uk/fast50/winners/2021/ (2021). Accessed 17 Sept 2022

Fabris, A., Mishler, A., Gottardi, S., Carletti, M., Daicampi, M., Susto, G.A., Silvello, G.: Algorithmic audit of italian car insurance: Evidence of unfairness in access and pricing. In: Proceedings of the 2021 AAAI/ACM Conference on AI, Ethics, and Society, pp. 458–468. ACM, Virtual Event (USA) (2021)

Huetter, J.: Mitchell partners again with A.I. photo estimating system Tractable. https://www.repairerdrivennews.com/2021/01/27/mitchell-partners-again-with-a-i-photo-estimating-system-tractable/ (2021a). Accessed 11 Nov 2022

Huetter, J.: Claim Genius newest addition to Mitchell photo estimating AI choices. https://www.repairerdrivennews.com/2021/03/05/claims-genius-newest-addition-to-mitchell-photo-estimating-ai-choices/ (2021b). Accessed 13 Nov 2021

Kumar, N., Srivastava, J.D., Bisht, H.: Artificial intelligence in insurance sector. J. Gujarat Res. Soc. **21**(7), 79–91 (2019)

Lutz, C.: Digital inequalities in the age of artificial intelligence and big data. Human Behavior Emerg. Technol. **1**(2), 141–148 (2019)

Mitchell: The Only Claims Automation Solution Powered by an Open Ecosystem. https://www.mitchell.com/solutions/auto-physical-damage/intelligent-solutions/estimating (2020). Accessed 4 Feb 2023

Nonninger, L.: Tractable hits unicorn status and takes on property damage appraisal. https://www.insiderintelligence.com/content/tractable-hits-unicorn-status (2021). Accessed 28 Sept 2022

Nonninger, L.: The AI in Insurance Report: how forward-thinking insurers are using AI to slash costs and boost customer satisfaction as disruption looms. https://www.businessinsider.com/the-ai-in-insurance-report-2019-6 (2019). Accessed 12 Oct 2022

Solera: Are we ready for artificial intelligence (AI) in automotive claims? https://www.qapter.com/wp-content/uploads/2021/02/EN-Qapter-AI-infographic-assessor-2022.pdf (2022). Accessed Oct 2022

Solera: Qapter. https://www.qapter.com/ (2022). Accessed 17 Feb 2023

Tractable: Put applied AI at the heart of your claims cycle. https://tractable.ai/en/industries/auto-claims (2022). Accessed 3 Feb 2023

WeProov: Car inspection app and body repair estimate: digitalize your vehicle tracking. https://www.weproov.com/en/vehicle-inspection-app-weproov (2022). Accessed 28 Nov 2022

WeProov: Artificial intelligence, from vehicle parts & damages detection to cost repair estimate. https://bit.ly/3xuObYG (2020). Accessed 1 Dec 2022

Zanzotto, F.M.: Human-in-the-loop artificial intelligence. J. Artif. Intell. Res. **64**, 243–252 (2019)

Preparing Students for Successful Business Careers in a Changing Society: Good Practices from a Private High School in Romania

Florentina Alina Grigorescu

Abstract The current paper aims to explore various methods and strategies for preparing students for successful careers in a changing business society, with a focus on students from a private high school in Bucharest, the capital of Romania. The literature review highlights the importance of equipping students with the necessary skills and knowledge in an increasingly complex and dynamic job market, and how traditional education methods may fall short in this regard. The research methodology used in this paper is based on the teachers' responses to an online questionnaire, on a quantitative research, with the aim of analyzing current practices and trends in education and career development, specifically in the context of a private high school from Bucharest. The main results of this study demonstrate that incorporating real-world experiences through internships, teaching critical thinking and problem-solving skills, and promoting lifelong learning and personal development are effective methods for preparing students for successful business careers. Additionally, the use of technology and digital tools in education is found to enhance students' career prospects in the business sector. The paper concludes by highlighting the significance of the research and its contributions to the fiels of education and career development, providing insights and recommendations for educators and policymakers to improve the preparation of students for success in a constantly changing society.

Keywords Preparing students · Successful business careers · Effective methods

1 Introduction

Todays jobs are constantly changing, developing and adapting to todays technology. It is imperative that students possess skills and knowledge that are more and more up-to-date and that will help them later adapt to the demands of the labor market. Traditional educational methods may become insufficient in the process of preparing

F. A. Grigorescu (✉)
University Politehnica, Bucharest, Romania
e-mail: alina.pirvu@lrfb.ro

M. Busu (ed.), *Rethinking Business for Sustainable Leadership in a VUCA World*, Springer Proceedings in Business and Economics,
https://doi.org/10.1007/978-3-031-50208-8_11

students for the jobs of the future. Therefore, there is a need to apply modern and effective methods in order to prepare students to become successful young employees in a changing society.

This paper explores various methods and strategies for preparing students for successful business careers, based on the primary school, middle school and high school students from a private school in Bucharest, Romania, through the comprehensive analysis of current practices and trends in education and development used by teachers today.

The research question for this paper is: What teaching methods and strategies are most effective in preparing students for successful business careers in a changing society?

To achieve this objective, the paper examines the benefits of incorporating real-world experiences through internships, teaching critical thinking and problem-solving skills, and promoting lifelong learning and personal development. Additionally, the use of technology and digital tools in education is explored as a means of enhancing students' career prospects.

The finding of this study will provide new perspectives and recommendations for educators to improve the preparation of students for successful careers in a constantly changing society.

2 Literature Review

2.1 Examining the Importance of Equipping Students with the Necessary Skills and Knowledge in an Increasingly Complex and Dynamic Job Market

The literature review of this paper aims to give an overview of the current state of knowledge and research on the topic of preparing students for successful business careers in a rapidly, changing and volatile, uncertain, complex, and ambiguous (VUCA) world. In recents years, there has been a growing recognition of the importance of equipping students with necessary skills and knowledge to thrive in an increasingly complex and dynamic job market.

In the current era, unemployment is increasingly prevalent, technology is advancing at a rapid pace, and we are moving towards an age in which robots will replace human workers. Furthermore, the Covid-19 pandemic has brought significant changes to many field of activity. The critical factor that can ensure individuals' success is education. Only through education can individuals achieve a successful business career in the future. This places a great deal of pressure on educators (Buchanan et al. 2023).

Mason (2007) posits that the failure in courses at the onset of university studies is largely due to students not having sufficiently developed necessary skills. Furthermore, most students do not possess sufficiently developed critical thinking and do not possess effective problem-solving strategies. Thus, it can be inferred that for students to be successful in the university environment, in addition to the knowledge and specific skills of the courses they are taking, they must also work on developing critical thinking and problem solving. This two skills need to be developed with the help and through education.

The educational community, along with parents, business leaders, and other educational partners, has the responsibility of preparing students for successful future business careers in a VUCA world. In addition to the content and skills developed in educational programs, students need real-world work experiences in order to be able to integrate into the workforce. This contact with the real world can be achieved through internships. Internship opportunities during school offer benefits for both students, as they provide the chance to learn and apply their knowledge and skills in real-world experiences, and employers, as they provide access to human resources at a reduced cost and the opportunity to mold the students to fit the organization (Galbraith and Mondal 2020).

Grande and Estebanez (2020) argue that students find internships highly valuable and perceive themselves to fully integrate into the company culture, learning to develop more transferable skills than cognitive ones. On the other hand, employers perceive that students who do internships have strong social skills and adapt easily to the company culture, but expected students to show more creativity and stronger cognitive skills. The authors also argue that through internships, employers (supervisors) should focus on developing students' critical thinking, creativity, problem-solving and aligning with labor market needs.

A number of studies have highlighted the importance of incorporating real-world experiences through internships in preparing students for successful business careers. However, while internships have become popular because students can gain experience and demonstrate skills related to the job, this type of practice has also raised concerns due to the potential for exploiting the vulnerability of young people (Costea et al. 2012). A study conducted at a public university in Malaysia presents the perspectives of students regarding internships. The researchers recorded a satisfaction level among the students about internship, with 61.1% of respondents stating that their experience in the internship program was a good one, 33.3.% choosing the moderate option, and 5.6% considering their experience as not positive all (Sadacharam 2021). Therefore, for the majority of students, experiences in internship programs are beneficial. These experiences have been found to provide students with valuable skills such as problem-solving, critical thinking, and communication, which are highly valued by employers.

Through participation in an internship, exposure to real-life scenarios helps to develop the knowledge acquired in school and helps students to determine if they are in the appropriate field of career. Internships provide students with real-world experience in the job market and assist them in applying their existing knowledge while also developing skills such as working under pressure and understanding the significance of deadlines (Vinay 2022).

Even during the pandemic, internships continued to be conducted in many educational organizations and companies. Despite being conducted online, the results of a study conducted in 18 Croatian IT companies indicate that virtual internships have numerous advantages, such as flexibility, time-saving, easy of work organization, comfort, the ability for self-organization, and the ability for interns to attend to their personal and academic obligations. However, respondents also identified some negative aspects of virtual internships, such as a lack of communication between students and mentors, limited opportunities for students to learn about the organization, and lack of understanding about human side of the company (Szombathelyi et al. 2022).

Other efficient methods of preparing students for a successful business career include developing digital skills among students. As technology advancements are rapidly increasing in our days, students need digital skills to meet the demand of the job market. Following the sudden shift to online activities in education due to Covid-19 pandemic, teaching at all levels has become mediated by technology (Pozo et al. 2021). Currently, students are exposed to different digital technologies and use various technology-based tools during lessons or in completing assignments. To enable students to integrate these resources into their practice and develop digital skills, digital literacy and ongoing training are needed to face future challenges.

Future professions are based on the use of digital technologies. The better students are prepared in this field, the more chances they will have for success in their future business careers. According to Vilbert (2019), both history and available data so far shows that technology does not eliminate jobs but it is a "great job-creating machine". He also argues that technology has taken over many repetitive or routine manual jobs but innovations will not replace human brains and hands.

According to Hetler (2022) the most demanded jobs based on technology are: software developer, data scientist, artificial intelligence specialist, IT manager, computer system analyst, computer network architect site reliability engineer, and DevOps engineer. For all these jobs, candidates need strong digital skills, general computer science knowledge, interpersonal and presentation skills, critical thinking skills, communication skills, ability to work well under high-pressure, leadership and many others. All these new jobs are in full development and their number increases as technology advances. Since there are not many qualified people for them, the offered salaries are represented by attractive sums.

In addition to the methods listed above, lifelong learning and personal development are also effective methods that assist in preparing young individuals for successful business careers. Lifelong learning helps individuals develop skills that enable them to meet job demands, increasing their employability. A study conducted by Jenkins et al. (2003) indicated that lifelong learning is perceived as a means of re-entry into labor market. Furthermore, the authors of the same study assert that learning leads to learning and not just obtaining qualifications is a motivation for lifelong learning.

Regarding personal development, students should participate in such activities from a very young age. Personal development should be prioritized because it enables the students to have a clear understanding of their strengths and weaknesses and how they can achieve success in life. According to Campbell (2019), personal development is the most important goal that we can strive to achieve. He also states that we live in constantly changing environments and that personal development and growth can take many forms in schools.

According to the results obtained from a personal development program addressed to first-year students from the Faculty of Electrical Engineering, Energy, and Applied Informatics at the Gheorghe Asachi Technical University of Iasi, after participating in the program, the students demonstrated the following behavioral acquisitions: flexible thinking, generative thinking in problem-solving, acquisition of coping strategies, optimization of personal and academic life (Jitaru et al. 2022). Thus, necessary skills to successfully face the challenges of student life and future employment.

Furthermore, as students know and learn more about available careers and make decisions about what they choose to do next, it is important that they plan and choose a program of study that is challenging to prepare them for the future career. The greater the challenges they face during high school are, the more they will improve their career development and cope successfully in an ever-changing workplace (Hughey and Hughey 1999).

Virtual Enterprises in collaboration with Deloitte have developed and implemented a career preparation program that identifies the abilities, knowledge and skills that students can develop and are in demand by employers. They are currently working to prepare youth, high school and middle school students for successful careers by bringing the workplace into the classroom and providing students with meaningful experiences to better understand the world of work. According to Fugazzie (2018), two key elements of successful student career preparation program are (1) connecting with real-world professionals and industry leaders who can communicate workplace expectations and knowledge of the latest innovations and trends and (2) providing ways for students to gain practical experience to develop and demonstrate their skills such as projects, workshops, scenarios or deliverable-based-assignments.

In conclusion, according to current knowledge and research on student preparation for successful business careers, the authors note that there has been a growing emphasis on developing necessary skills and knowledge in recent years. Furthermore, the literature review highlights the significant importance of internships in preparing students as they provide real-world work experiences and assist students in applying their existing knowledge while also developing skills such as problem-solving, critical thinking, and communication. Additionally, researchers also emphasize the importance of developing digital skills as future careers are increasingly reliant on technology. In conclusion, the literature suggests a combination of internships, digital skills training, lifelong learning, and personal development for preparing students for successful business careers.

Analyzing the shortcomings of traditional education methods in preparing students for success in a changing society. Currently, career education and guidance has begun to take a more prominent role in the lives of students, teachers, and educational decision-makers. In order for students to make informed decisions about continuing their studies and choosing a professional career, they must have a thorough understanding of their interests, abilities, and personality profile. Beginning in secondary school, students should have access to academic and vocational counseling. This counseling is provided by the school counselor and the student's class advisor. It involves a detailed analysis of the student's family, school, and peer environment, as well as a series of interviews and the use of specific methods and tools for investigating the student's personality and identifying their abilities and interests (Mihailescu 2021).

In Romania, career guidance for students is carried out in two ways. First, within the educational system, it is done through the county's Centers for Resource and Psychological Assistance, through counseling and career guidance classes (class advisor classes), through optional counseling and guidance courses, and through counseling provided by teachers of other subjects, as well as extracurricular activities, programs, and projects. Second, outside the education system, career education and guidance can be provided through individual psychology offices or through EU-funded projects.

In summary, the literature review highlights the importance of equipping students with the necessary skills and knowledge in an increasingly complex and dynamic job market, and how traditional education methods may fall short in this regard. The study argues that incorporating real-world experiences through internships, teaching critical thinking and problem-solving skills, and promoting lifelong learning and personal development are effective methods for preparing students for success. Additionally, the use of technology and digital tools in education is found to enhance students' career prospects.

3 Methodology

The method used in this quantitative study is a survey carried out through an online questionnaire. The questionnaire was sent to teachers by email and was created in the Microsoft Forms application. A number of 42 teachers from the Romanian-Finish High School in Bucharest completed the questionnaire, in January 2023. Among the respondents, 30 of them are teaching at the middle and high school level and 12 at the primary level.

The questionnaire was composed of 15 questions. The first two questions were closed questions, with a single answer, and were intended to identify the level where they teach and whether they also fulfill the role of class master. The next four questions were also closed questions, and the remaining 9 were open questions all about the methods, strategies and activities carried out by teachers in order to prepare and guide students in their future careers. The data provided by the teachers was thoroughly processed and analyzed based on the Excel report generated by the Microsoft Forms application.

The main research hypothesis of the present study are: H1: Integrating real-life experiences through internships or activities that put students in real-life situations is an effective method for preparing students for successful business careers. H2: Developing critical thinking and problem-solving skills is an effective method for preparing students for successful business careers. H3: Personal development, including developing digital skills and promoting lifelong learning, is an effective method for preparing students for successful business careers. The argument for this hypothesis is the fact that jobs of the future require employees who are able to cope with the changes that have occurred and the increasing advance of technology.

The research question is: What teaching methods and strategies are most effective in preparing students for successful business careers in a changing society?

4 Results and Discussions

Following the analysis of the answers provided by the teaching staff to the online questionnaire, the results obtained support the research hypothesis and answer the research question.

First of all, the integration of activities that involve contact with the real-world is a practice often used by teachers at the Romanian-Finish High School. Almost all the teachers stated that they carried out activities that put students in real-world situations either through visits to different institutions, internships carried out within the school or within other organizations or visits to centers that carry out activities where students can use their knowledge and skills by taking on real or fictional jobs.

All teachers concur that internships assist students in preparing for a successful future business career. In particular, given that at the high school level there are classes with a pedagogical profile, students benefit from a day of practice per week where they have the opportunity to teach and learn from their own experience alongside mentor teachers. Furthermore, over half of them assert that they have implemented projects and/or programs aimed at educating and guiding students in their career choices. In addition to physical activities, teachers also listed virtual activities such as presentations and simulations of various professions as methods for career development.

In the following table are presented activities carried out by teachers in order to prepare students for integration into real jobs, putting them in real life situations. Most of the activities are extracurricular activities, conducted by primary, secondary or high school teachers (Table 1).

The following questions included in the questionnaire referred to the subjects that the teachers consider important in terms of education and business career guidance for students. A number of only two teachers stated that the subject they taught does not directly influence the career guidance process. On the other hand, more than half of the teachers said that career education and guidance activities can be carried out in all subjects, and the rest listed a number of subjects they consider important in this process such as: Civic Education, Psychology, Social Education, Entrepreneurial Education, Personal Development and Leadership classes.

The teachers listed a variety of methods and strategies used in the career guidance process, including those used during regular class hours such as conversation, problematization, and debate, as well as activities conducted during guidance and

Table 1 Activities that put students in real-world situations

Activity	The organization	The position/job
Internship	Romanian finish high school	Teacher, assistant teacher, secretary, accountant, receptionist, librarian, sound assistant, event organizer, team coordinator
Internship	Destiny park	Doctor, fireman, policeman, pharmacist, cashier, IT-ist
Visits	Court of appeal bucharest	Lawyer
	The hills of Arges	Baker
	Magurele Science park	Scientist

Source Authors' own research

personal development classes such as presentations on different professions and successful individuals from various fields, interviews with notable figures, special guests speaking about their profession and education and career path, discovering students' strengths through various activities or tools, and simple discussions with the school psychologist, teachers, counselor, or other individuals who can guide students in the career guidance process.

In terms of the skills that teachers consider necessary for students to develop during their school years, the top three essential skills selected by teachers were problem-solving, emotional intelligence, and critical thinking (Fig. 1).

Analyzing these responses and comparing them to the research hypothesis, it can be observed that problem-solving and critical thinking are skills that students must have for a successful integration into the workforce. Although personal development

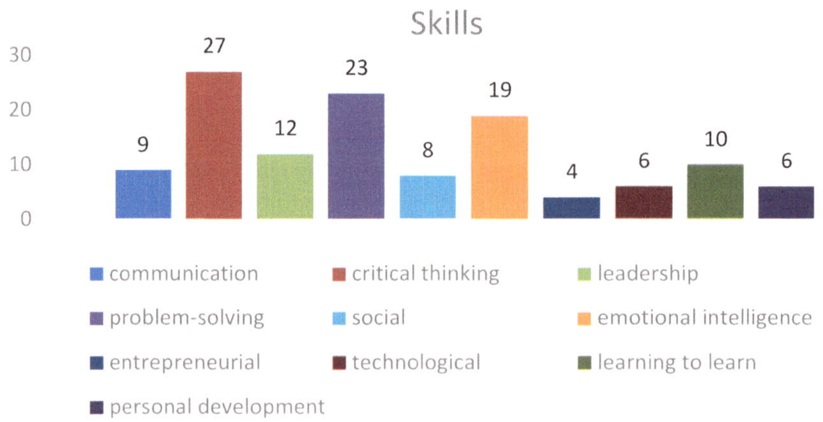

Fig. 1 Selected skills by teachers. *Source* Authors' own research

was identified as an essential skill in the research hypothesis, few teachers selected it as a necessary skill for students to develop. However, teachers did select skills such as emotional intelligence, learning how to learn, and communication, all of which can be developed through specific activities for students' personal development (Table 2).

Regarding the digital skills necessary for a successful future business career, all teachers stated that these skills are necessary and, from the perspective of the discipline they teach, each one contributes to the development of these skills among students. Additionally, teachers listed a series of applications, platforms, websites, and virtual labs used during lessons and by students for homework. The most used applications according to the answers given by the teachers who completed the survey are: Canva, Wordwall, Padlet, Kahoot, Bamboozle and Mentimeter. In terms of platforms used by teachers, they stated that they only use Microsoft Teams and Moodle. In addition to these, the educators listed a series of websites such as: Twinkl, Livresq, and other discipline-specific websites. Also, for students' homework, they stated that the Google extensions (Docs, Slides, Sheets, Forms) are the most used because they allow for collaborative work, but depending on the specific homework, teachers offer students the opportunity to use any other tools they consider useful for completing homework.

The following questions in the questionnaire referred to the professional orientation projects carried out by teachers in class. Firstly, at all levels of education, the Young Leadership project is carried out in high school. Within this project, students from each class are divided into several teams, and in turn, each member receives the role of leader. Teams periodically receive different projects, challenges, or tasks that they must complete, and the leader is responsible for his team throughout the activities. Students develop their teamwork, communication, leadership, problem-solving, critical thinking and other skills specific to the activity carried out.

Another project that is being carried out in several classes within the high school is the "I Want to Become" project. The project involves several stages, each aimed at helping students in their professional orientation process. First, students go through several tests to identify their strengths and weaknesses. Then, each student will gather

Table 2 Activities for the development of critical thinking and problem-solving skills

Critical thinking	Problem-solving
Alternative teaching methods (Thinking Hats, cube method, 6/3/6 technique)	Conflict resolution
Role-playing	Problem-solving (in real or fictional situations)
Case study	Event planning
Debate	Assumption of responsibilities in the classroom/school
Analysis of situations/characters in literary texts	Involvement in projects
Scaffolded questioning	Virtual/video simulation

Source Authors' own research

information about the profession they desire and analyze if their own knowledge and skills match it. Finally, students will perform and analyze the resources they need to meet the requirement of the profession and will seek people who work in the field to offer advice and guidance to help them in their educational and professional journey. This project takes place throughout a school year and is repeated multiple times throughout the primary, middle, or high school cycle so that students can observe and analyze how their desires have evolved, how their skills have developed, and how they have succeeded in improving their plan.

Other examples of activities through which teachers contribute to the education and professional orientation of students are: Teacher for a day, Learning to Learn, interdisciplinary projects on themes chosen based on the needs of students or charitable projects. All of these contribute to the development of necessary skills and knowledge that will help students meet the requirements and challenges of their future chosen workplaces.

The final questions in the questionnaire refer to the importance of contact with individuals who are already working in fields of interest to the students. All the teachers who responded to the questionnaire stated that they periodically invite special guests to their classes to speak to scholars about the job or profession they chosen, how they obtained it, their educational journey, the challenges they face, and other information the students would like to hear. These guests may include parents, teachers, writers, business people, athletes, political figures, singers, scientists, or other well-known individuals who can assist the students in their educational and career orientation process.

Another important aspect highlighted by teachers is the exceptional influence of psychologists and school counselors. In addition to guidance, professional development or psychology classes, school psychology offices play a very important role in students' lives. School psychologists regularly carry out activities that contribute to students' professional development and support them in their education and career orientation process. These activities are held in class, at the school level, or through meetings with students in small groups or individually.

The final essential aspect, with regard to the education and professional orientation of students, is the meetings between teachers, students and their families at the beginning and end of each school year. During these meetings, important issues discussed by students regarding their desires, plans, and objectives for the next period, feedback regarding their progress to that point, an analysis of their school situation, and the creation of an individualized plan together with the family and teachers to improve results and achieve desired success are discussed.

5 Conclusion

In conclusion, this study highlights the importance of education and guidance for students in regard to their effective integration into the labor market in the context of a constantly changing society. The results of this study support the notion that integrating real-world experiences through internships fostering critical thinking and problem-solving skills, promoting personal development, and promoting lifelong learning are effective methods to prepare students for successful business careers. Furthermore, Xuereb (2006) argues that authentic partnerships between the education and business sector are necessary to adequately prepare students for workforce, benefiting both parties and effectively equipping students for success in the job market.

Additionally, the use of technology and digital tools in education also plays a significant role in enabling students to develop the necessary digital skills for future business careers. Ore et al. (2022), also conclude that students must acquire 21st-century digital skills as they are essential for high performance in the business world and for success in life.

This study contributes to the field of education and professional development, within the context of a VUCA world, by providing a series of guidelines and recommendations for teachers and policy makers to enhance students preparation for success in a constantly changing society.

However, this research is limited as it was conducted only in a private high school from Bucharest, Romania and its results cannot be generalized to a larger extent. Further studies are needed to generalize these findings and apply them at national or even international level.

Finally, the results of this study provide practical recommendations for teachers and education decision-makers to improve the career guidance process for primary, secondary, and high school students. This research provides a foundation for future research and practical applications in the field of education.

References

Buchanan, J., Allais, S., Anderson, M., Calvo, R. A., Peter, S., Pietsch, T.: The futures of work: what education can and can't do. https://unesdoc.unesco.org/ark:/48223/pf0000374435. Accessed 2 Apr 2023 (2023)

Campbell, R.: Self-growth is a key to students' success, Self-growth is a key to students' success (theavtimes.com). Accessed 5 Feb 2023 (2019)

Costea, B., Amiridis, K., Crump, N.: Graduate employability and the principle of potentiality: an aspect of the ethics of HRM. J. Bus. Ethics **111**(1), 25–36 (2012)

Fugazzie, T.: Preparing students for Career Success, Preparing Students for Career Success | Getting Smart. Accessed 21 Feb 2023 (2018)

Galbraith, D., Mondal, S.: The potential power of internships and the impact on career preparation. High. Educ. J. **38**, 1–15 (2020)

Grande, R., Estebanez, J.: The role of internships in the development of transversal skills. J. High. Educ. **71**(5), 708–723 (2020)

Hetler, A.: Top 9 most in-demand tech jobs of 2022. Top 9 most in-demand tech jobs of 2022 (techtarget.com). Accessed 27 Mar 2023 (2022)

Hughey, K.K., Hughey, J.K.: Preparing students for the future: making career development a priority. (PDF) Preparing Students for the Future: Making Career Development a Priority (researchgate.net) (1999). Accessed 2 Mar 2023

Jenkins, A., Vignoles, A., Wolf, A., Galindo-Ruenda, F.: The determinants and labour market effects of lifelong learning. Appl. Econ. **35**, 1711–1721 (2003)

Jitaru, O., Axinte, R.B., Vatavu, C.: The role of personal development in students adapting to the academic life. In: Scientific Research and Education in the Air Force Conference (2022)

Mason, M.: Critical thinking and learning. Educ. Philos. Theory **39**(4), 339–349 (2007)

Mihailescu, T.: From academic and professional guidance to career guidance. De la Orientarea Şcolară şi Profesională la Orientarea în Carieră (teodoramihailescu.net). Accessed 1 Mar 2023 (2021)

Ore, E.R., Hassan, A.M., Ogungbo, M.O.: Business education and digital skills: a conceptual review. Int. J. Adv. Res. Multidiscip. Stud. **2**(1), 41–45 (2022)

Pozo, J.-I., Echeverria, M.-P., Cabellos, B., Sanchez, D.L.: Teaching and learning in times of COVID-19: uses of digital technologies during school lockdowns. Front. Psychol. **12** (2021)

Sadacharam, J.: The Importance of Internship for Public University Students in Malaysia. Universiti Saints Malaysia, Apex (2021)

Szombathelyi, D., Jakopec, T., Aparac Jelusic, T.: Student internships in the time of crisis: how IT companies cope with new challenges and opportunities. Proc. Assoc. Inf. Sci. Technol. **59**(1), 718–719 (2022)

Vilbert, J.: Technology creates more jobs than it destroys. In: Technology Creates More Jobs than it Destroys—Foundation for Economic Education (fee.org). Accessed 13 Feb 2023 (2019)

Vinay, S.: The importance of an internship: top 5 reasons why internships are critical (capital-placement.com). Accessed 17 Apr 2023 (2022)

Xuereb, P.: Preparing students for the world of work: the benefits of education—Business partnerships. Symposia Melitnesia **3**, 50–64 (2006)

Technological, Physical and Cultural Strategies for Generation X and Y Toward Their Performance

Jefta Harlianto, Harjanto Prabowo, Mohammad Hamsal, and Boto Simatupang

Abstract In dynamic business environments and with the growing influence of multiple generations, employees expect a valuable, attractive, and satisfying work experience based on their preferences. This expectation makes the way employees experience work more important in many companies. Companies are developing integrated strategies to enhance their employees' experience in the workplace. Although much research states the positive effect of employee experience, there is a lack of focus on the effect toward employee performance. This study compared the impact of employee experience strategy (consisting of technological environment, physical environment, and cultural environment) on Generation X and Generation Y toward their performance. The study sample size was 194; the sampling method was simple random sampling. The collected data was examined using Structural Equation Modeling (SEM). The empirical findings demonstrated that employee experience had a significant impact on employee performance for both generations while Generation X had a higher significant impact compared with Generation Y. The findings also showed that the higher value of cultural environment will impact the higher value of adaptive performance for both Generation X and Generation Y. This study contributes new findings about the positive impact of employee experience on employee performance.

Keywords Employee experience · Employee performance · Generation X · Generation Y · Human resources

---

J. Harlianto (✉) · H. Prabowo · M. Hamsal · B. Simatupang
Management Department, BINUS Business School Doctor of Research in Management, Bina Nusantara University, Jakarta, Indonesia
e-mail: jefta@binus.edu

H. Prabowo
e-mail: harprabowo@binus.edu

M. Hamsal
e-mail: mhamsal@binus.edu

B. Simatupang
e-mail: bsimatupang@binus.edu

M. Busu (ed.), *Rethinking Business for Sustainable Leadership in a VUCA World*, Springer Proceedings in Business and Economics,
https://doi.org/10.1007/978-3-031-50208-8_12

1 Introduction

The business world is facing an unexpected situation that organizations must deal if they want to survive amongst fierce competition. As important as it is to a company ability to navigate disruption, change and business uncertainty, research from Deloitte reveals that only nine percent of business leaders believe they are very ready to address the issue (Verasai 2021).

For the first time, organizations have four generations working together: traditionalists, baby boomers, Generation X and Generation Y. Each generation has job expectations shaped by their own cohort's characters. Organizations should redesign their methods to how they work together with their employees. The new multi-generational workplace requires a new way of thinking about human resource management strategy in many organizations. Prioritizing the employee journey in a company can produce vastly different methods to improve high employee performance in the organization.

Rather than focus narrowly on employee commitment, organizations should develop an integrated focus on the entire employee experience, bringing together involvement in all HR and management systems that impact an employee in an organization. Through employee experience, human resources divisions are now focusing on identifying and enhancing this journey experience to facilitate greater employee achievement.

This shift is becoming so common that we are even noticing the urgent emergence of entire roles dedicated to employee experience strategy in organizations. We have confidence in the employee experience, and its relationship with commitment and accomplishment is critical to recognize and prioritize (Plaskoff 2017). If organizations can provide the right employee experience strategy, they can achieve double the customer satisfaction and innovation, and create 25% higher profits than those that do not (Verasai 2021).

Deloitte Global Human Capital Trends (2017) research shows that organizations' ability to address this issue of engagement and culture has dropped by 14% since 2016. Employee experience ranks as a major trend again; nearly 80 percent of executives rated employee experience as very important (42%) or important (38%), but only 22% reported that their companies were excellent at building differentiated employee experiences, and fifty-nine percent of survey respondents reported they were not ready or only somewhat ready to address the employee experience challenges.

Meanwhile, research from IBM and Globoforce (2017) shows optimistic employee experience can help retain talent and improve performance. Workers in the top 25% of the Employee Experience Index reported they were 52% less likely to leave, 73% more likely to go above and beyond, and 32% more likely to achieve higher performance. It also reports that employee experience varies by country and Indonesia is ranked fifth compared with other countries in the Asia Pacific region.

XYZ University is one of the leading higher education institutions in Indonesia. It was established in 1981 and its main campus is located in the capital city, Jakarta. Most of its campuses are located within the area of Greater Jakarta. It also has campuses in other cities in Indonesia (Bandung, Malang, and Semarang). XYZ University faces similar global employment trends in which it has four generations working together. The percentage of employee composition in XYZ University is mostly Millennials (55%), followed by Gen X (29%), Generation Z (15%), and baby boomers (1%). Currently there is no formal survey related to employee experience while there are many programs that have been implemented to support the employee experience in this university.

Acknowledging that most employees are composed of Gen X and Millennials as well as fluctuating employee performance in the organization for the last three years, researchers proposed evidence for generational differences in employee performance. Therefore, this research focuses on the comparison between Generation X and Generation Y on the impact of employee experience on their performance.

The factors that are used in this study are based on existing literature by Morgan (2017), and Pradhan and Jena (2021). The unit of analysis are Generation X and Generation Y employees in XYZ University. The results should provide a detailed comparison about Generation X and Generation Y on the impact of employee experience on their performance.

2 Literature Review

2.1 Employee Experience

The business world is changing, and many companies are now prioritizing their employee journeys to have outstanding experiences in their organizations. Many HR professionals have created employee experiences in the organization similar with how they created customer experience within their company. It is in accordance with (Harris 2007) theory which posits that employee experience is based on customer experience management. Although there is a lack of study around the topic of employee experience, we want to expand on the research that provides the basic foundations of this research.

One article states that the definition of employee experience is "The employee journey has many milestones and interactions (or touchpoints), and the quality of employee experiences has a direct influence on employee satisfaction, engagement, commitment and, in the end, performance." (Plaskoff 2017, p. 2). Quantitative research has been conducted in assessing the employee experience, and in particular the human resources process. One of the studies in employee appraisal identified that there was a significant impact of line manager actions on employee-level outcomes but this was also constrained by the organizational climate (Farndale and Kelliher 2013).

Another study also identified the evolution of employee experience. It starts with the phase of utility (what do employees need to work); productivity (what do employees need to work better and faster); engagement (how can we make employees happy so they perform better); and experience (how can we create a company where people want vs. need to show up). Many companies today use employee engagement and employee experience interchangeably without any differentiation, which is inaccurate. Employee engagement is the short-term adjustment, while employee experience is the long-term redesign of the company. The research also defines employee experience as "the intersection of employee expectations, needs, and wants and the organizational design of those expectations, needs, and wants" (Morgan 2017). We can conclude the employee experience is beyond employee engagement.

HR in many companies has begun to implement outstanding journeys for their employees, starting from recruitment, onboarding, performance management, training development, career and succession management, working environment, compensation, benefits, safety, and the off boarding process. HR believe it will result in building emotional and greater employee connections with the company.

2.2 *Critical Drivers of Employee Experience*

Employee experience provides a comprehensive method that can assist many companies to build the best workplace practices. Many researchers have conducted studies to identify the important drivers of employee experience. One study identified an employee experience framework that consisted of physical environment, cultural environment, and technological environment (Morgan 2017).

Deloitte's employee experience framework discovered the factors that promoted a positive employee experience which were: meaningful work, supportive management, positive environment, growth opportunity, trust in leadership, collaboration, and connection (Abbatiello et al. 2017).

IBM and Globoforce (2017) designed the employee experience index (EXI) to measure employees' personal experience at work in terms of belonging, purpose, achievement, happiness, and vigor.

Another study contributed in finding the factors of employee experience: career; personal; business strategy; technology; cultural; physical environment; brand; and leadership (Gheidar and Zanjani 2021). A study gathered the critical drivers of employee experience that are: employee's experiential needs and desires; embracing holistic thinking; radical holistic thinking; radical participation; experimenting and iterating; trusting and appreciating the process; making intangible experiences visible tangibly; better workplace practices; and leadership and management practices (Itam and Ghosh 2020).

2.3 Employee Performance

Employee performance has been studied by researchers from different perspectives. Many studies focus on the factors impacting employee performance; others are focused on the dimensions of employee performance.

One study also revealed that employee performance had three dimensions: task, adaptive, and contextual performance. Performance in the form of task performance consisted of job explicit behaviors which included fundamental job responsibilities assigned as a part of an employee's job description. An adaptive performance involves employees' ability to efficiently deal with unpredictable working situations. Along with task and adaptability, employee also needed a kind of prosocial behavior demonstrated by employees in a workplace which is called contextual performance (Pradhan and Jena 2021).

Another study also intended to discover the concept of employee performance; they found that Organization Citizenship Behavior (OCB), team learning, LMX, individual learning, and innovative behavior were factors affecting performance (Atatsi et al. 2019).

In an empirical study conducted by Diamantidis and Chatzoglou (2019), the results revealed that job environment and management support had the strongest influences (direct and indirect) on job performance, while adaptability and intrinsic motivation directly affected job performance.

2.4 Generation X and Y

The multigeneration model has been studied by researchers from various viewpoints. Currently, there are several employee generations in companies. Each generation will have job expectations shaped by their own cohort's characters, which makes it is interesting for company to manage the needs of different individuals in the same organization. In this study, the age groups are established based on the categories by Kapoor and Solomon (2011) (Table 1).

Experts expect that there will be a shift in the composition of employee generations in workplaces in the future. Generation X and Y will occupy the workplace, as the two most populous cohorts in many companies. The changing in workforce generational groups indicates that organizations should recognize the personalities

Table 1 Age groups

No	Generation	Years born
1	Generation X	Between 1965 and 1980
2	Generation Y	Between 1980 and 1999

Source Kapoor and Solomon (2011)

of each age group (Harlianto 2021). Organizations should reframe their methods to how employees work together.

Organizations should recognize the attributes of Generation X and Y. One study found that the work style attributes of Gen X are emphasizing the great expectations of their needs, while Gen Y has high expectations technology in the company (Calloway 2018). The attributes of their job expectations are also different; Gen X needs more prospects for growth, autonomy, advancement, and flexibility in working schedules, whereas Gen Y needs the developments and advancement that provide the meaningfulness of the job.

2.5 Employee Experience and Employee Performance

A study shows that employee experience is the greatest level of fulfilment that an employee can feel and gain from their relations with work and the work environment in the company. It would lead to the highest levels of empowerment, engagement, and commitment leading to employee satisfaction and customer delight. Their study also provided the employee experience method; it described that top management action and workplace practices will support employee experience in the organization. The employee experience would develop employee engagement and performance in the organization (Itam and Ghosh 2020).

This study will focus on the relationship between various critical drivers of the employee experience which support improving the employee performance. This study combines ACE technology, COOL physical spaces, and CELEBRATED culture as the three dimensions of employee experience that was constructed by Morgan (2017), while the three dimensions of employee performance (task, adaptive, and contextual performance) are based on Pradhan and Jena (2021). Further, this paper also emphasized the employee experience strategy that impacted upon employee performance based on a comparison of Generation X and Generation Y.

3 Methodology

3.1 Research Design

This study uses a quantitative method. There are two types of research approaches used here, namely: (1) descriptive research, describing the characteristics of the respondents and its rating scale from the survey results; and (2) comparative research, comparing one or more variables on two or more different samples to investigate the difference between those samples. This study compares between Generation X and Generation Y on the impact of employee experience toward their performance. The research analysis unit is knowledge workers in a higher education institution.

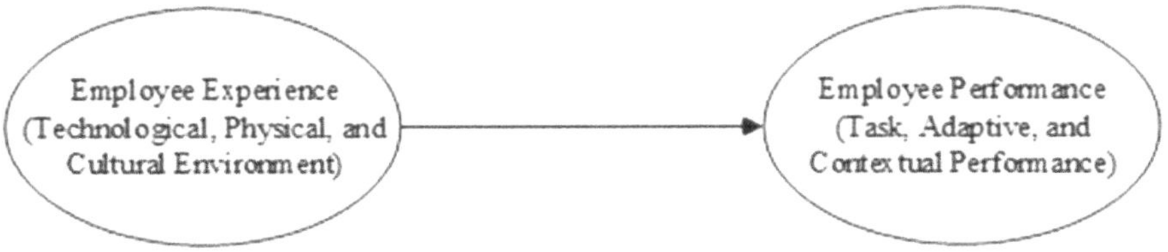

Fig. 1 Research model *Source* Author (2022)

3.2 Measurement of Variables

All items used to measure the constructs in this study were adapted from previous studies.

Employee experience was treated as the independent variable and was measured as described by Morgan (2017). In evaluating employee experience, respondents were asked to provide their assessment of employee experience in terms of technology, physical spaces, and culture as the three categories of employee experience based on a five-point Likert scale (1 = "strongly disagree" to 5 = "strongly agree").

Employee performance was treated as the dependent variable and was measured as described by Pradhan and Jena (2021). In evaluating employee performance, respondents were asked to provide a self-assessment of their performance in terms of task, adaptive, and contextual performance as the three categories of employee performance based on a five-point Likert scale (1 = "strongly disagree" to 5 = "strongly agree"). Employee performance was based on the self-evaluation of the employees.

3.3 Research Framework

Referring to the problem statement and literature review, the research framework is illustrated in the following Fig. 1.

3.4 Research Hypothesis

The research hypothesis is as follows:

H1: Employee experience has a significant impact on employee performance for Gen X and Gen Y.

3.5 Research Methodology

The statistic method used the structural equation modeling (SEM) techniques to analyze the standardized regression weights between two employee groups (Generation X and Y) on the impact of employee experience on their performance. This study was conducted in several steps, namely: (1) designing questionnaire instrument based on the research framework, (2) deciding the respondent sample, (3) conducting online survey, (4) conducting Validity and Reliability Test, (5) processing data with structural equation modeling using SPSS-Amos V26, and (6) interpreting and analyzing the data.

3.6 Data Collection

The data collection in this study used a survey with a Likert scale (1–5). The questions were designed based on the research framework and literature review. The research was conducted at XYZ University in Greater Jakarta, Indonesia. In response to the results of the literature study, a questionnaire was created. The study collected the primary data from 194 knowledge workers. The results of the current study were not published before in their current form.

4 Results and Discussions

This research utilized a quantitative research method to empirically test the variables under study. This part describes the analysis result and interpretation of the data processing.

4.1 Respondents' Profile

The survey results show that there were 72 respondents of Generation X employees and 122 respondents of Generation Y employees at XYZ University. The following table shows the respondent's profile based on several characteristics (Table 2).

4.2 Hypothesis Testing

This research aimed to examine the significant relationship between the independent variables and the dependent variable by comparing employee generations (Gen X

Table 2 Respondents' profile

		Generation Y		Generation X				Generation Y		Generation X	
		N	%	N	%			N	%	N	%
Gender	Male	50	51	48	49	Job function	Finance	3	75	1	25
	Female	72	75	24	25		Information technology	14	78	4	22
Education	High school equivalent	7	44	9	56		Marketing	35	85	6	15
	Diplomas	5	71	2	29		Human resources	23	62	14	38
	Undergraduate	84	68	39	32		Academic	14	74	5	26
	Master	24	55	20	45		Operations and support	32	48	35	52
	Doctoral	2	50	2	50		General affairs	1	13	7	88
Tenure	<1 year	18	90	2	10						
	1–3 years	32	89	4	11						
	4–6 years	37	88	5	12						
	7–10 years	23	66	12	34						
	>10 years	12	20	49	80						

Source Author (2022)

and Gen Y). The independent variable, employee experience, consisted of technological, physical, and cultural dimensions, while the dependent variable, employee performance, consisted of task, adaptive, and contextual dimensions. The statistic method used the structural equation modeling (SEM) techniques to analyze the standardized regression weights between two employee groups (Generation X and Y) on the impact of employee experience on their performance.

The following analysis aimed to strength the research framework by examining the research hypothesis using structural equation modeling (Table 3).

Based on the results of regression analysis, to identify the significant effect, it compared the value of the Critical Ratio (CR) > 1.96 with probability (P) < 0.05 and if there was significance of *** then it meant P < 0.001. If the CR value was above 1.96 and P < 0.01 then it could be concluded as a significant effect.

The results showed that the CR value for Generation Y was 5.879 > 1.96 with a P value of <0.001 or significant with ***, while the CR value for Generation X was 5.988 > 1.96 with a P value < 0.001 or significant with ***. It showed that employee experience had a significant impact on employee performance for both generations while Generation X had the higher significant impact compared to Generation Y.

The percentage of the impact of the employee experience on employee performance can be seen with the details for each indicator as follows (Table 4)

It can be seen also for the impact estimation from each indicator based on structural equation modeling (Figs. 2 and 3).

Based on the result, it states that the impact of employee experience toward employee performance for Generation Y was 0.69 or 69%, while for Generation X it was 0.86 or 86%. It meant that Generation X had a higher impact compared

Table 3 Result comparison of regression weights analysis

				Estimate	S.E.	C.R.	P
Generation Y	EP	←	EE	1.595	0.271	5.879	***
	Technology	←	EE	1			
	Physical	←	EE	2.267	0.361	6.287	***
	Cultural	←	EE	5.53	0.828	6.678	***
	Task	←	EP	1			
	Adaptive	←	EP	1	0.078	12.844	***
	Contextual	←	EP	1.585	0.134	11.87	***
Generation X	EP	←	EE	1.77	0.296	5.988	***
	Technology	←	EE	1			
	Physical	←	EE	3.257	0.485	6.719	***
	Cultural	←	EE	1			
	Task	←	EP	1.697	0.133	12.723	***
	Adaptive	←	EP	1.546	0.288	5.36	***
	Contextual	←	EP	1.122	0.081	13.9	***

Source Author (2022)

Table 4 Analysis result of standardized regression weights

			Estimate	
			Generation Y	Generation X
EP	←	EE	0.688	0.857
Technology	←	EE	0.617	0.720
Physical	←	EE	0.682	0.660
Cultural	←	EE	**0.961**	**0.881**
Task	←	EP	0.864	0.880
Adaptive	←	EP	**0.895**	**0.980**
Contextual	←	EP	0.862	0.940

Source Author (2022)

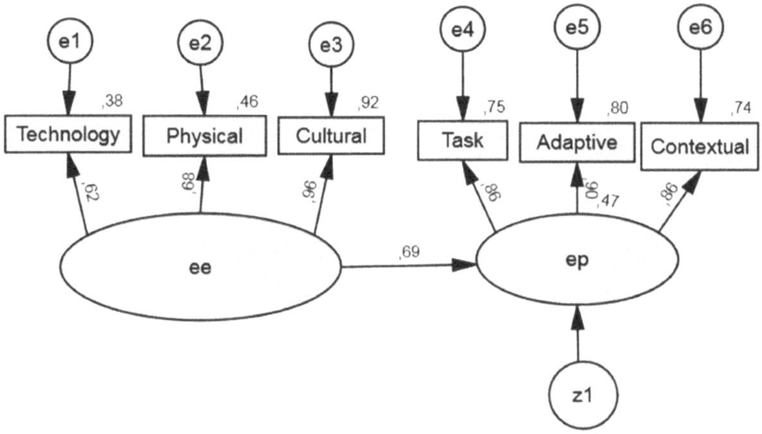

Fig. 2 Employee experience toward employee performance (Generation Y) *Source* Author (2022)

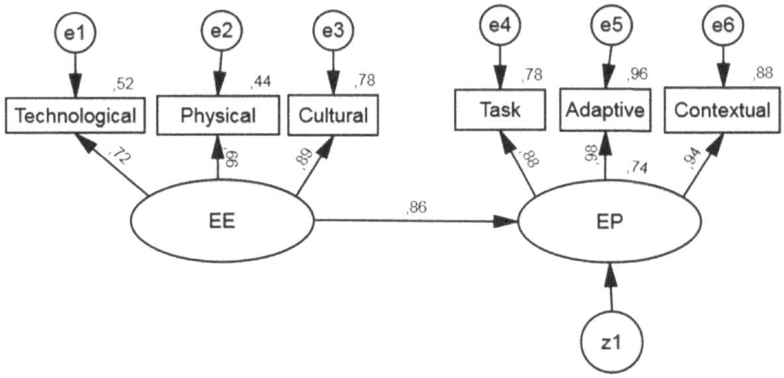

Fig. 3 Employee experience toward employee performance (Generation X) *Source* Author (2022)

Table 5 Goodness of fit

Goodness of fit index	Cut-off value	Analysis result Gen X	Analysis result Gen Y	Model
Chi square df 8	7.344	12.695	11.617	Good of fit
Df	<5.99	8	8	Good of fit
RMSEA	>0.08	0.091	0.061	Moderate fit
GFI	Close to 1	0.943	0.970	Good of fit
AGFI	Close to 1	0.850	0.921	Moderate fit
TLI	Close to 1	0.976	0.984	Good of fit
CFI	Close to 1	0.987	0.991	Good of fit

Source Author (2022)

with Generation Y. Other findings were the higher value of cultural environment would impact the higher value of adaptive performance for both Generation X and Generation Y (Table 5).

For both Generation X and Y, the findings were that the model was fit or in accordance with the expected. The model that shows the Goodness of fit measurement is a good model for the data. In sum, the more criteria met by the model, the more suitable the model is for this data.

5 Discussion

The result findings showed that the hypothesis, 'Employee experience has a significant impact on employee performance' was also statistically supported.

These results provided evidence that employee experience leads to employee performance. In general, employee experience was regarding employee judgments about their journey through all the touchpoints at their company, starting as a job applicant through to the exit process from the organization. Many organizations believe that providing the best employee experience will give the best impact for the organization and employees. Based on the results, it implied that the best employee experience can lead to an impact on employees, especially on their performance.

There is not much previous research that explore the impact of employee experience on employee performance, although much of the literature states the positive impact for organizations, such as employee productivity, engaging the multi-generational personnel, work satisfaction, competitive advantage, and organization growth (Itam and Ghosh 2020). This study contributed new findings about another positive impact of employee experience which is employee performance.

According to these findings, there is also difference in the result of the impact if we compare between multiple generations in a company. The two biggest populations nowadays in many companies are Generation X and Generation Y which have different results from the statistical calculation. The result reveals that Generation

X has a higher impact of employee experience on employee performance compared with generation Y.

Generation X and Y will occupy the workplace, meaning that organizations should understand their characteristics. Their characteristics also contribute to deference to their workstyles and job expectations within the company. Companies should reframe their approaches in how these multiple generations work together. Based on the findings, it implied that organizations should focus on the strategy in designing and implementing the employee experience according to the characters of each cohort of employees within a company.

The results of this study also indicated that specific dimensions of employee experience could impact the dimension of employee performance. Multiple generations also indicated a difference in impact results. The findings showed that higher levels of cultural environment would impact the higher value of adaptive performance for both Generations X and Y.

Every organization should focus on cultural environment that was proposed by Morgan (2017). An organization needs to be viewed positively by employees; this could be achieved if every employee feels valued and it also provides genuine sense of purpose that makes employees feel like they are part of the organization. This culture environment design should be appropriate for both Generation X and Generation Y.

Generation X need more fair treatment, commitment to employee health, wellness, diversity, and inclusion, while Gen Y need growth and advancement, and the organization should provide the ability to learn new things and offer the resources. Executives and managers are coaches and mentors for employees. Those implementations should produce a positive experience and referrals can come from employees.

The company's physical workspace, culture and technology are all important components of the employee experience according to Morgan (2017). The cultural environment is not the one that everyone can see but it is the only environment that everyone feels within the organization. Based on the findings, it implied that a component of employee experience should be focused on the specific multiple generations the company currently has; each generation has a different perspective on employee experience components, and it can lead to different results on specific types of employee performance (e.g., task, adaptive, or contextual performance).

5.1 Managerial Implications

The empirical research findings of this study also have managerial implications for higher educational institutions. Higher educational institutions need to understand the important of employee experience in an organization. They need to recognize the impact of employee experience that can contribute to employee performance.

It is important to design employee experience in at every touchpoint within an organization, but the company also needs to consider the multi-generational characteristics. Each generation should be assessed about their needs in undergoing their journey as employees in the organization, whether they emphasize more the physical

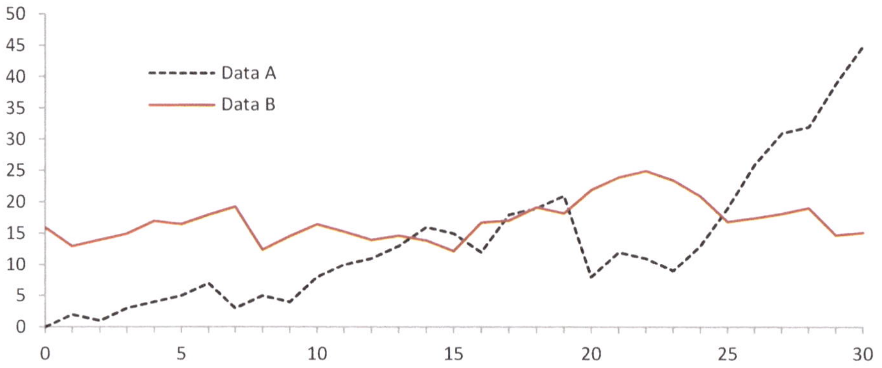

Fig. 4 A figure caption is always placed below the illustration. Short captions are centered, while long ones are justified. The macro button chooses the correct format automatically

workspace, culture, or technology. Organizations should have strategy in designing, implementing, and evaluating their employee experience within every touchpoint, since employee experience will have a significant impact on their performance in the organization (Fig. 4).

6 Conclusion

The empirical findings have demonstrated that the proposed research framework showed that employee experience had a significant impact on employee performance for both generations while Generation X had a higher significant impact compared with Generation Y. The findings also show that a higher value of cultural environment will impact the higher value of adaptive performance for both Generation X and Generation Y. Companies should have strategy in designing, implementing, and evaluating their employee experience within every touchpoint; they also need to consider the multi-generational characteristics since employee experience will have significant impact toward their performance in the organization.

Although much of the literatures states the positive impact of employee experience, few focus on the impact on employee performance. This study contributes new findings about another positive impact of employee experience which is employee performance. We expect that future research will continue to combine other variables of human resources practices, while adding Gen Z which will also be highly significant for the labor market. Its idiosyncrasies should be also considered to improve our understanding of the impact of employee experience toward employee experience in HRM implementation.

References

Abbatiello, A., Knight, M., Philpot, S., Roy, I.: Leadership disrupted: pushing the boundaries. Deloitte Glob. Hum. Capital Trends, 77–85 (2017). https://www2.deloitte.com/content/dam/Deloitte/global/Documents/About-Deloitte/central-europe/ce-global-human-capital-trends.pdf

Atatsi, E.A., Stoffers, J., Kil, A.: Factors affecting employee performance: a systematic literature review. J. Adv. Manag. Res. **16**(3) (2019). https://doi.org/10.1108/JAMR-06-2018-0052

Calloway, C., Calloway Recruiting, C., Generations, R.X. (n.d.). Recruiting and Retaining Generations X and Y in Organizations Recruiting and Retaining Generations X and Y in Organizations Recruiting and Retaining Generations X and Y Recruiting and Retaining Generations X and Y in Organizations. https://corescholar.libraries.wright.edu/honors

Diamantidis, A.D., Chatzoglou, P.: Factors affecting employee performance: an empirical approach. Int. J. Product. Perform. Manag. **68**(1) (2019). https://doi.org/10.1108/IJPPM-01-2018-0012

Farndale, E., Kelliher, C.: Implementing performance appraisal: exploring the employee experience. Hum. Resour. Manag. **52**(6) (2013). https://doi.org/10.1002/hrm.21575

Gheidar, Y., Zanjani, M.: Designing a conceptual framework for digital employee experience. Iran. J. Manag. Stud. **14**(4) (2021). https://doi.org/10.22059/IJMS.2020.305871.674131

Harlianto, J.: Factors influencing the employee retention of millennial generation (Case study: XYZ Foundation). Int. J. Manag. (IJM) **12**(1), 1115–1126 (2021). https://doi.org/10.34218/IJM.12.1.2021.097

Harris, P.: We the people: the importance of employees in the process of building customer experience. J. Brand Manag. **15**(2), 102–114 (2007). https://doi.org/10.1057/palgrave.bm.2550123

IBM and Globoforce.: The Employee experience index around the globe: how countries measure up and create human workplaces introduction **15** (2017)

Itam, U., Ghosh, N.: Employee experience management: a new paradigm shift in HR thinking. Int. J. Hum. Capital Inf. Technol. Profession. **11**(2) (2020). https://doi.org/10.4018/IJHCITP.2020040103

Kapoor, C., Solomon, N.: Understanding and managing generational differences in the workplace. Worldw. Hospital. Tour. Themes **3**(4) (2011). https://doi.org/10.1108/17554211111162435

Morgan, J.: The Employee Experience Advantage How to Win the War for Talent by Giving Employees the Workspaces they Want, the Tools they Need, and a Culture they Can Celebrate. Wiley (2017)

Plaskoff, J.: Employee experience: the new human resource management approach. Strateg. HR Rev. **16**(3) (2017). https://doi.org/10.1108/shr-12-2016-0108

Pradhan, R.K., Jena, L.K.: Employee performance at workplace: conceptual model and empirical validation. Bus. Perspect. Res. **5**(1) (2021). https://doi.org/10.1177/2278533716671630

Verasai, A.: Employee experience: how to influence and win over workers. The HR Digest (2021). https://www.thehrdigest.com/employee-experience-how-to-influence-and-win-over-workers/

The Lagging of HR Analytics Adoption Rate: Understanding from Individual Attitude of HR Professionals

Jefta Harlianto, Rano Kartono Rahim, Mohammad Hamsal, and Asnan Furinto

Abstract HR functions also have become progressively data-centric, while organizations have higher expectation for HR in providing important insights to improve overall productivity. This expectation should be shown in higher adoption HR Analytics by HR professionals. Organization should prioritize HR analytics adoption but there was slow progress towards adopting the technology. The objective of this paper is to determine the factors influencing attitude of HR professionals towards Human Resources analytics in Jakarta. Partial least squares structural equation modeling (PLS-SEM) was used to confirm the model based on data compiled from HR professionals. This study examines the effect of technology self-efficacy and tool trialability on attitude towards Human Resources Analytics. The result shows a significant effect of technology self-efficacy on attitude towards Human Resources analytics. However, tool trialability does not have effect on attitude towards Human Resources analytics. The findings of this study suggest an important message for organizations to focus on technology self-efficacy in increasing attitude towards Human Resources analytics, this would strengthen the implement HR Analytics successfully. It also concludes that internal motivation should drive the personal intention on adoption of HR Analytics. This empirical study contributes to a better understanding about the factors of HR professionals in adopting HR analytics as needed by their organizations and essential for competitiveness.

Keywords Technology self-efficacy · Tool trialability · Attitude towards human resources analytics · HR analytics adoption

---

J. Harlianto (✉) · R. K. Rahim · M. Hamsal · A. Furinto
Bina Nusantara University, Jakarta, Indonesia
e-mail: jefta@binus.edu

R. K. Rahim
e-mail: rano.kartono@binus.ac.id

M. Hamsal
e-mail: mhamsal@binus.edu

A. Furinto
e-mail: afurinto@binus.edu

M. Busu (ed.), *Rethinking Business for Sustainable Leadership in a VUCA World*, Springer Proceedings in Business and Economics,
https://doi.org/10.1007/978-3-031-50208-8_13

1 Introduction

The business world is going digital, and we cannot neglect it since it's one of the major trends changing society and business. Human resources (HR) function currently is increasing important in many organizations, HR has role as strategic business partner for top management in company. HR functions also have become progressively data-centric, while organizations are now expecting to HR in pursuing valuable insights to improve overall productivity. HR data and the integration of other functions company data will be beneficial to support decision-making in company (e.g., retain high potential employees, design benefits that match with employee needs, etc.). This will be possible by having availability of HR data with the support of technological developments. However, Human resources (HR) function is lagging form other functional fields of management in the adoption of data analytics technology (Angrave et al. 2016).

Several surveys have been conducted regarding Asia-Pacific adoption of HR analytics. In 2017, one survey by Deloitte discovered that 71% of companies extremely prioritize HR analytics within their organization but there was slow progress towards adopting the technology (Abbatiello et al. 2017). Meanwhile, KPMG in 2019 revealed that plans to implement people analytics is a low importance among companies including the HR executives, even though more than 80% of the respondents agreed with the statement that HR could add more benefit to their organization through analytics. While other data showed that there were slight changes in the number of organizations that linked people data to business results.

Many studies also want to contribute to the growing argument on analytics as a strategic ability within many companies. Recently, HR analytics is promising as a significant field in both academic research and in practitioner-related implementation (Margherita 2022). HR analytics adoption is described as the method through which company invests in, implements, and integrates HRA into the workforce's decision-making process in organization. However, the adoption of HR analytics by company is flawed by some challenges. A major challenge that adversely influences the implementation of HR analytics is the differing levels of process maturity within the HR function across its organizations (Shet et al. 2021).

Other study found that the adoption of HR analytics is exposed to the emergence of some obstacles. It identified 14 obstacles which are grouped into four main categories: data and models, software and technology, people and, management (Fernandez and Gallardo-Gallardo 2021). This is also in line with one research show that HR professionals also need to develop their skills and knowledge to become HR analytics champions and become more strategic and board wide impact within organization (Angrave et al. 2016). One recent study has evaluated the adoption of HR analytics giving more importance to personal factors, although technology might be a challenge in its implementation (Fernandez and Gallardo-Gallardo 2021). Other study also revealed that even in today's technology-driven work environment, HR practitioners continue to rely on spreadsheets for data analytics rather than on human resources management platforms (Shet et al. 2021).

The objective of this paper is to determine the factors influencing attitude of HR professionals towards Human Resources analytics in Jakarta. It focusses on the investigation the predictor of intention to adopt HR analytics. This study examines the effect of technology self-efficacy and tool trialability on attitude towards Human Resources Analytics. This empirical study contributes to a better understanding about the factors of HR professionals in adopting HR analytics as needed by their organizations and essential for competitiveness.

2 Literature Review

2.1 Human Resources Analytics

Analytics is a relatively emerging discipline that is growing in terms of methods, stakeholders, and impact (Huselid 2018). In the last ten years, many literatures have provided contribution and address many critical perspectives about Human resources (HR) analytics. Human resources (HR) analytics concept is growing with different naming, such as people analytics, workforce analytics, talent analytics, HR big data analytics and human capital analytics.

Basically, the ideas refer to the method to managing individuals within companies and creating more objective, rational and effective judgments about employees based on the data analysis (Margherita 2022). All HR Analytics definitions mentioned in many literatures, those are about collecting, analyzing, and reporting data to improve people-related decisions in organization and, as result to enhance employee and business results. Therefore, HR Analytics can be found as a strategic method that depend on data and statistics simulations, and it is more than just HR measures and metrics (Fernandez and Gallardo-Gallardo 2021).

Previous literature also provides the objective of HR Analytics is to deliver reliable fundamentals for employee data-related decisions that impact the personal and organizational outcomes. HR Analytics clearly refer to enhance or improve organization performance outcomes or determine business goals (Fernandez and Gallardo-Gallardo 2021).

One literature revealed that the characteristics of HR Analytics are mostly as an analysis process or decision-making process, as a method and as an HR practice (Fernandez and Gallardo-Gallardo 2021). While other study also highlighted number of main elements of HR analytics which are (a) an evidence-based approach to people-related decision-making; (b) it implements methodical approaches of HR data analysis and visualization (c) it assists the requirements of leader and top management decision making; (d) a multi-procedure and multi-application effort with various potential effects (Margherita 2022).

2.2 Attitude Towards Human Resources Analytics

The attitudes formed towards the innovation under consideration will determine the persuasion stage of the innovation-decision process. An organization which has adopted analytics will expect this knowledge will promote a positive attitude towards analytics toward individual employees. The HR professionals' commitment and willingness to learn and intent in adopting the HR Analytics determine the readiness level to move towards HR Analytics (Vargas et al. 2018).

Previous studies shows that behavioral intention can be explained as individual willingness towards any aspect, and it reflects person behavior. It has documented the connection between intention and behavior (Venkatesh et al. 2000). While other study shows that an individual's willingness to use a technology depends on their acceptance and intention to use it (Lin and Chang 2011). One literature show that there is significant positive relationship between adoption Intention of HRA and HRA adoption behavior (Ekka and Singh 2022).

HR Leaders who have been focused on the soft skills development related to analytics within the organization seem to lack of reassurance in dealing with HR analytics. Previous literature found that HR analytics is exposed to the emergence of some obstacles and one category of the obstacles is people. It shows the lack of knowledge, skills and competences related to analytics (Fernandez and Gallardo-Gallardo 2021). It is becoming progressively visible that HR professionals need training in understanding analytics.

2.3 Hypothesis Development

Technology self-efficacy

The Fundamental theory for technology self-efficacy, we should refer to self-efficacy according to Bandura. It is the principle that an individual's abilities in terms of performing certain tasks required to generate given attainments reflects on individual abilities due to the skills they own (Bandura 1997). The Technology Self-Efficacy framework is also recognized as Computer Self-Efficacy, and it shows an essential part in the adoption of innovation at personal level (Ejaz et al. 2020). Similarly, Technology Self-Efficacy also impacts on the behavioral intention to use according to Hu et al. (2003); as cited evidence by Perera and Jayawardana (2022).

Specific topic in HR analytics, one literature show that technological factors are critical for the adoption of HRA in company (Shet et al. 2021). Previous study shows that technology Self Efficacy influences the HRIS Use Behavior and Behavioral Intention HRIS use (Perera and Jayawardana 2022). Technology Self-Efficacy also have a positive relationship among the UTAUT model's HRIS Use Behavior-related variables (Perera and Jayawardana 2022). Other study in healthcare also found that technology self-efficacy had effect on perceived usefulness, perceived ease of use, and attitude toward using personal health record (Dutta et al. 2018). While earlier literature has provided empirical evidence about the positive effect of self-efficacy on the intention behavior to use towards the actual usage for the technology (Al-Haderi 2013). The technology self-efficacy also found that it can determine the individual level of HR analytics adoption (Vargas et al. 2018). Based on previous literature, we hypothesize the following:

H1: Technology Self-efficacy has significant effect on Attitude towards Human Resources Analytics.

Tool trialability

Trialability is the degree to which an innovation may be investigated with on a limited basis (Roger 2003). The five attributes of innovations are: (1) relative advantage, (2) compatibility, (3) complexity, (4) trialability, and (5) observability. One literature revealed that relatively earlier adopters perceive trialability as more important than do later adopters (Gross 1942; Ryan 1948). The level of adoption is also explained by the five attributes (relative advantage, compatibility, complexity, trialability, and observability) (Roger 2003).

Previous study found that trialability had a significant influence on intention to use (Ullah et al. 2021), mean while other literature shows that trialability does not have a positive influence on the adoption process (Indra 2022). Other studies related to trialability; it shows that trialability is one of the success factors for successful adoption. While trialability determine the factors influencing customer adoption (Ali et al. 2019). Trialability was also found impacting the intention of m-banking adoption both positively and significantly (Changchun et al. 2017). Based on previous literature, we hypothesize the following:

H2: Tool Trialability has significant effect on Attitude towards Human Resources Analytics.

3 Methodology

3.1 Research Design

This study uses a quantitative method. There are two types of research approaches used here, namely: (1) descriptive research, describing the characteristics of the respondents and its rating scale from the survey result; and (2) The type of research used is associative (describing causal relationships). This study examines the effect of technology self-efficacy and tool trialability on attitude towards Human Resources Analytics.

3.2 Measurement of Variables

All instruments used to measure the constructs in this study were appropriately adapted from previous studies. The technology self-efficacy and tool trialability were treated as the independent variables. The technology self-efficacy (e.g., "I could complete a job or task using technology if I had training materials readily available") was measured through the instrument of Lee and Huang (2014) and Chau (2001). The instrument of Moore and Benbasat (1991) was used to measure HR Analytics tool 'trialability' (e.g., "I have had a great deal of opportunity to try various HR Analytics applications"). In evaluating technology self-efficacy and tool trialability, respondents were asked to provide their assessment of instruments based on a five-point Likert scale (1 = "strongly disagree" to 5 = "strongly agree").

The attitude towards analytics was treated as the dependent variable. The Johnston and Warkentin (2010) through a four-item scale was used to assess the attitude towards analytics (e.g., "I like working with HR Analytics"). In evaluating technology self-efficacy and tool trialability, respondents were asked to provide their assessment of instruments based on a five-point Likert scale (1 = "strongly disagree" to 5 = "strongly agree").

3.3 Research Model

Referring to the problem statement and literature review, the research framework is illustrated in the following figure:

3.4 Research Methodology

This study was designed to use quantitative approach. The examination of causal relationship across variables was done with Structural Equation Modeling-Partial Least Squares (SEM-PLS). Two types of variables were involved, latent (construct) variable also known as unobserved variable, and indicator variable also known as observed variable of each latent variable. Latent variable is divided into exogenous latent variable and endogenous latent variable. In this study, the exogenous latent variable refers to technology self-efficacy and tool trialability while the endogenous latent variable is represented by attitude towards human resources analytics. The model of hypotheses is shown in Fig. 1.

This study is conducted in several steps, namely: (1) designing questionnaire instrument based on the research framework, (2) deciding the respondent sample, (3) conducting online survey, (4) processing data using SMART-PLS software, and (5) interpreting and analyzing the data.

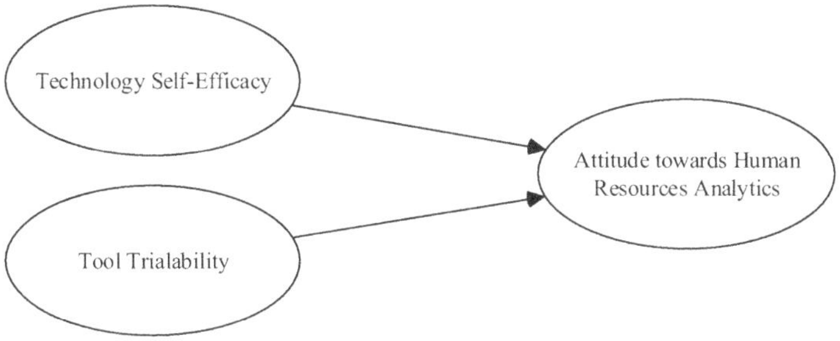

Fig. 1 Research model

3.5 Data Collection

Data collection was done with survey method that using questionnaire with Likert Scale (1–5), where the questions designed based on the research framework and literature review mentioned before. The targeted respondent for this research consisted of HR professionals who are familiar with HR Analytics, or currently working in organizations adopted it. The respondents' responses were requested through online survey to ensure confidentiality and privacy. The study collected the primary data from 71 h professionals in Jakarta.

4 Results

This part describes the analysis result and interpretation of the data processing. The Structural Equation Modeling-Partial Least Squares (SEM-PLS) is used to examine the causal relationship of technology self-efficacy and tool trialability on attitude towards Human Resources Analytics.

4.1 Respondents' Profile

The online survey was conducted to gather the result from HR professionals. The following table shows the respondent's profile based on several characteristics (Table 1)

Table 1 Respondents profile

Demographic	Categories	%	Demographic	Categories	%
Gender	Male	47	Tenure	<1 year	4
	Female	53		>10 years	33
Age	≤20 years	0		1–3 years	9
	21–37 years	62		4–6 years	13
	38–55 years	38		7–10 years	40
	≥56 years	0	Industry	Banking	27
Education	High school equivalent	2		Telecommunication	9
	Diplomas	0		Finance technology	2
	Undergraduate	60		Education	53
	Master	38		Information technology	4
	Doctoral	0		Multi industry	2
				Consulting	2

Source Author (2022)

4.2 Evaluation of Measurement Model (Outer Model)

This part describes about assessment of measurement model. For the assessment of measurement model, there was conducted convergent Validity, discriminant validity, and composite reliability.

Convergent validity is determined based on the principle that the metrics of a construct should be highly correlated. It should be done based on value of loading factor and AVE. Outer loading Factor > 0.7 (Hair et al. 2017: 102), and Average Variance Extracted (AVE) > 0.705 (Hair et al. 2017: 115). As shown in Table 2, it indicates the values of convergent validity.

Discriminant validity intends to define whether a reflective indicator is really a good measure of its construct based on the assumption that each indicator must be highly correlated with its construct only. Measures of different constructs should not be highly correlated (Ghozali and Latan 2015). In the SmartPLS application, the discriminant validity test uses cross loadings values and the Fornell-Larcker Criterion, (Henseler et al. 2015).

If the square root value of the AVE for each construct is greater than the correlation value between constructs and other constructs in the model, then the model is said to have a good discriminant validity value (Fornell and Larcker 1981; Wong 2013). Cross Loading is based on the factor loading of all indicators in one latent variable must be greater than those in other latent variable. As shown in Tables 3 and 4, it indicates Fornel Larcker and cross loading value.

Composite Reliability measured the reliability of the reflective constructs. Composite Reliability must be >0.6 and Cronbach's Alpha > 0.7 (Ghozali and Latan 2015). As shown in Table 2, it indicates the Cronbach's Alpha and Composite Reliability value.

Table 2 Outer loading factors, AVE, Cronbach's alpha and composite reliability

Latent variable	Indicators	Outer loading factors	Average variance extracted (AVE)	Cronbach's alpha	Composite reliability
Attitude towards human resource analytics	HRA1	0,868	0,78	0,906	0,934
	HRA2	0,87			
	HRA3	0,884			
	HRA4	0,909			
Tool trialability	TR1	0,874	0,739	0,825	0,894
	TR2	0,831			
	TR3	0,873			
Technology self-efficacy	TSE1	0,905	0,775	0,713	0,873
	TSE2	0,856			

Source Author (2022)

Table 3 Fornel lacker

Latent variable	HRA (Attitude towards human resource analytics)	TR (Tool trialability)	TSE (Technology self-efficacy)
HRA (Attitude towards human resource analytics)	0,883		
TR (Tool trialability)	0,411	0,859	
TSE (Technology self-efficacy)	0,766	0,591	0,881

Source Author (2022)

Table 4 Cross loading

Latent variable	HRA (Attitude towards human resource analytics)	TR (Tool trialability)	TSE (Technology self-efficacy)
HRA1	0,868	0,31	0,643
HRA2	0,87	0,456	0,676
HRA3	0,884	0,257	0,666
HRA4	0,909	0,423	0,72
TR1	0,406	0,874	0,622
TR2	0,299	0,831	0,459
TR3	0,34	0,873	0,416
TSE1	0,736	0,533	0,905
TSE2	0,605	0,507	0,856

Source Author (2022)

Table 5 R^2

	R^2	R^2 adjusted
HRA (Attitude towards human resource analytics)	0,59	0,571

Source Author (2022)

4.3 Evaluation of Structural Model (Inner Model)

Evaluation of structural model is used to predict the causal relationship between latent variables based on the hypothesis (Ghozali and Latan 2015). In the evaluation of the structural model, an evaluation is carried out by testing the value of R-square (R2), Q-square (Q2) test, and NFI.

The following is a description of each stage carried out in the evaluation of the structural model and the criteria for the evaluation.

4.4 R-Square

The R-square (R^2) test is intended to measure the proportion of changes in exogenous variables to endogenous variables. The condition is that if the R-square (R^2) value of 0.67 means that the model is good, 0.33 means the moderate model, and 0.19 means the weak model (Chin 1998) (Table 5).

The R Square value of the joint effect of technology self-efficacy and tool trialability on attitude towards Human Resources Analytics is 0.59 with an adjusted r square value of 0.571 (Table 7), it can be explained that all independent variables (technology self-efficacy and tool trialability) simultaneously effect attitude towards Human Resources Analytics by 0.59 or 59%. Because Adjusted R^2 57.1 < 67%, the effect of technology self-efficacy and tool trialability on attitude towards Human Resources Analytics is moderate.

4.5 T-Statistic (Bootstrapping)

The results of the PLS SEM direct effects bootstrapping analysis are as follows (Table 8).

4.6 Direct Effects of Technology Self-efficacy on Attitude Towards Human Resources Analytics

Based on calculations using bootstrap, where the test results of the estimated coefficient of Technology Self-Efficacy against Attitude towards Human Resources Analytics bootstrap results are 0,805 with T Statistics value of 8,225 and standard deviation of 0,098. The P value is 0.00 < 0.05 so accept H1, it means that direct effect of Technology Self-Efficacy on Attitude towards Human Resources Analytics is significant.

4.7 Direct Effects of Tool Trialability on Attitude Towards Human Resources Analytics

Based on calculations using bootstrap, where the test results of the estimated coefficient of Tool Trialability against Attitude towards Human Resources Analytics bootstrap results are −0,065 with T Statistics value of 0,521 and standard deviation of 0,124. The p value is 0,603 > 0.05 so reject H2, it means that direct effect of Tool Trialability on Attitude towards Human Resources Analytics is not significant (Table 6).

Table 6 T-statistic

	Original sample (O)	Sample mean (M)	Standard deviation (STDEV)	T statistics (|O/STDEV|)	P values
TR (Tool trialability) → HRA (Attitude towards human resource analytics)	−0,064	−0,065	0,124	0,521	0,603
TSE (Technology self-efficacy) → HRA (Attitude towards human resource analytics)	0,804	0,805	0,098	8,225	0

Source Author (2022)

Table 7 Predictive relevance

	SSO	SSE	Q^2 (=1-SSE/SSO)
HRA (Attitude towards human resource analytics)	180	99,571	0,447
TR (Tool trialability)	135	135	
TSE (Technology self-efficacy)	90	90	

Source Author (2022)

Table 8 Model fit

	Saturated model	Estimated model
SRMR	0,079	0,079
d_ULS	0,284	0,284
d_G	0,231	0,231
Chi-square	59,962	59,962
NFI	0,773	0,773

Source Author (2022)

4.8 Predictive Relevance

Q-Square (Q2) testing to measure how well are the observed values generated by the estimated model and parameter. The requirement is that if the Q-square value is more than 0 (zero) it means that the model has a good predictive relevance value, whereas if the Q-square value is less than 0 (zero) it means that the model lacks good predictive relevance. Based on calculations using Q-Square (Q2), the value is 0.447. It is more than 0 (zero) it means that the model has a good predictive relevance value.

4.9 Model Fit

NFI value to measure how good is the research model. The requirement is that if NFI value produces a value from 0 to 1, a good NFI value is a value close to value 1. Based on calculations using NFI, the value is 0.773 (value close to value 1). It means is a good research model to estimate the effect of technology self-efficacy and tool trialability on attitude towards Human Resources Analytics.

Based on this result, the author considers research model as fit and perceives that it can be used for hypothesis testing (Fig. 2).

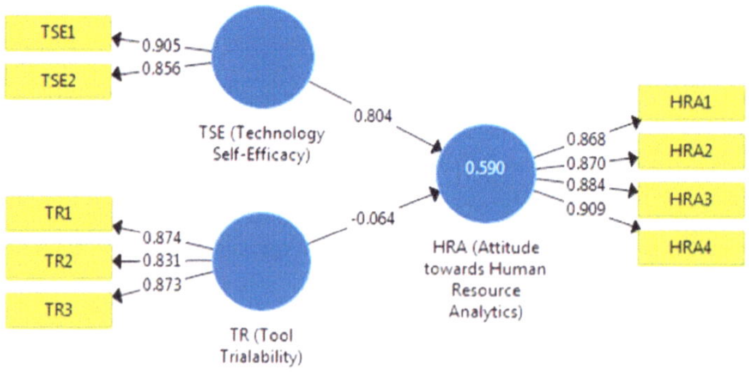

Fig. 2 Hypothesis model result

5 Discussion

This study examines the effect of technology self-efficacy and tool trialability on attitude towards Human Resources Analytics. Previous study shows that findings have mainly focused on personal obstacles in adopting HR Analytics (Vargas et al. 2018). One recent literature also has evaluated the adoption of HR analytics giving more importance to personal factors, although technology might be a challenge in its implementation (Fernandez and Gallardo-Gallardo 2021).

The survey result show that the direct effect of Technology Self-Efficacy on Attitude towards Human Resources Analytics is significant. This is in line with previous study stating that Previous study revealed that the individual level of adoption is driven predominantly by attitude towards analytics followed closely by technology and quantitative self-efficacy (Vargas et al. 2018). Other literature also shows that technology Self Efficacy influences the HRIS Use Behavior and Behavioral Intention HRIS use (Perera and Jayawardana 2022). Technology Self-Efficacy also have a positive relationship among the UTAUT model's HRIS Use Behavior-related variables (Perera and Jayawardana 2022). It concludes the increase of technology self-efficacy provide greater adoption intention of Human Resources Analytics.

The survey result also shows that the direct effect of Tool Trialability on Attitude towards Human Resources Analytics is not significant. This is in line with previous literature shows that trialability does not have a positive influence on the adoption process (Indra 2022). One literature shows that trialability might play an important role for organizations that have adopted HR Analytics, however, the total effect on the individual level of adoption is lesser (Vargas et al. 2018). We assume tool trialability might need or role-as a mediating variable for effecting the attitude towards Human Resources analytics, and previous study also examined the roles of trialability as mediator (Changchun et al. 2017).

The finding of this research can be summarized that there is a significant effect of technology self-efficacy on attitude towards Human Resources analytics. However, tool trialability does not have effect on attitude towards Human Resources analytics. In this study, we divide the factors into two categories which are individual's perception and external motivation. Technology self-efficacies appears to be associated to the individual's perception, and this variable have the significant impact on HR Analytics adoption. The increase of technology self-efficacy provides greater adoption intention of Human Resources Analytics. This assume that each HR professional could complete a job / task using technology which makes their job more interesting and enjoyable. While external motivation related to tool trialability, the study implies that the variable act as non-directly lead driver of HR analytics adoption intention process. This should have further study to examine tool trialability whether it "need" or "role-as" a mediating variable in effecting the attitude towards Human Resources analytics.

In other words, we can conclude that internal motivation should drive the personal intention on adoption of HR Analytics. Therefore, the findings of this study suggest an important message for organizations in Jakarta to focus on technology self-efficacy in increasing attitude towards Human Resources analytics, this would strengthen the implement HR Analytics successfully.

6 Conclusion

This paper is expected to determine the factors influencing attitude of HR professionals towards Human Resources analytics in Jakarta. It focusses on the investigation the predictor of intention to adopt HR analytics. It examines the effect of technology self-efficacy and tool trialability on attitude towards Human Resources Analytics. The finding of this research shows a significant effect of technology self-efficacy on attitude towards Human Resources analytics. However, tool trialability does not have effect on attitude towards Human Resources analytics.

This empirical study contributes to a better understanding about the factors of HR professionals in adopting HR analytics as needed by their organizations and essential for competitiveness. This study should be beneficial for the Human Resources (HR) leaders to understand about individual intention of HR Professionals in adopting HR analytics. They can focus on the technology self-efficacy to increase the intention and take action to eliminate the barriers to the personal's decision.

The study still has limitations. For further study, we suggest using the research instruments in wider range of industry, while it could be developed by expanding the model and including other predictors of intention to adopt HR analytics should also be included. The existing instruments can also be utilized to analyze in other functional areas (e.g., marketing analytics etc.).

References

Abbatiello, A., Knight, M., Philpot, S., Roy, I.: Leadership disrupted: pushing the boundaries. Deloitte Global Human Capital Trends, 77–85 (2017). https://www2.deloitte.com/content/dam/Deloitte/global/Documents/About-Deloitte/central-europe/ce-global-human-capital-trends.pdf

Al-Haderi, S.M.S.: The effect of self-efficacy in the acceptance of information technology in the Public Sector Sami Mohamed Saleh Al-Haderi School of Business, Univeristy Utara Malaysia UUM, sintok 06010 Kedah, Malaysia. Int. J. Bus. Soc. Sci. **4**(9), 188–198 (2013)

Ali, M., Raza, S.A., Puah, C.H., Amin, H.: Consumer acceptance toward takaful in Pakistan: an application of diffusion of innovation theory. Int. J. Emerg. Mark. **14**(4), 620–638 (2019). https://doi.org/10.1108/IJOEM-08-2017-0275

Angrave, D., Charlwood, A., Kirkpatrick, I., Lawrence, M., Stuart, M.: HR and analytics: why HR is set to fail the big data challenge. Hum. Resour. Manag. J. **26**(1), 1–11 (2016). https://doi.org/10.1111/1748-8583.12090

Bandura, A.: Self-efficacy: toward a unifying theory of behavioral change. Psychol. Rev. **84**(2), 191–215 (1977). https://doi.org/10.1037/0033-295X.84.2.191

Changchun, G., Haider, M.J., Akram, T.: Investigation of the effects of task technology fit, attitude and trust on intention to adopt mobile banking: placing the mediating role of trialability. Int. Bus. Res. **10**(4), 77 (2017). https://doi.org/10.5539/ibr.v10n4p77

Chau, P.Y.K.: Influence of computer attitude and self-efficacy on IT usage behavior. J. End User Comput. **13**(1), 26 (2001)
Chin, W.W.: Commentary: issues and opinion on structural equation modeling. MIS Q. **22**(1), 7–17 (1998)
Dutta, B., Peng, M.H., Sun, S.L.: Modeling the adoption of personal health record (PHR) among individual: the effect of health-care technology self-efficacy and gender concern. Libyan J. Med. **13**(1) (2018). https://doi.org/10.1080/19932820.2018.1500349
Ejaz, S., Akbar, W., Shaikh, M.: Slow adoption of HR analytics: understanding from the lens of innovation diffusion theory. Int. J. Manag. **11**(February), 2090–2101 (2020). https://doi.org/10.34218/IJM.11.11.2020.198
Ekka, S., Singh, P.: Predicting HR professionals' adoption of HR analytics: an extension of UTAUT model. Organizacija **55**(1), 77–93 (2022). https://doi.org/10.2478/orga-2022-0006
Fernandez, V., Gallardo-Gallardo, E.: Tackling the HR digitalization challenge: key factors and barriers to HR analytics adoption. Compet. Rev. **31**(1), 162–187 (2021). https://doi.org/10.1108/CR-12-2019-0163
Fornell, C., Larcker, D.F.: Evaluating structural equation models with unobservable variables and measurement error. J. Mark. Res. **18**(1), 39–50 (1981). https://doi.org/10.2307/3151312
Ghozali, I., Latan, H.: Partial least squares: Konsep, Teknik dan Aplikasi Menggunakan Program SmartPLS 3.0. Semarang: Badan Penerbit Universitas Diponegoro (2015)
Gross, N.C.: The Diffusion of a Culture Trait in Two Iowa Townships, M.S. thesis, Ames, Iowa State College (1942)
Hair, J.F., Hult, G.T.M., Ringle, C.M., Sarstedt, M.: A Primer on Partial Least Squares Structural Equation Modeling (PLS-SEM), 2nd edn. Sage Publications Inc., Thousand Oaks, CA (2017)
Henseler, J., Ringle, C.M., Sarstedt, M.: A new criterion for assessing discriminant validity in variance-based structural equation modeling. J. of the Acad. Mark. Sci. **43**, 115–135 (2015). https://doi.org/10.1007/s11747-014-0403-8
Hu, P.J.-H., Clark, T.H.K., Ma, W.W.: Examining technology acceptance by school teachers: a longitudinal study. Inf. Manag. **41**(2), 227–241 (2023). https://doi.org/10.1016/s0378-7206(03)00050-8
Huselid, M.A.: The science and practice of workforce analytics: introduction to the HRM special issue. Hum. Resour. Manag. **57**(3), 679–684 (2018). https://doi.org/10.1002/hrm.21916
Indra, R.H.: Determinant factors of voluntary adoption of mobile money in Indonesia. J. Ilmiah Indonesia **7**(8.5.2017), 2003–2005 (2022)
Johnston, A.C., Warkentin, M.:. Fear appeals and information security behaviors: an empirical study fear appeals and information security behaviors: an empirical study1. Source: MIS Q. **34**(3), 549–566 (2010). http://www.jstor.org/stable/25750691%5Cn, http://about.jstor.org/terms
KPMG.: The future of HR 2019: In the know or in the No (2018)
Lee, C.L., Huang, M.K.: The influence of computer literacy and computer anxiety on computer self-efficacy: the moderating effect of gender. Cyberpsychol. Behav. Soc. Netw. **17**(3), 172–180 (2014). https://doi.org/10.1089/cyber.2012.0029
Lin, J.S.C., Chang, H.C.: The role of technology readiness in self-service technology acceptance. Manag. Serv. Qual. **21**(4), 424–444 (2011). https://doi.org/10.1108/09604521111146289
Margherita, A.: Human resources analytics: a systematization of research topics and directions for future research. Hum. Resour. Manag. Rev. **32**(2) (2022). https://doi.org/10.1016/j.hrmr.2020.100795
Moore, G.C., Benbasat, I.: Development of an instrument to measure the perceptions of adopting an information technology innovation. Inf. Syst. Res. **2**(3), 192–222 (1991). https://doi.org/10.1287/isre.2.3.192
Perera, A.H.U., Jayawardana, A.K.L.: Role of charismatic leadership and technology self-efficacy in HRIS use behavior: a conceptual study. J. Manag. Res. **9**(1) (2022). https://doi.org/10.29145/jmr/91/04
Rogers, E.M.: Diffusion of Innovations, 5th edn. Free Press, New York, NY (2003)
Ryan, B.: A study in technological diffusion. Rural. Sociol. **13**, 273–285 (1948)

Shet, S.V., Poddar, T., Wamba Samuel, F., Dwivedi, Y.K.: Examining the determinants of successful adoption of data analytics in human resource management—A framework for implications. J. Bus. Res. **131**(March), 311–326 (2021). https://doi.org/10.1016/j.jbusres.2021.03.054
Ullah, N., Al-Rahmi, W.M., Alzahrani, A.I., Alfarraj, O., Alblehai, F.M.: Blockchain technology adoption in smart learning environments. Sustainability (Switzerland) **13**(4), 1–18 (2021). https://doi.org/10.3390/su13041801
Vargas, R., Yurova, Y.V., Ruppel, C.P., Tworoger, L.C., Greenwood, R.: Individual adoption of HR analytics: a fine grained view of the early stages leading to adoption. Int. J. Hum. Resour. Manag. **29**(22), 3046–3067 (2018). https://doi.org/10.1080/09585192.2018.1446181
Venkatesh, V., Davis, F.D.: A theoretical extension of the technology acceptance model: four longitudinal field studies. Manag. Sci. **46**(2), 186–204 (2000). https://www.jstor.org/stable/pdf/2634758.pdf
Wong, K.K.K.: Partial least squares structural equation modeling (PLS-SEM) techniques using SmartPLS. Mark. Bull. **24**(1), 1–32 (2013)

The Pension Systems Reform as a Symptom of Capitalism Financialization

Antonia Ioana Ion

Abstract The pension systems are an important object of public policy for many reasons, first of all because they concern in each country millions of retired people who's well-being depends on it. In the last 3–4 decades the interest for the reform, especially for a structural one, is growing having as main motivation the sustainability risks implied by the system of defined benefits which is still dominant, at least, in the European Union. By analyzing the drivers of the structural reform from the perspective of the capitalist economy evolution I found that the interest of main players-as World Bank, and International Monetary Fund-promoting the reform concern not only the sustainability of the pension systems, but also the attraction in the financial circuit of the pensions' money deposits. We consider this phenomenon to be a symptom of the capitalism's financialization. Backed by a larger analysis of such financialization symptoms, I consider the international financial institutions' strive for a structural reform of the pension systems as a particular, specific symptom of the capitalist financialization present in all the contemporary varieties of the capitalism. One doesn't make value judgements, adopting a neutral position and even avoiding causal explanations; we only signal a characteristic of the present financial capitalism and put it in the framework of the problems concerning retirement compensations.

Keywords Capitalism · Financialization · Defined benefits · Defined contributions · Pension system · Pension system reform

1 Introduction

From a point of view the economy it's like nature, it has its processes, laws, etc. objective, compelling. From another point of view economy is different from nature, because its law, processes, etc. are directly embodied by man. We can intervene in the nature, but we are not so intricated with the nature as we are with the economy. This

A. I. Ion (✉)
Università degli Studi di Milano Statale, Milano, Italia
e-mail: ionioanaantonia@gmail.com

M. Busu (ed.), *Rethinking Business for Sustainable Leadership in a VUCA World*, Springer Proceedings in Business and Economics,
https://doi.org/10.1007/978-3-031-50208-8_14

position of ours is the source of the complexity and risks in analyzing the economy; it is necessary for us to take a position in the nexus between the economy as such and our interventions—mainly through public policies, especially through regulation—on it. There are privileged areas and problems which permit us to analyze this profound interconnection between economy and the society at large. Such problems allowed a much more informed look on the economy itself and as a consequence of that look, a better understanding of important aspects of the society. By choosing to study the pension systems reform as a symptom of the capitalism financialization, we make in the first place, a foray in that complexity nexus between economy and public policy with the aim to better characterize the current form of the capitalist economy, the financial one. In the second place, but also of relevant importance is the attempt to better understand the implications of the proposed reform. In short, to follow the money, the pensions contributions and where they are going. That journey of the money is the connection between financial capitalism and a very important societal problem that of pensions.

This for the relevance of the study. Now, my aim. By making a parallel with health approaches, one can tell that my study is not an etiological one, but a symptomatic one. Certainly, if it looks like a duck, it walks like a duck, it is a duck. But I don't have any ambition concerning causal explanations, being satisfied with the success in proving that the recent impetus of the pension systems reform is a symptom of the capitalism financialization. Certainly, I didn't intend to deny the fact that such a reform is motivated by sustainability needs, by social commends etc., I only want to prove that there is an important financial interest in introducing the reform. Therefore, I don't make value judgements; I hope to describe and argue in a convincing manner that the reform is also drove by the evolution of the financial capitalism. Therefore, my mean research question is: How to interpret the strive for a structural reform of pension systems?; accordingly my hypothesis is: to interpret it as a symptom of capitalism financialization.

In order to make my contribution clear it is useful to already introduce some remarks on the main terms of the study. Concerning the capitalism and its evolution, one can consider the search for profit the main driver of the economy and use it as a criterion to make distinction between the stages of the capitalism; I adopt the traditional taxonomy, commercial, industrial, financial. It is also useful to consider the varieties of financial capitalism/ recent capitalism, in order to see that the financial aspects are determinant in any variety (another known taxonomy is considered: state, big firms, oligarchic, entrepreneurial capitalism). This characterization of capitalism and each evolution allows me to make a taxonomy of the symptoms of the capitalism financialization: first of all there are general symptoms as globalization and delocalization characteristic for the evolution of the capitalism to the financial stage; then, there are particular or specific symptoms, like the reform of pension systems or the profusion of financial instruments, characteristic for the financialization of the financial capitalism (this is not a linguistic barbarism, it expresses the development of the financial capitalism on its own base to the extreme); finally, there are mixed symptoms, like the public policies adopted (e.g. Glass Steagall Act and Dodd Frank Act), characteristic for the conjugated vertical evolution to the financial capitalism with horizontal financialization of the financial capitalism. I will briefly characterize them, and insist on the pension systems reform as specific symptom. Concerning the pension system, I will mainly discuss the Bismarckian system which is with defined benefits and the structural reform proposed by the World Bank which introduces a defined contributions system; the core of the debate is the introduction of a mandatory privately administrated pillar with defined contribution without guaranteed benefits; the most salient feature is the mandatory participation imposed by the state without state guarantees for the benefits; interesting are the positions of the European state in this regard.

Specifying my aim and terms conducts us to the explanation of my approach, my method. My entire research is larger than the section presented here. My approach is like this: introduction of a model which has as main structure the taxonomy of capitalism (stages and varieties) and the associated symptoms for the financial capitalism (general, particular and mixed), then the construction of a few case studies in order to see if the phenomena analyzed in the case studies are fitted in the model; if they are fitted, one can consider that phenomena as being one type of symptom; the article exposes a case study, the one of the pension systems reform, which is contrasted with the model and I conclude that this phenomenon, the reform, is a particular symptom of the capitalism financialization.

The exposition of my research follows the traditional way: introduction, literature review, methodology, results and discussions, conclusion.

2 Literature Review

Beyond being a professional exigency the literature review of a problem can orientate the research and highlight its importance; one has to study the most relevant contributions in order to have a referential for its own efforts. In the present case the supposed fields implied by the research are the financialization of capitalism, the pension systems and their reform. Here are some remarks on the important literature in these fields and in the contributions reviewed here one can found many references to an extended literature which I can't discuss for obvious reasons.

The literature of the financialization must be divided in one branch consecrated to the taxonomy of capitalism with its stages and varieties and another branch consecrated to the financialization of the last stage of capitalism, the absolute dominance of the financial sector on all the economy and even the society at large. In the first branch were useful: for the stage structure of capitalism the ideas of Hilferding on commercial, industrial and financial capitalism (Hilferding 1981, passim); for the varieties of contemporary capitalism, we referred to the distinction between entrepreneurial, big firms, state-guided and oligarchic capitalism (Baumol et al. 2007, p. 60–92); for a general discussion of the different perspective on capitalism it can be consulted a recent synthesis (Aligica and Tarko 2015, p. 116–117). The ideas advanced in these contributions were the base for the model sketched for the capitalism especially for the financial one and for the typology of its symptoms. The second branch of the literature on capitalism financialization is developed especially in the last two decades. It is a literature where we find contributions on a large range, from strictly analyzing the financial aspects in major corporations (do Carmo et al. 2023, p. 196–212) to making a link between growing financialization and contraction of the welfare state (Scarano 2023, p. 191–196). A good outlook on the main problems and evolutions in the field of financialization study (approaches, actors, domains, techniques, implications and risks), useful for my research, we found in a recent handbook (Mader et al. 2020, passim). I was more interested in the relatively reduced literature which made a link between the financialization and the pension systems considering relevant the problem of attracting in the financial circuit of the money resources of the households with lower income through instruments like mortgage, pension systems, insurance, credit (van der Zwan 2014, p. 111–114). But this process and its results are hardly dependent on the institutional setting in various countries as comparative studies on European cases prove (Dixon and Sorsa 2009, p. 360–361) and also on U.S.A. and Netherlands comparative case (van der Zwan 2017, p. 555).

Despite a larger set of pension systems, my research is concentrated on the structural reform of Bismarckian pension system (pay as you go or defined benefits) because the proponents of the reform themselves concentrate on this system; it was the pension system in Romania and in most of the European countries and it still is in the most developed western European countries. Very instructive for the evolution of German pension system is a book which studies its root from Bismarck until the present pressure on Germany to deepen the reform of pensions (Mierzejewski 2016, p. 14–17). Fundamental for the standard conception on the structural pension systems reform as it is proposed and supported by the World Bank and IMF are the contributions of a team leaded by Robert Holzman, the author or editor of almost 10 titles on the pension systems reform, including in East Europe (Holzman et al. 2008, passim). One can find revealing data on the reform of pension systems in E.U. in the publications of the European Commission (2018, p. 175, 2021, p. 178).

3 Methodology

In few words one can draw a model for the evolution of the capitalism; a model which included on the vertical three stages: commercial, industrial and financial; each one is delimitated by the most important way to make profit, because profit is the driver of the capitalist enterprise; the model included on the horizontal of the last stage, the financial one, four varieties characterized by the way to do business; this horizontal, no matter the varieties, is dominated by the financial activity; by including in the model the idea of financialization of the financial capitalism—because in the financial activity where made the biggest profits and the financial organizations took the most important part of the total corporate profit—one can introduce in the model also the idea of the symptoms of the capitalism financialization. There are general symptoms which characterize the financial capitalism as an entire stage like globalization or delocalization; then, there are particular, specific symptoms for the recent evolution of the financial capitalism—for its intense financialization—symptoms like the structural reform of pension systems (see below); finally, mixed symptoms expressing both the stadial evolution of capitalism and the specific financialization of it, like the legislation on financial capitalism crises (the regulatory difference between Glass Steagall Act and Dodd Frank Act). This model can be used in order to analyze various case studies. Despite the fact that the case study is considered a qualitative research method, in reality a case study supposes a set of techniques and methods which can include many quantitative data (Schrank 2006, p. 171); my case study—the reform of pension system—has as object data from the European Union offered under its authority. By comparing the results of the case study with my theoretical model I try to answer to the research question and to prove my hypothesis that the structural reform of pension systems is a symptom of the capitalism financialization; this is a result which confirms the above-mentioned results of van der Zwan, Dixon and Sorsa.

4 Results and Discussions

We can affirm that "capitalism is essentially the investment of money in the expectation of making a profit" (Fulcher 2004, p. 2); therefore, as Adam Smith says "it is not from the benevolence of the Butcher, the brewer or the baker that we expect our dinner, but from their gaze to their own interest", (Smith 1998, p. 30); the rate, the mass, and the sustainability of profit is paramount in allocating capital and taking risks. For this reason, we can consider the profit under its different aspects as the main criterion in analyzing the stage of the capitalist evolution; the economic sector were in a lapse of time is realized a greater rate of profit is dominant; obviously the other sectors are present, but in a subordinated role (Vaggi and Groenewegen 2006, p. 20). Considering the profit as a criterion we adopt the taxonomy for the stages of the capitalism theorized by Rudolf Hilferding and accepted even by Hayek: commercial, industrial and financial capitalism (Hilferding 1981, pp. 208–223 and 239–267). For example, the concentration of the capital in the commercial sector determined the fact that "the Dutch merchant fleet around 1670 was larger than that of the rest of Europe put together" (van Zanden 2001, p. 79). We can see an analogous example in the financial stage of the capitalism, when capital is invested in finance in order to obtain more profit (rate and mass): "in 2007 the United States financial sector generated 8% of GDP and accounted for 40% of total corporate profits" (UNCTAD Task Force 2009, p. 21–22). This situation didn't change much in the following period; there were variations of a few percents. Certainly, the situation is different in countries in lower development, but the main idea concerning the three stages of capitalism is correct. If we consider the various taxonomies of current varieties of capitalism we adopt the one proposed by Baumol and colleagues: entrepreneurial, big firms, state-guided and oligarchic capitalism. One can exemplify: entrepreneurial,

U.S.A. economy in the first half of the 20th century; big firms, current U.S.A., U.K. economies; state-guided, China's Economy; oligarchic, Russia's economy (Baumol et al. 2007, p. 60–92). It is worth to mention that Baumol can't offer a current example of entrepreneurial economy; he affirms that the current blending of entrepreneurial and big firm capitalism in U.S.A. has opportunities to foster innovation, but also risk to become defensive and to retard the development (Baumol et al. 2007, p. 90). My main observation in accord with the analysis of Baumol and colleagues is that in all these varieties of capitalism the financial sector is essential and paramount with all its positive and negative implications; the fact that by total assets the first four places in the ranking of banks are Chinese and the fact that Evergrande Group has liabilities of more than 300 billion dollar (Jim and Yu 2023) show the importance of financial sector even in the state-guided economy; this dominance is even clearer in the case of all other mentioned varieties. In short, I can say that despite the difference between capitalism varieties in the last decades the financial capitalism is more and more financialized.

In order with that simple taxonomy, one can sketch a taxonomy for the symptoms of the capitalism financialization. First of all, there are general symptoms of the evolution of the capitalism from its commercial through its industrial to its financial stage. Such general symptoms are globalization, relocation, etc. I will say a few words only on globalization which is a term with many meanings, therefore it is useful to restrain the use of the term at its quantitative and verifiable aspects. We can define the globalization in an accountancy manner as the tendency to strong growth of the income realized abroad, or by considering the trade flows and the ones of portfolio and direct investment. So understood the globalization can be measured e.g. by KOF Globalization Index which show starting with 80's a sustained growth of the globalization *pari passu* with the financialization of the economy; they are strongly statistically correlated (Savina et al. 2019, p. 572–573). Then, there are particular, specific symptoms to the more profound financialization of the capitalism in the less decades like the diversification and massification of financial instruments (in particular of derivatives) or the attraction in the financial circuit of the money of the lower income households (through pension systems reform, mortgages, etc.). I will insist on pension systems reform in the following. For the moment, only a few words on the last category of symptoms of capitalism financialization, the mixed ones; i.e. symptoms which connect the vertical general evolution of capitalism to the financial stage with the horizontal financialization of the current financial capitalism. Such a mixed symptom is the legislation adopted to tackle the crisis; I describe the symptom as the differences between the provisos of Glass Steagall Act which forbid the merger of banks with insurance firms and forbid to such entities to own productive companies (Glass and Steagall 1933, Sections 16, 20, 21, 32) and the provisos of Dodd Frank Act which forbid only the bail-out for the swaps operated over the counter (Dodd and Frank 2010, Section 616). Clearly put, the difference is between Glass Steagall Act which intend to suppress the causes which lead to the Great Depression, while Dodd Frank Act address only to the results—and that only partially—of the financial activity that generated the Great Recession. The difference is motivated with the great complexity of the financial capitalism and the so called too big to fail;

in other terms, if you regulate the finance industry you risk to kill it and with it the economy with catastrophic consequences to the society.

An important figure in the German research establishment observed that "motivating non-owners…to work hard and diligently in the interest of the owners…requires artful devices" (Streeck 2016, p. 2). One of these devices is pension system. Certainly, there were pension schemes—in a private form—before the introduction of the two main pension systems, the Bismarckian, and the Beveridgean ones. Here is considered only the first one because it is still the most widespread and the first introduced; it is sufficient for the needs of my research. In the ninth decade of the 19th century Chancellor Bismarck introduced the legislation for social security in Germany, the first one in the world; the most important and the last adopted, in 1889 was the law of pensions. Bismarck, in a discourse in the Reichstag highlighted exactly the idea of Streeck, that the measures are not driven by socialist ideas, but by the need to motivate the workers which is in the interest of the capitalist, the economy and the country (Bismarck 1884). The Bismarckian pension system included: the funding by three parts of the pensions (employer, employee, state); the standard and the minimal contribution period and the quantum of the contribution (which determines the rate of replacement of the wage by the pension; nowadays with a great variation from around 80% to around 30%); the benefits and the beneficiaries of the pensions (the retired workers and his family in case of decease); the pension contributions are paid by the actual employees and the beneficiaries are the retired employees which satisfy the conditions of the law; the contributions are collected in a national fund publicly managed by a House of pensions as a budget separated from the state budget. Despite some important variations (concerning the contributors, the periods of contributions, the quantum of the benefits, etc.) that system of defined benefits (the employees when they start working and contribute now which will be their benefits 30 or 40 years later) is the most widespread in the European Union and all around the globe. This pension system has an important contribution to the growth of the welfare in the European countries and it enforced the intergenerational solidarity and it has the advantage of a public an accountable public management without the fear that the pension budget will be dilapidated by its managers or by the state. However, an important characteristic is the ratio between the actual contributing employees and the actual benefiting retired. At the beginning of the pay as you go system there where not problems with the pension funding, because they were many employees and few retirees; but the success of the social security legislation which generated welfare produced an important growth of the life span; therefore the ratio between employees and retirees diminished and the sustainability of the Bismarckian defined benefits pension system becomes a problem. This is, starting with the '80s, a veritable mantra of the proponents of the Payg pension system reform. The international financial organizations were the most important proponents of the structural reform with the specialists of the World Bank in the forefront of this effort. Until today their production, *Averting the Old Age Crisis*, (World Bank 1994) is an important reference in the strive for the reform. It is very significant the fact that in the following year the main specialist behind that book was endorsed by the IMF which publish in one of its reviews an article summarizing and supporting that book (James 1995).

The Payg pension system, as said above, included more parameters; the proponents of the structural reform argue that the evolution of this parameters conduces to the non-sustainability of the system on long term in the conditions of old age population growing. Accordingly, initially they proposed a pension system with three pillars: the first one continues the publicly managed fund in the Bismarckian setting; the second is a privately managed fund with compulsory defined contributions and aleatory benefits; the third is privately managed fund with voluntary contributions (World Bank 1994, p. 238–239). In the next 15 years W.B., IMF and various specialists elaborated many works arguing that the structural reform is necessary in order to assure the sustainability of the pension system; in parallel in these articles more and more took place the idea that the reform has a collateral benefit, the introduction of the financial circuit of more and more money. The main argument was that the existing voluntary pension funds—already present in the U.S.A., U.K. and other countries—are important players on the financial markets. In 2008 the World Bank, under the lead of Robert Holzmann, produced a text which sets once and for all the main positions and arguments of the strive for the reform. Now it must be five pillars: the first, a general net of safety funded through general taxation; the second, the continuation of the Bismarckian system with contributions publicly managed; the third a compulsory contributions privately managed pillar; the fourth a voluntary contributions privately managed pillar; the fifth a non-monetary support from the state and family pillar (Holzmann et al. 2008, p. 5–6). The authors, in order to support the structural reform, introduced to sets of criteria for evaluation of a pension system; they argue that their system satisfy the criteria better than the old one. The first set of criteria is intrinsic to the pension systems; the criteria are adequacy to the needs of pensioners, affordability, the capacity to finance the system, sustainability i.e., affordability on long term, predictability and robustness to the shocks (ibidem, p. 8). The criteria in the second set are the development of financial markets, the increase of saving and the reduction of labor market distortions (ibidem, p. 9). Obviously, the international financial organization are most interested in the performance evaluated through this second set of criteria, an extrinsic one to the pension systems itself.

There are two main observations to be made. The first one concerns the arguments on the sustainability of the Payg pension system and the necessity of the structural reform. If the growing ageing of the population and the decreasing ratio between employees and retirees are uncontestable realities, it is questionable the fact that the only solution is the structural reform. There is another important perspective, that of a parametric reform consisting in such a tuning of the above-mentioned parameters of the Payg system in order to assure its sustainability. It is interesting that Holzmann in the name of World Bank together with an important economist, Joseph Stiglitz have also written a book on a parametric reform of pension systems (Holzmann and Stiglitz 2001) showing that this path of reform is also possible. The fact that in the last three decades, in the conditions of ageing population, almost all the Western European countries didn't introduce the mandatory pillar and acted through parametric interventions supported the idea that the structural reform is not a must, especially when we consider its difficulties (like compelling the same generation of employees to pay from the past retirees and for funding their own pensions). The second observation concerns the fact that their structural reform proposes the introduction of a compulsory contributions pillar privately administered. Here is a manifest disequilibrium: the state with its coercive force makes compulsory the contributions without any guarantee for the risks of the private management, i.e. against the risk of the introduction of pensions money in the financial circuit. If the first interest of the state is for the retirees and their pensions, such a disequilibrium doesn't exist. Here we can already conclude that an important driver of the structural reform of pension systems is of a financial nature, not only an answer to the sustainability need of the pension system.

One can say more, if one looks to the answers of different European countries, to the pressure for a structural reform of the pension system. The European Commission publish every three years a report on the ageing population and the problem of the pension system. I will present two tables, one for 2018 and the other for 2021, which show the situation of the compulsory pillar in selected E.U. countries (Tables 1 and 2).

By confronting the two tables we have a clear image about the negotiation force of E.U. countries and the importance of the strongly settled institutions; the important western countries refused the introduction of a mandatory pillar while the new E.U. countries accepted this conditionality. Neither the last three years and the growing pressure of the ageing population didn't change the situation. These findings are in order with the above-mentioned results of van der Zwan, Dixon and Sorsa which highlight the different position of the pension systems reform determined by the institutions of each country. Now, one can say not only that the pension system reform is a symptom of capitalism financialization, but can add that the symptom manifestation depends on characteristics of various countries, despite the pressure of global international financial organizations. We consider that we tested and confirmed the hypothesis that structural reform of pension systems is a symptom of capitalism financialization.

Table 1 Presence of mandatory pillar in selected E.U. countries 2018

Countries	Compulsory pillar
France	No
Germany	No
Italy	No
Spain	No
United Kingdom	No
Bulgaria	Yes
Croatia	Yes
Polonia	Yes
Romania	Yes

Source European Commission (2018, p. 175)

Table 2 Presence of mandatory pillar in selected E.U. countries 2021

Countries	Compulsory pillar
France	No
Germany	No
Italy	No
Spain	No
United Kingdom	No more E.U
Bulgaria	Yes
Croatia	Yes
Polonia	Yes
Romania	Yes

Source European Commission (2021, p. 178)

5 Conclusion

Now if one contrasts the model of the capitalism evolution that we have drawn above with its stages and symptoms of the financial stage with my case study of pension systems reform, one can say that the main results are as it follows: we proved that the financial motivation is a strong driver of the pension systems structural reform; accordingly we confirmed my hypothesis reading that reform as a specific symptom of the capitalism financialization; by doing so we consolidated the theoretical model proving that it is relevant, useful by its applications. I want to say a few words on two possible developments on the research. The first one refers to an exercise of testing other symptoms of capitalism financialization (globalization, legislation, etc.) starting from the model. The second one refers to comparative research of the specific capitalism symptoms as pension systems reforms, mortgages etc. with the proliferation of financial instruments (e.g. derivatives).

References

Aligica, P.D., Tarko, V.: Capitalist Alternatives, Models, Taxonomies, and Scenarios. Routledge, New York (2015)

Baumol, W.J., Litan, R.E., Schramm, C.J.: Good Capitalism, Bad Capitalism, and the Economics of Growth and Prosperity. Yale University Press, New Haven (2007)

Bismarck von, O.: Reichstag speech (1884). http://ghdi.ghidc.org/sub_document.cfm?document_id=1809. Accessed 02 May 2023

Carmo do, M.J., Neto, M.S., Donadone, J.C.: Corporate financialization. In: An Economic Sociology Perspective. Routledge, New York (2023)

Commission, E.: The 2018 Ageing Report. Publications Office of the European Union, Luxembourg (2018)

Commission, E.: The 2021 Ageing Report. Publications Office of the European Union, Luxembourg (2021)

Dixon, A.D., Sorsa, V.-P.: Institutional change and the financialization of pensions in Europe. Compet. Change **13**(4), 347–367 (2009)
Dodd, C., Frank, B.: Wall street reform and consumer protection act (2010). https://www.congress.gov/bill/111th-congress/house-bill/4173/text. Accessed 02 May 2023
Fulcher, J.: Capitalism: a Very Short Introduction. Oxford University Press, Oxford (2004)
Glass, C., Steagall, H.B.: Banking act of 1933 (1933). https://fraser.stlouisfed.org/title/federal-reserve-bank-new-york-circulars-466/1248-banking-act-1933-15952. Accessed 02 May 2023
Hilferding, R.: Finance Capital. Routledge, London (1981)
Holzmann, R., Stiglitz, J.: New Ideas About Old Age Security. World Bank, Washington (2001)
Holzmann, R., Hinz, R., Dorfman, M.: Pension Systems and Reform Conceptual Framework. World Bank, Washington (2008)
James, E.: Averting the old age crisis. In: Finance & Development, pp. 4–7 (1995)
Jim, C., Yu, X.: China Evergrande races court deadline on debt restructuring terms-sources (2023). https://www.reuters.com/world/china/china-evergrande-races-court-deadline-debt-restructuring-terms-sources-2023-03-09/. Accessed 02 May 2023
Mader, Ph., Mertens, D., van der Zwan, N. (eds.): The Routledge International Handbook of Financialization. Routledge, London (2020)
Mierzejewski, A.C.: A History of the German Public Pension System: continuity amid Change. Lexington Books, Lanham (2016)
Savina, G., Haelg, F., Potrafke, N., Sturm, J.-E.: The KOF globalisation index—Revisited. Rev. Int. Organiz. **14**(3), 543–574 (2019)
Scarano, G.: Financialisation and Macroeconomics. The Impact on Social Welfare in Advanced Economies. Routledge, London (2023)
Schrank, A.: The case study and causal inference. In: Perecman, E., Curran, S.R., (eds.) A Handbook for Social Science Field Research, Thousand Oaks, Sage Publications (2006)
Smith, A.: An Inquiry into the Nature and Causes of the Wealth of Nations. Elecbook Classics, London (1998)
Streeck, W.: How Will Capitalism End? Essays on a Failing System, London, Verso (2016)
UNCTAD Task Force The Global Economic Crisis: Systemic Failures and Multilateral Remedies ,United Nations, New York (2009)
Vaggi, G., Groenewegen, P.: A Concise History of Economics Thought from Mercantilism to Monetarism. Palgrave-Mcmillan, London (2006)
van der Zwan, N.: Making sense of financialization. Soc. Econ. Rev. **2014**(12), 99–129 (2014)
World Bank: Averting the Old Age Crisis. Oxford University Press, New York (1994)
Zanden van, J.L., Early Modern Economic Growth. A Survey of the European Economy 1500–1800. In: Prak, M., (ed.) Early Modern Capitalism Economic and Social Change in Europe, 1400–1800. Routledge, London (2001)
Zwan van der, N.: Financialization and the pension system: lessons from the United States and the Netherlands. J. Modern Euro. History **15**, 4(2017), pp. 554–578 (2017)

Game Theory and Its Usefulness in the Development of Innovative Strategies

Denisa Ligia Matei

Abstract Companies, as well as individuals, believe that they should be in competition with others around them and rarely consider cooperation. We constantly feel that we have to outdo the other player and that the credit is ours. When a company develops its strategy, it must consider what decisions and strategies its competitor is also initiating. Until a black swan appears in the market, companies often do not initiate strategies for a good position in the market. Large companies around the world manage to make a real impact in communities and reduce poverty across vast areas. It is assumed that over time leadership has been greatly undervalued, but at the same time it has now come to be overvalued, considering that it can bring about real change in the organization. The reality is that without a sufficiently good strategy, no leader will be able to bring added value to the organization. In this lesson we will address some theoretical aspects of strategy, as well as game theory. Also, in the second part we will analyze the impact it can have for players who are going to develop a strategy, but there are also some examples and simulations done hypothetically with two players. In the last part of the paper, we will present a series of conclusions and recommendations in what game theory means in the practical life of large companies.

Keywords Game theory · Strategy · Players · Cooperation · Change

This paper was co-financed by The Bucharest University of Economic Studies during the Ph.D. program.

D. L. Matei (✉)
Bucharest University of Economic Studies, Bucharest, Romania
e-mail: Mateidenisa16@stud.ase.ro

Bucharest, Romania

M. Busu (ed.), *Rethinking Business for Sustainable Leadership in a VUCA World*, Springer Proceedings in Business and Economics,
https://doi.org/10.1007/978-3-031-50208-8_15

1 Introduction to Concepts

Every day we interact to cooperate either with other individuals or with other legal entities. Often from the purpose of a possible cooperation we can end up in conflict due to some factors that can intervene. To be able to analyze the behaviors that influence the final result we want from an approach, we have a number of tools at hand.

Among them, we highlight game theory, which has multiple practical applications both in negotiation, international relations, and in critical moments of decision-making and defining a strategy. The field of game theory takes the form of strategy games useful in large companies and beyond.

Frederic Laloux, in his book "The Reinvented Organization" presents a series of conditions necessary for a company to reinvent itself. We will associate how these points can have a conclusion using game theory.

Top management is the first essential element for an organization to reinvent itself. They must be strategic thinking and success oriented (Laloux 2017).

The authors of the book "Thinking Strategically" proposed as a model of analysis the fact that we live in turbulent times. Through their research, both the increase in the pace of change and the uncertainty or dynamics of organizational environments are captured. They say that turbulence occurs in two worlds: in the world of chaos, where we know that there is order, but we also understand that we cannot discern that state "ordered from within the organizational environment and the world of the complex", limited by stability on the one hand and chaos on the other the other, where the general directional order is identifiable, but the organization is continually faced with unexpected "lightning strikes" or "stochastic shocks" from the environment (Glassman et al. 2005).

Managers need guidance on how to deal with turbulent environments to improve corporate performance. Research on environmental turbulence has suggested that firms adopt a less centralized, more organic structure in dynamic and uncertain environments.

The ultimate test of a manager is when he leads people in times of crisis: how do they manage the challenges that arise? How does it come up with solutions?

George (2009) believes that "leaders appear after crises". As the leadership theories say, not every manager is also a leader and not always a leader is also the manager. Returning to the way a leader is formed, we can liken him to the sailor who will never specialize on a calm sea.

2 Concepts and Approaches of Game Theory and Strategy

Before defining game theory, we will dwell a little on a few approaches to strategy in order to be able to connect the data of the problem more easily.

From the perspective of Eugen Burduș, Ghoerghița Căprărescu and Tiberiu Zorlentan, strategy represents "the science, but also the art of establishing the general objectives of the organization, whether we are talking about them in the medium or long term. Strategy is also defined by those previously mentioned as the science and art of formulating action options to achieve them, in order to effectively adapt the organization to the requirements of the environment in which it operates".

On the other hand, Corneliu Russu defines strategy as "the product of the strategic management process which is based on an exhaustive, integrative and unitary plan of managerial action. The strategy is present in order to achieve the established objectives, which specify how the company will be run and how it will act, as well as what actions will be carried out to ensure the organization fulfills its mission" (Russu 1999).

McKinsey defines strategy as "the set of decisions and relative actions for choosing means and for articulating resources in order to achieve the objective" (McKinsey 1984).

Among the major components of an organization's strategy, we mention the following: the company's mission: the philosophy of the company's evolution, but also a differentiating factor compared to competitors in the same market or field of activity; the mission of the company is at the same time an assumption of the activity, as it is necessary to establish the objective of the company, the customers, but also the purpose of the company. Customers, but also customers' needs, justify the company's reason for existing.

Drucker (1990) said that "a business is not defined only by its name, logo or other visual elements, but is defined by the need that the customer can satisfy when accessing the company's service or good".

In addition to the company's mission, we also mention the fundamental objectives, strategic options and resources. While objectives help implement the mission so that it does not remain just an intention, strategic options show how the strategic objectives will be achieved. Resources are gaining their status as the main element of the strategy concept, whether we are talking about human, material, financial or informational resources.

Game theory appears wherever there are strategic type interactions. Although the necessary documentation showing tested theories appears later, in the nineteenth or twentieth century, it is assumed that there have been analyzes of strategy games since the Renaissance.

According to Jean Blaise Mimbang, "game theory analyzes what are the consequences of the interaction between rational players, given the fact that they pursue their own goals, within a well-defined framework" (Mimbang 2023). In order for an organization to define an innovative strategy, it is important for managers to know game theory and to understand possible consequences of decisions taken by other players in the market.

The theory not only analyzes interactions, but aims to highlight the interdependence that exists between individuals and companies. At the geo-political level including countries must understand that they are interdependent. In this sense, it is necessary to find a point of balance in all socio-economic relations and not only.

A game can exist if there are at least two players or agents, each of whom has a number of possible strategies. The outcome of the game will be directed by the choices made by each player, by the strategies they choose. The most famous example for game theory is the prisoners' dilemma. They have the option of admitting the deed or denying it. And depending on the chosen strategy, the number of years of imprisonment will be defined. Of course, the lower the number, the bigger the win for the defendants.

There are several types of games. The ones with zero amount do not bring profit for any of the players of the game. When we talk about dominance it is important to know that a strategy T is dominated by a strategy S. If every player decided to choose a dominant strategy, although the recommendation theories say that they should never play such a strategy, then s- would reach the equilibrium of the dominant strategies.

In game theory we have motion diagrams as a variant. It is when one player wants to maximize his own payoff while minimizing the other player's payoff. At the same time. The saddle point or pure minimax equilibrium is the result of a matrix game. This is the column maximum and the row minimum, being a special case of the Nash equilibrium. For matrices that have a minimax equilibrium, players should ideally play the strategy indicated by the matrix, where one player has a strategy that guarantees him a win, but the second player has a strategy that shows him that the initially it will not earn more than the anticipated value. When fixed probabilities come into play we speak of mixed strategy, where there is therefore a mixed minimax equilibrium. In the case of matrix games, the games can be transposed to the simple cases discussed previously. In these game theories, the Nash equilibrium is discussed which guarantees that no player can obtain additional gains if they unilaterally change their strategy, thus being an ensemble of strategies, one strategy for each player (Nash equilibrium). The Nash equilibrium is one of the basic elements of game theory precisely to ensure an optimal outcome for the players.

These types of games can be made using classic methods, matrices or they can use automatic tools that contribute to the efficiency and accuracy of the result, such as Excel.

For behavior and strategic thinking, the backward method is known, the one by which forecasting and looking into the future is essential for decision-making.

3 How Can Game Theory and Innovative Strategy Development

Coming into practice with the Nash equilibrium from game theory, we will thus present the following analysis.

We assume two big players in the IT&C industry, a market that is becoming increasingly competitive for attracting candidates. Everyone needs innovative strategies to attract their workforce, especially now in the new socio-economic context of being able to work from anywhere, for anyone.

By assumption and simulation of a case study: we think Microsoft and Google, two giant players in IT, could have a talent war or come to an agreement.

We will thus list a series of strategies:

Microsoft decides to offer higher salaries while Google doesn't announce anything about it, surely according to the game theory, Microsoft will win all the potential candidates in the market who are looking for opportunities.

If, on the other hand, Google decides to raise wages and Microsoft does not, Google will in this case succeed in attracting all the candidates.

The third strategy is in the context where both Microsoft and Google are raising wages, and in this case neither will manage to have a better position than the other. If neither raises wages, neither has anything to gain in this situation either.

Let's consider the two players mentioned previously, each of them choosing between a high salary level and some low ones. Given their choices about how many employees they can bring on, their number of employees brought on will be (Table 1).

According to game theory, does Microsoft have a dominant strategy? What about Google?

- If we are in Microsoft's shoes:
  - if the Google company has high salaries, the Microsoft company should have low salaries (30 employees > 20 employees);
  - if the Google company has low wages, the Microsoft company should have low wages (20 employees > 10 employees);
  - thus, the Microsoft company has a dominant strategy, because the decision it takes, regardless of that of the Google company, does not change for the Microsoft company; choose a low production so as not to be at a disadvantage;
- If we are Google:
  - if the Microsoft company has high salaries, the Google company must have high salaries (20 employees > 10 employees);
  - if the Microsoft company has low production, the Google company must have low production (40 employees > 30 employees);
  - thus, the Google firm does not have a dominant strategy, because it changes its decisions according to the variation of the Microsoft firm's decision;

Table 1 Microsoft versus google salary

	Google		
Microsoft		High salaries	Low salaries
	High salaries	Google takes 20 employee	Google takes 10 employee
		Microsoft takes 20 employee	Microsoft takes 10 employee
	Low salaries	Google takes 30 employee	Google takes 40 employee
		Microsoft takes 30 employee	Microsoft takes 20 employee

Source Author

The Nash equilibrium is given if both Microsoft and Google choose to have low wages; it is not favorable for either to take another unilateral decision;

How could the understanding between the two help them in this case? Microsoft would gain 20 employees instead of 10, and Google would gain 40 employees instead of 10 if the Nash equilibrium is considered.

From another perspective of the game theory approach, we assume that Microsoft and Google have a controversy regarding alignment on a subject of legislative decision in the field in which they operate. Depending on the decision they make, they may or may not win votes from the fans of the two companies. The voting matrix is as follows (Table 2).

- If we are the Microsoft company:
  - if Google supports, Microsoft should oppose (70 > 60 > 30);
  - if Google opposes, Microsoft must support (85 > 80 > 35);
  - if Google abstains, Microsoft must oppose (40 > 25 > 20);
  - thus, Microsoft does not have a dominant strategy, because it changes its decisions according to the variation of Google's decision;
- If we are Google:
  - if Microsoft supports, Google must abstain (75 > 70 > 15);
  - if Microsoft opposes, Google must abstain (60 > 30 > 20);
  - if Microsoft abstains, Google must abstain (80 > 65 > 40);
  - thus, Google has a dominant strategy, because the decision it makes, regardless of Microsoft's, does not change for Google.

They do not both have dominant strategies, as detailed in the previous point; only Google has a dominant strategy, while Microsoft does not have a dominant one; in this case we have only one dominant strategy;

The equilibrium is given, if Microsoft opposes and Google abstains, according to the Nash equilibrium; it is not favorable for either to take another unilateral decision;

The minimax algorithm can be applied to ensure that we minimize the other's maximum.

Let's assume that Microsoft is a monopoly in the field of gaming, it makes consoles, and Google does not. Microsoft is threatening a price war if any potential rival enters the market. Let's assume that the Google company intends to enter the

Table 2 Microsoft versus google decisions

		Google		
		Support (%)	Oppose (%)	Abstain (%)
Microsoft	Support	30, 70	85, 15	25, 75
	Oppose	70, 30	80, 20	40, 60
	Seabţine	60, 40	35, 65	20, 80

Source Author

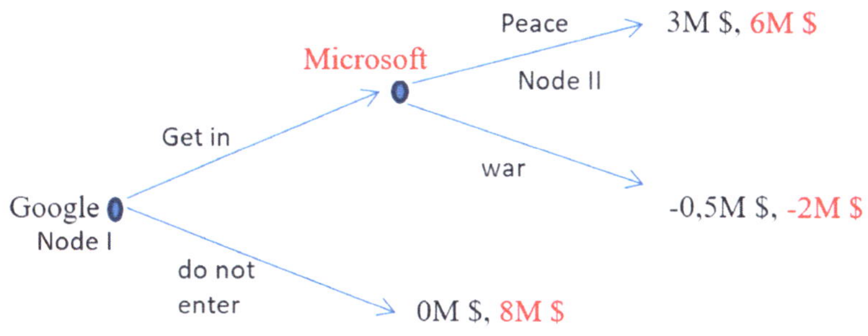

Fig. 1 Google versus microsoft decision for profit

market, to produce gaming consoles as well. The decision tree looks like this (the results of the company Google are highlighted in the first row; the results of the company Microsoft are highlighted in the second row) (Fig. 1).

Should Google enter the market? Is Microsoft's threat credible? Google should enter the market because it has a chance to earn 3 M rather than 0 M and leave it, if Google does not enter, to Microsoft to earn 8 M; thus, if the Google company enters the market, which it is recommended to do, the Microsoft company will take into account the peace anyway, because the profit for the Microsoft company will be 6 M instead of losing −2 M; It can be a threat to Microsoft, if Google doesn't want to enter the market, they don't earn anything and let Microsoft earn 8 M. Regardless of Google's decisions, Microsoft will win in favor of Google, especially choosing peace.

4 Conclusions

Even if it is difficult to put the Nash equilibrium into practice, because not all the conditions necessary for an equilibrium can exist, the concept and the possibility of their application in the solution of some economic problems, represents a significant contribution. At the same time, this concept shows that there is more to be gained through cooperation than through competition.

From strategy games companies can learn that who will always win the game and cooperation and alliance are important. After a longer sequence of the same games, winning strategies tend to repeat themselves and be copied. When agents have the same strategy, in most situations both lose the game. The winner in this situation may be a competitor who at the time was not fighting a direct battle.

Companies must consider their own resources and use them effectively, without stealing successful strategies from competitors.

Strategy is therefore the element that connects the present to the future, the facts of today that will create impact in the future. A company's strategy must, in the new context, be as innovative as possible and anticipate changes of a technological nature, customer tastes and needs that are continuously developing.

References

Drucker, P.: Lessons for Successful Nonprofit Governance. Wiley Online Library (1990)
George, B.: Lessons for Leading in Crisis. Meteora, Bucureşti (2009)
Glassman, A., Zell, D., Duron, S., Posner, B.: Thinking Strategically in Turbulent Times: an Inside View of Strategy Making. M.E. Sharpe, London (2005)
Laloux, F.: Organizaţia Reinventată. Vellant, Bucureşti (2017)
McKinsey.: Strategic Management. Prentince-Hall, London (1984)
Mimbang, J. B. (2023). *Teoria jocurilor: Arta de a gândi strategic*. 50Minutes.com.
Russu, C.: Management Strategic. ALL Beck, Bucureşti (1999)

Energy Transition: How Solar PV Would Shape the Final Household Electricity Consumption. An Economic Analysis on Sizing, Integration and Risk of Prosumer Energy Systems

Lucian V. Pamfile and Miruna E. Proşcanu

Abstract The paper explores the potential impact of prosumers on household energy consumption in Europe and at global level until 2030 and provides a centralised summary of their current status. Also, it analyses the official forecasts of international organisations for residential consumption, with a focus on Europe and beyond, and examines the trend of becoming a prosumer and its potential impact on future trends of consumption, production and prices. According to the results, the number of prosumers worldwide is expected to grow rapidly and, by 2030, they could potentially account for a significant share of electricity production (5% in the European Union and 14% globally). Using applied research and statistical analysis, the paper also finds that while the number of prosumers in Europe is expected to continue growing, the rate of growth is impacted by economic incentives. Financial advantages are one of the main drivers of prosumption, followed by environmental and social benefits—and, following on this, prosumers could help reduce carbon emissions and improve energy security, with a high degree of flexibility. Overall, the paper concludes that while the impact of prosumers on household energy consumption may be limited in the short term, the trend towards prosumption could have significant long-term implications for the way in which we consume, produce and pay for energy. It is thus important to carefully monitor and regulate the trend towards prosumption to ensure a smooth transition to a more sustainable energy system.

Keywords Prosumers · Electricity · Households · Photovoltaics · Energy transition

L. V. Pamfile (✉) · M. E. Proşcanu
Bucharest University of Economic Studies, 010731 Bucharest, Romania
e-mail: pamfilelucian17@stud.ase.ro

M. E. Proşcanu
e-mail: miruna.proscanu@csie.ase.ro

M. Busu (ed.), *Rethinking Business for Sustainable Leadership in a VUCA World*,
Springer Proceedings in Business and Economics,
https://doi.org/10.1007/978-3-031-50208-8_16

1 Introduction

Energy is an essential resource in everyday life, for people and for economic operators, but also a resource with a strong strategic dimension, highlighted in particular by the difficult security and geopolitical context, as the war started in Ukraine in 2022 showed and also by the recovery efforts following the COVID-19 pandemic. In recent years, the significant rise in electricity prices has triggered a domino effect: in a deeply interconnected economic and social system, this has led to a general jump in inflation as well as in the prices of all consumer goods.

Therefore, prosumers are transforming the way energy is produced and consumed around the world, while also impacting the electricity price by reducing the demand and increasing the generation of electricity. With the increasing affordability of photovoltaic (PV) panels and other renewable energy technologies, more and more households are choosing to generate their own electricity and even sell any excess back to the grid. While this trend has been particularly prominent in Europe, it is now spreading rapidly to other regions of the world and is now part of the energy transition.

Energy transition is the process by which the global energy sector is switching from energy production and consumption systems based on fossil fuels (i.e. oil, natural gas and coal) to technologies based on renewable energy sources. It thus represents a significant structural change in the system because, in order to meet the goal of reducing carbon dioxide (CO_2) emissions, as called for in the Paris Agreement, the entire energy value chain needs to change: from production, transport, distribution to consumption, and even storage.

Changing the way electricity is consumed is therefore part of this process, as part of the energy value chain, and one of the impacts is the emergence and growth of prosumers. They are individuals or households that both consume and produce energy and they are becoming an increasingly important part of the energy system. Prosumers are playing an increasingly important role in the global energy landscape, as more households around the world are generating their own electricity using PV panels on their rooftops.

Prosumers can be divided into two areas of activity, domestic and industrial, and this paper focuses solely on the first, respectively the residential prosumer. Reducing consumption and changing the way it is generated also requires the application of energy efficiency measures, thus one of the ways to accomplish a consumption decrease is using PV panels and becoming a prosumer (see Fig. 1).

As described by the European Environment Agency (EEA), the possibility of becoming a prosumer and the appropriate type of prosumption depend on various factors such as the type of dwelling, neighbourhood, geographical and climatic conditions, financial considerations and neighbours' interest. Homeowners can invest in their own photovoltaic rooftop system, while renters can participate in an energy community to become active prosumers (https://www.eea.europa.eu/publications/energy-prosumers-and-cities). Therefore, the rise of prosumers has the potential to significantly impact energy consumption, production and prices worldwide.

Fig. 1 Schematic diagram of a possible prosumer household (https://www.eea.europa.eu/publications/energy-prosumers-and-cities)

As a resource, solar energy is an essential component of residential and non-residential energy supplies. With the exception of tidal, geothermal and nuclear power, the majority of the energy sources used by societies to generate electricity depend on the Sun. Hydroelectricity is dependent on the Sun due to the fact that rivers would not flow without the Sun driving the water cycle, evaporating water and precipitating water at higher elevations. Wind is primarily caused by the uneven heating of the Earth's surface by the sun. Nonetheless, despite the fact that solar energy is capable of directly converting the sun's rays into electricity, the amount of power that is currently being generated is insufficient, and the world still requires additional sources of energy.

The majority of the world's energy is currently generated through the combustion of fossil fuels like coal, natural gas and oil. These fuels are the product of about 500 million years of living matter being subjected to the heat and pressure of the Earth's crust. Prior to human extraction, the entire energy supply of fossil fuels in the Earth's crust was approximately 2×10^{23}/J, but for a sustainable annual usage rate of fossil fuel it is now 2×10^{14}/J (BP Statistical Review of World Energy 2019). Even with this knowledge, total world energy consumption is 10^{20} J per year which will lead to run out of gas in 49 years, oil in 54 years and coal in 139 years (Our World in Data 2023).

Solar energy is the most plentiful source of clean energy and the quantity of solar power that penetrates the surface of the Earth exceeds the current global energy demand by orders of magnitude. The Sun gives 2.3×10^{16} J of energy to Earth annually, which is equivalent to 23.000 terawatt/years, while Earth's total yearly energy need is 17 TW/year. Notwithstanding the accessibility of solar energy worldwide, just a small part of the electricity in the world is currently generated by solar power (Wang et al. 2022).

PV panels are the most used method for capturing solar energy and are commonly found on the rooftops of residential and commercial properties. In photovoltaics, sunlight striking a particular substance known as a semiconductor stimulates electrons, which are then collected to produce electrical current. PV allows for the direct conversion of sunlight to electricity (Kroposki et al. 2009).

In Europe, the energy transition has been gaining momentum in recent years as countries strive to meet their climate goals, but also driven by the high increase of electricity prices. EEA noted that the amount of electricity produced by prosumers in eastern Europe was still rather modest due to lack of encouraging policies and generally low electricity prices in the past years. However, their number seems to increase quickly in several nations, once the electricity prices increased significantly (https://data.europa.eu/doi/10.2800/030218), as in Romania and in Europe in general. Also, as part of this transition, there has been a growing trend towards the installation of PV panels on rooftops by households, which has recently led to the emergence of prosumers.

This paper investigates the potential impact of prosumers on household energy consumption, starting from official forecasts of international organisations as a basis for analysis. It also examines the trend of becoming a prosumer and its potential risks for the energy system. However, the dilemma that has arisen is how much this production will influence consumption, in figures, given the data currently available.

2 Literature Review

Sustainability, energy efficiency and renewable sources of energy are frequently listed as viable strategies for accomplishing climate as well as energy security objectives and transition. Energy is a crucial component of economic progress, which can be hampered by the negative externalities associated with the combustion of fossil fuels. Renewable energy sources have the potential to supplant fossil fuels, but their costs and technological challenges to widespread deployment are still substantial (Verdolini and Galeotti 2011). To achieve this level of decarbonization, a variety of policies will be required, including the replacement of conventional power stations with a mix of gas, wind farms and solar thermal plants, the recognition of significant energy efficiency improvements in electrical products, housing developments, transportation and urban design and the inculcation of pervasive energy conservation practises among the general public (Deng and Newton 2017).

Solar energy solutions, such solar photovoltaics, are commonly viewed as essential possibilities for reducing society's dependence on fossil-fuel-generated electricity, the accompanying carbon emissions and other environmental issues (Marszal et al. 2011).

There are also arguments regarding potential adverse effects. Electric utilities must collect enough income to recoup the up-front cost of infrastructure investments. The transmission and distribution network is designed to meet the greatest demand of customers (McLaren et al. 2015). Consumers that install solar panels can cut their energy costs and their effective average power rates. When the price declines, a solar user may consume more power than before—a solar rebound effect, since solar adoption reduces consumers' energy bills and consequently the average price they pay for electricity (Qui et al. 2019).

In addition to the obvious impact that the deployment of photovoltaic panel technology has on reducing end-user costs and relieving pressure on conventional generation units and the distribution system, access to electrification in underserved rural areas is of equal importance in accomplishing the decarbonisation objectives of the energy transition (Buragohain 2012).

However, it should be also taken into account that manufacturing and assembly of clean energy technologies are likewise very concentrated, with three-quarters of solar PV modules and EV batteries being produced and assembled in China. China is responsible for 75% of the worldwide production capacity for battery cells and up to 97% of the capacity for wafer manufacturing for photovoltaic cells, which are the two main components of today's clean technology supply chains (International Energy Agency 2023c). Thus, manufacturing tends to be controlled by a limited number of businesses due to its capital-intensive nature and technological complexity. For instance, in 2021, only three companies produced 65% of the world's battery cells (International Energy Agency 2023a).

3 Methodology

In this study, applied research was used, as the analysis focuses on applying scientific knowledge to solve practical problems: how much of global and European consumption is accounted for by domestic consumers, how much of this is accounted for by consumers who are and can become prosumers, how the consumption curve will change and what the potential energy produced and injected into the electricity grid will be at the peak of production. The data used are provided as a result of collaboration between researchers and industry, government or other organisations that are interested in reporting global data on the energy system and its status. This type of research is aimed at solving specific practical problems and developing innovative technology, in this case finding the impact of prosumers on the electricity consumption.

In terms of qualitative and quantitative methods, statistical analysis was used in the analysis because it involves analysing numerical data using statistical methods to identify patterns or relationships. A consumption and production forecasting simulation was created based on household consumers and prosumers numbers. A historical consumption hourly curve was analysed, based on official data from organisations, as well as consumption forecasts for the coming years, until 2030. The two curves of estimated consumption and estimated production were joined together to see the impact of prosumers in the decrease of overall consumption.

The Spearman's rank correlation was also used in this particular research. The Spearman correlation examines the direction and strength of the relationship among two variables. It is considered a non-parametric measure, indicating that it makes no assumptions about the data's distribution. The Spearman's correlation coefficient, indicated by the symbol rho (), falls between the range of -1 to 1 (Spearman 1904). This type of correlation was considered because the available data are limited, do not fit a standard Gaussian distribution and the relationship between them is not linear.

Therefore, the paper presents an inductive research, as it is a way of gathering data and constructing ideas based on observations gained from that data and it is frequently used in exploratory research when there is little prior knowledge about the issue under investigation. In the case of analysing the impact of prosumers on electricity consumption, there were multiple steps taken to apply the inductive research method such as collecting data, identifying patterns, developing theories, testing those theories and refining them based on the results of testing. In theory, this approach can help to uncover new insights into the ways that prosumers are affecting the electricity industry and may inform future policy decisions or investment strategies.

In order to process statistical data, the authors used Microsoft Excel and SPSS as software programs and part of the conclusions are the result of the analysis done with these programs.

4 Results and Discussions

4.1 Impact on Electricity Consumption

The number of prosumers worldwide has been growing rapidly in recent years and several international organisations have made forecasts about the potential impact of prosumers on household energy consumption. According to the IEA (https://www.iea.org/reports/world-energy-outlook-2021), the total residential electricity consumption in 2021 worldwide was 6072 TWh (representing around 26% of total world consumption) (https://www.iea.org/reports/electricity-information-overview/electricity-consumption), while in the European Union (EU), households, or the residential sector consumed 747 TWh (representing around 27% of total consumption). At the same time, the organisation says the number of prosumers worldwide could reach over 100 million by 2030, up from around 25 million today (https://www.iea.org/reports/approximately-100-million-households-rely-on-rooftop-solar-pv-by-2030). New production capacities are deployed every year and this figure will continue to rise due to greater PV competitiveness, a growing need for sustainable energy sources and a need to reduce electricity bill costs in the residential area. This number could potentially reach several hundred million, accounting for a significant share of electricity production worldwide, IEA also noted (https://www.iea.org/reports/net-zero-by-2050).

This growth is expected to have a significant impact on household energy consumption, with some estimates suggesting that prosumers could generate part of the world's electricity demand by 2030. For example, the IEA predicts that by 2040, up to 240 million households worldwide could be prosumers, representing 14% of global electricity generation (International Energy Agency 2023b). Similarly, the European Commission predicts that by 2050, up to 98 million EU citizens could become prosumers by 2050 (https://sustainable-energy-week.ec.europa.eu/news/lets-boost-prosumerism-models-and-state-play-2022-09-02_en) and up to 83% of households in the EU could have solar panels on their roofs, demand response and/or energy storage solutions installed (Kampman et al. 2016).

In its World Energy Outlook 2020 (https://iea.blob.core.windows.net/assets/a72d8abf-de08-4385-8711-b8a062d6124a/WEO2020.pdf), IEA noted that the increase in the number of large household appliances and air conditioners in emerging and developing nations will account for more than 80% of the 700 TWh increase in worldwide residential power demand through 2025. At the same time, demand for domestic space cooling alone is expected to rise by 260 TWh by 2025, while according to the Stated Policies Scenario (STEPS), an additional 285 million people with access to power by 2030, meaning an addition of 80 TWh to demand in 2030. In total, based on the current worldwide consumption and on the above estimations, the total residential electricity consumption in 2030 worldwide would be 7112 TWh—this being a 14% increase compared with today.

Globally, in total, about three-quarters of the worldwide rise in power consumption in 2021 occurred in emerging markets and developing economies and the largest electricity consumers were China (which alone accounted for 50% of the global increase in 2021), United States and Europe. The three accounted for over 60% of global electricity demand (https://www.iea.org/news/global-electricity-demand-growth-is-slowing-weighed-down-by-economic-weakness-and-high-prices). By 2030, the global electricity will increase between 2.4 and 2.7%, the same source indicated, the residual consumption being one of the sectors which would add pressure on the demand side.

Therefore, in terms of household energy consumption, the impact of prosumers will depend on the balance between consumption and production. In some cases, prosumers may produce more electricity than they consume, resulting in a net injection of electricity into the grid. This surplus electricity can help reduce carbon emissions and improve energy security. However, in other cases, prosumers may consume more electricity than they produce, resulting in a net increase in energy consumption (see Figs. 2 and 3).

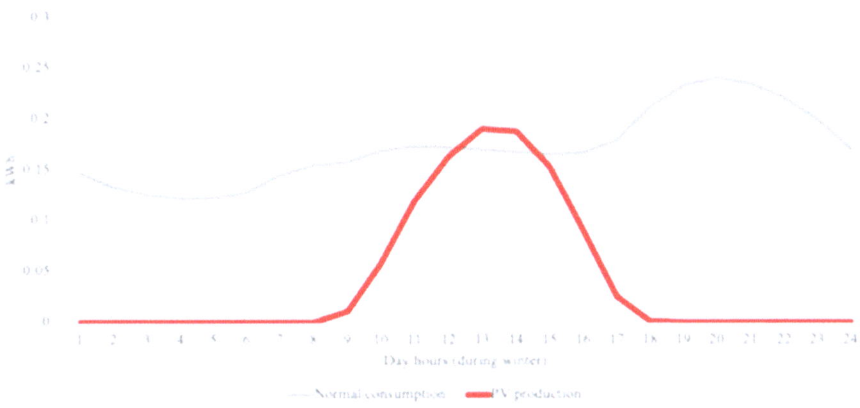

Fig. 2 Prosumer consumption and production profiles during the winter. *Source* Authors' own research results/contribution

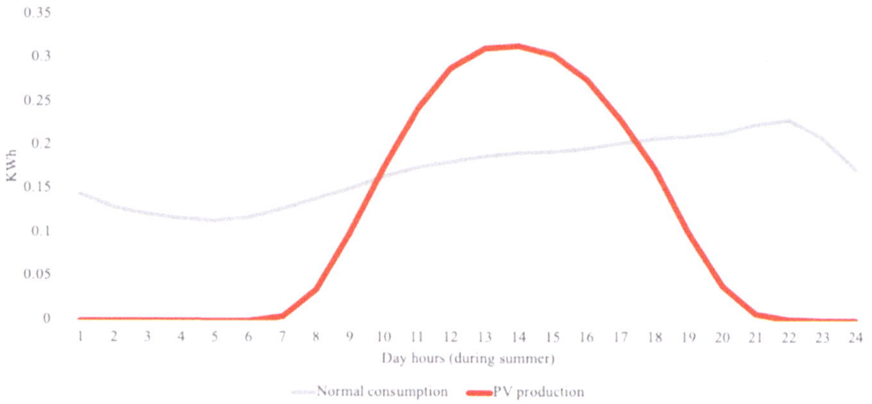

Fig. 3 Prosumer consumption and production profiles during the summer. *Source* Authors' own research results/contribution

The impact of prosumers on household energy consumption will vary depending on the country and region. In countries with high electricity prices and low PV installation costs, prosumers can potentially reduce their electricity bills and even earn income by selling surplus electricity to the grid. In contrast, in countries with low electricity prices and high installation costs, the economic incentives for becoming a prosumer are lower. But the rising electricity prices in recent years have led to increased interest in becoming a prosumer, particularly in Europe. For example, in Romania, the number of household prosumers increased three times in just one year, in 2022—from 13,109 at the beginning of the year to over 39,000 at the end of it, according to the national energy regulator (https://www.anre.ro/ro/presa/comunicate/informare-de-presa-09-02-2023-anre-incurajeaza-producerea-de-energie-electrica-de-catre-prosumator)—representing approximately 0.4% of the total number of final consumers in the country, which is 8.738.575 (https://www.anre.ro/ro/print?modul=stiri&link=comunicat-de-presa-06-01-2021).

Thus, delving into the impact of prosumers, it is essential to understand the household consumption trends. According to the European Commission, residential energy consumption in the EU is expected to remain relatively stable until 2030, with a slight decrease in 2030 (https://ec.europa.eu/eurostat/statistics-explained/index.php?title=Electricity_production,_consumption_and_market_overview). This decrease is due to improvements in energy efficiency and the adoption of energy-saving technologies. But the EEA warns that these savings could be offset by population growth, economic growth and an increase in the number of households (European Environment Agency 2023), and the IEA predicts that the number of prosumers in Europe will continue to grow, but at a slower rate than in recent years due to the decreasing economic (https://www.iea.org/reports/european-union-2020). On the other hand, high electricity prices in 2022, caused by the war in Ukraine, have prompted many consumers to consider becoming prosumers in order to reduce their electricity bills.

Therefore, it is worth noting that the impact of prosumers on household consumption will depend on the number of prosumers in the future, yet unknown.

4.2 *Impact on Electricity Production*

The rise of prosumers is also expected to have a significant impact on energy production, as mentioned in the introduction part. As more households generate their own electricity using photovoltaic panels, they are able to sell any excess electricity back to the grid, which can help to reduce the need for traditional power plants. There will be a difference between winter and summer periods, as PV production during summer will be substantially higher than in winter period, but the benefit of this category of consumers is very high, since they are able to cover some of their own demand, reducing the electricity extracted from the grid, and also are able to inject their surplus into the grid (see Figs. 4 and 5). In addition, if more prosumers would have batteries installed in their homes—to use the excess energy produced in the middle of the day at night—the better would be for the energy system (Fig. 5).

This shift towards prosumption could help to reduce carbon emissions (by lowering the greenhouse gas emissions) and improve the security of energy supply (by reducing the reliance on traditional energy sources). Also, this shift towards distributed generation could help to improve the efficiency and reliability of the electricity grid, as well as reduce the need for new transmission and distribution infrastructure.

The growth of prosumers could potentially lead to a shift in the way energy is produced and distributed. As more households generate their own electricity, the role of centralised power plants could decrease and the grid could become more decentralised. This could lead to challenges in managing the grid and maintaining

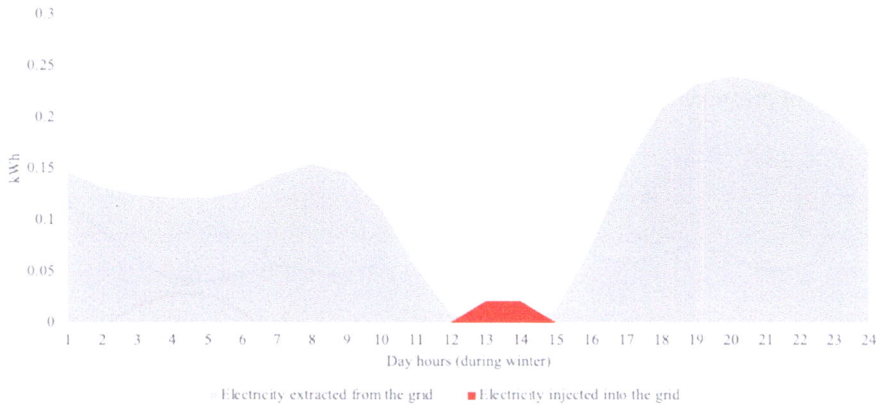

Fig. 4 Prosumer's electricity injection and extraction from the grid during the winter. *Source* Authors' own research results/contribution

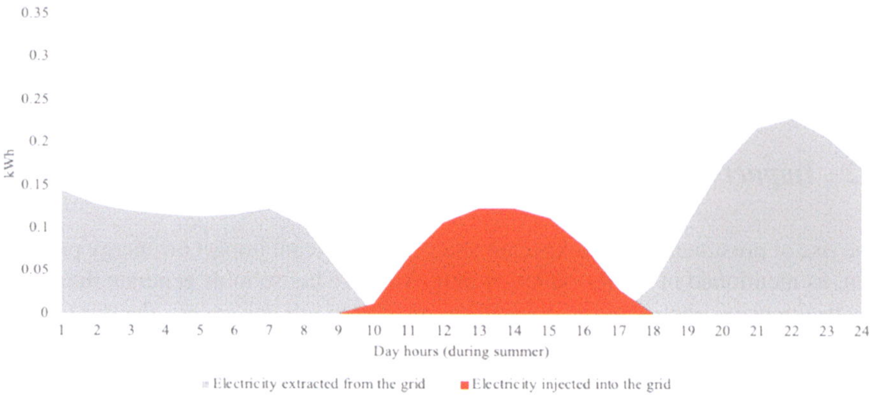

Fig. 5 Prosumer's electricity injection and extraction from the grid during the summer. *Source* Authors' own research results/contribution

grid stability, but it could also create opportunities for new business models and technologies.

At EU level, prosumers could produce and cover around 5% of Europe's electricity demand by 2030 and almost a quarter of EU's electricity consumption, based on existing building stock (Bódis et al. 2019), which will help to reduce carbon emissions and improve the security of energy supply. The Commission estimates that, by 2030, prosumers will produce around 83 TWh of electricity annually, which is equivalent to the annual electricity consumption of approximately 20 million households. In percentage terms, this electricity injected into the grid will represent about 11% reduction in total household consumption. In terms of production, the electricity injected into the grid would be almost at the same level as the EU's residential consumption during the summer (see Fig. 6). This suggests again that prosumers will play an increasingly important role in meeting the energy needs of households in Europe. At the same time, the IEA expects the total capacity of installed rooftop solar in Europe to increase to 133 GW in 2040 (https://www.iea.org/reports/approximately-100-million-households-rely-on-rooftop-solar-pv-by-2030), also emphasising the role of prosumers in the energy system.

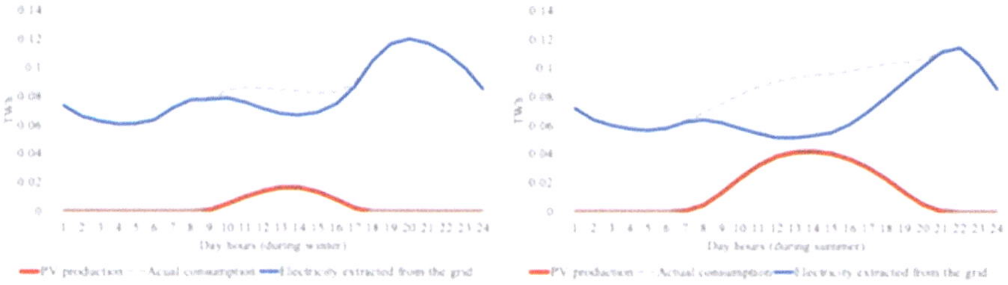

Fig. 6 Consumption and production profiles by 2030 at EU level (winter and summer scenarios). *Source* Authors' own research results/contribution

However, because of a lack of statistics and monitoring of prosumption, the overall number of prosumers in EU Member States and the EU as a whole is unknown (Doračić et al. 2020) and there are various reasons for this: Member States have not been compelled to monitor prosumption and many various forms of prosumption notions use a variety of technology and ownership structures. As mentioned by the EEA (https://data.europa.eu/doi/https://doi.org/10.2800/030218), this makes it very difficult to define uniform categories for EU-wide monitoring. And the situation is the same globally. The only estimated number of prosumers is given by the IEA, which reported that there are about 25 million today and they could increase to 100 million by 2030 (https://www.iea.org/reports/approximately-100-million-households-rely-on-rooftop-solar-pv-by-2030).

In its World Energy Outlook 2022, IEA noted that renewables are on track to become the world's leading source of power in the long term. Total renewable electricity output climbed by more than 500 TWh from 2020 (https://www.iea.org/reports/global-energy-review-2020) to a record high of over 8000 TWh in 2021, driven mostly by expanding solar PV and wind generation (International Energy Agency 2023a). According to the same source, renewables will continue to expand and their share of generation will increase to 43% by 2030—of which 14% will represent electricity generated and injected into the grid by prosumers—and 65% by 2050.

Taking into account the numbers provided by IEA for 2030 (International Energy Agency 2023a), if current forecasts turn out to be true, it could be seen an increased quantity of electricity injected into the grid coming from prosumers which, during the summer, would even cover all household demand globally (see Fig. 7).

In the case of Romania, as example, a huge increase in roof-mounted PV panels could be seen in 2022, as data from the country's national energy regulator shows. According to the source, the installed capacity of electricity production from renewable sources increased substantially, from 84 MW (of which 56 MW residential) and 13,109 household prosumers (96% of total number) in 2021 to 423 MW and 40,171 prosumers in total at national level in 2022 (https://www.anre.ro/ro/legislatie/prosumatori/prosumatori-rapoarte). The main drivers were high electricity prices, consumers' desire to lower their energy costs and to become more energy efficient, thus accelerating also the energy transition process (Figs. 8 and 9).

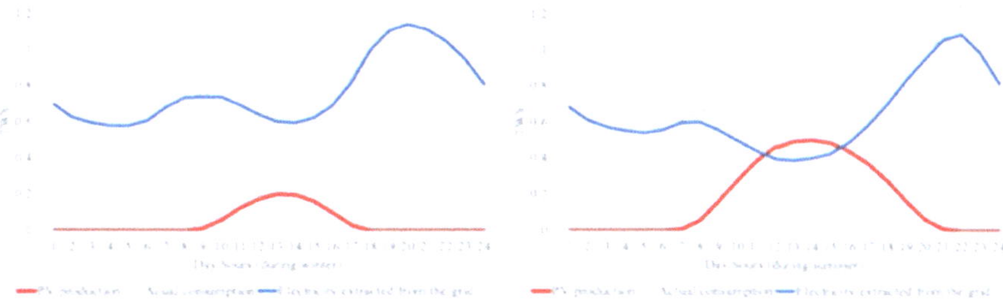

Fig. 7 Consumption and production profiles by 2030 globally (winter and summer scenarios). *Source* Authors' own research results/contribution

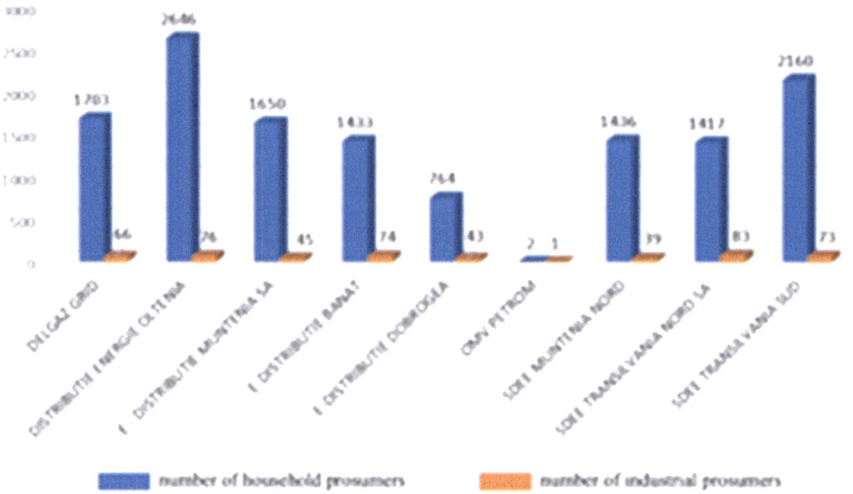

Fig. 8 Number of consumers connected to own network (per distribution operator and type of consumer) in 2021 in Romania (https://www.anre.ro/ro/legislatie/prosumatori/prosumatori-rapoarte)

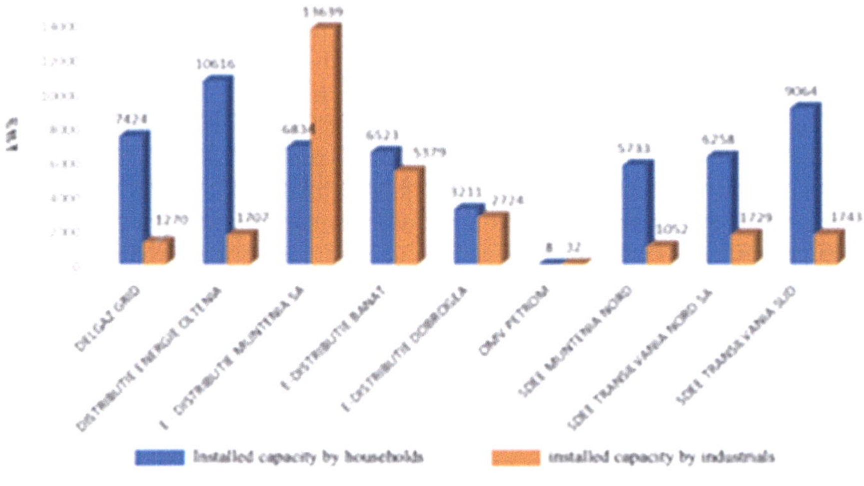

Fig. 9 Installed capacity for prosumers connected to the distribution operators' own network (by form of organisation) in 2021 in Romania (https://www.anre.ro/ro/legislatie/prosumatori/prosumatori-rapoarte)

4.3 Impact on Electricity Prices

The growing number of prosumers worldwide will have significant implications for energy prices as well. In countries with a high penetration of prosumers, such as

Germany (Moser et al. 2020), the growth of prosumers has already had a significant impact on the electricity market. This has led to a decrease in wholesale electricity prices and an increase in price volatility.

The impact on energy prices is more complex and depends on the specific market and regulatory framework. On the one hand, in some markets, increased prosumption could lead to lower prices as more households generate their own electricity and sell any excess back to the grid. On the other hand, prosumers still rely on the grid for backup power when their PV panels are not producing electricity, which means they still need to pay grid fees and other charges.

According to the European Commission (https://ec.europa.eu/eurostat/statistics-explained/index.php?title=Electricity_production,_consumption_and_market_overview#Household_electricity_consumption), several prosumer models have emerged in recent years, aided by market, local, or community strategies. There are numerous prosumer models available, including small and large businesses, and households which have solar panels, storage, vehicle-to-grid injections or opting for a "plunge pricing tariff" (when a supplier pays its customers to take excess energy off the grid) or demand-side flexibility. Therefore, the potential value they bring for people and the electric system is enormous.

For example, a balcony solar panel could potentially reduce household utility expenses by 25% over the course of 20 years (https://www.eea.europa.eu/publications/energy-prosumers-and-cities), considerably assisting individuals living in energy poverty. At the same time, community models have shown a wide range of positive environmental and social consequences. These imply direct citizen participation in the decision-making required for project implementation and governance processes and it can also increase social cohesion within a community and assist in the reduction of energy poverty.

There will also be a financial impact on the price of electricity (especially that traded on the spot and balancing markets) at peak consumption and production (during the day) compared to off-peak consumption (night-time).

Nonetheless, households have the most considerable flexibility potential. Flexibility provides for better network management and increased usage of renewable energy—by disconnecting prosumers from the grid when they inject too much electricity—and smart technology and storage solutions allow consumers to tap into this potential without losing comfort of money. For example, a prosumer without batteries may sell excess electricity during the middle parts of the day, when residential consumption is low, and hence receive a lower payment per kilowatt-hour than at peak times. In contrast, they may purchase electricity from the grid in the evening, when the sun does not shine but demand is high and pay a higher price. In other words, prosumers have the potential to reduce household energy consumption, as they are incentivized to use energy efficiently in order to maximise their self-consumption and minimise the amount of energy they need to purchase from the grid. In addition, prosumers can contribute excess energy back to the grid, which can help to stabilise the grid and reduce the need for expensive peaking power plants.

Moreover, in other markets, increased prosumption could lead to higher prices, as traditional utilities seek to recover their fixed costs from a smaller customer base.

The impact on prices will also depend on the specific regulatory framework, as some countries provide more favourable incentives and support for prosumers than others. Therefore, the impact of prosumers on energy prices will depend on the balance between their contribution to reducing demand and their reliance on the grid.

However, the trend towards becoming a prosumer also presents potential risks for the energy system. For example, if the number of prosumers grows too rapidly, it could lead to grid instability and a reduction in the revenue that utilities collect from selling electricity to households. Furthermore, there are concerns that prosumers may not be paying their fair share of grid maintenance and other fixed costs, which could lead to higher costs for non-prosuming households.

When it comes to drivers to become a prosumer, the price level in the market is also a very important aspect, as the following analysis made on Romania shows. Two variables were analysed: the monthly average price of the Romanian electricity spot market and the monthly number of prosumers between January 1, 2020, and January 31, 2022. And the results of the Spearman correlation indicated that there is a significant large positive relationship between them, ($r(34) = 0.933$, $p < 0.001$). This demonstrates that throughout the examined period of 2020–2022, the increase in energy costs had a positive effect on the number of prosumers. This investigation is limited, therefore it is possible that factors other than price rises have contributed to the rapid increase in the number of prosumers, like regulation, incentives and PV panels price decrease (Fig. 10).

Therefore, based on this result, as electricity prices are likely to remain at current levels, according to energy experts (https://www.investenergy.ro/george-andrei-tecusan-si-lucian-pamfile-fel-romania-pretul-energiei-electrice-in-viitor-vom-mai-reveni-vreodata-la-preturile-de-dinainte-de-pandemie/), there should continue to be a large increase in the number of prosumers. For a country like Romania, this would mean a 6-fold increase compared to the current level, which could lead to an installed production capacity of about 1.7 TWh by 2030, just in the residential area.

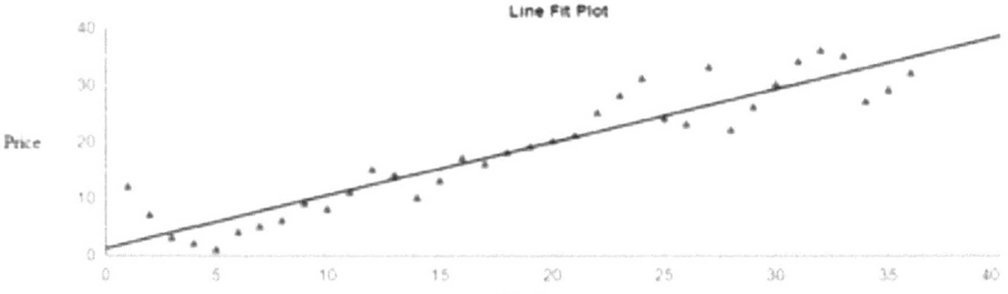

Fig. 10 Spearman correlation between prosumers' number and price evolution. *Source* Authors' own research results/contribution

5 Conclusion

The study represents an overview of the prosumers situation globally and its aim is to show a centralised summary of their status, as information. At the same time, the paper also highlights the growth potential of this category of energy market participants and their impact on the energy system in the future. However, the presented study and the results obtained are subject to certain limitations. For example, in the research process, it was impossible to collect data for variables indicated in the literature as potentially necessary, e.g., the number of total household prosumers today and the share of electricity produced by prosumers at present, due to lack of statistics and monitoring of prosumption. Therefore, the results are less detailed and relate to the assessment based on forecasts for 2030 and an overview of the sizing, integration and risk of prosumer energy systems in the near future. In addition, the data is scattered between multiple reporting organisations, which makes it difficult to create an overall picture of prosumers.

Prosumers are transforming the way electricity is produced and consumed around the world and their rising number will have a significant impact for the system until 2030. As noted in this paper, they have multiple roles: first, to reduce electricity consumption; second, to increase the electricity production; third, to reduce the CO_2 emissions (https://www.iea.org/reports/global-energy-co2-status-report-2019), since the surplus of electricity they inject in the grid is produced from a renewable source.

While the impact of prosumers on household energy consumption may vary depending on the country and region, the overall trend towards prosumption is expected to continue growing rapidly. The trend towards prosumption is expected to have significant implications for energy consumption, production and prices worldwide. The number of prosumers is expected to grow significantly over the next decade—a trend already started, as can be seen from the example of Romania—and this shift towards distributed generation could help reduce the demand, while also contributing to the reduction of carbon emissions and improvement of the security of energy supply.

In addition to environmental and social benefits, one of the important advantages of prosumption is financial rewards. Individual citizens can profit financially from prosumption models under the right circumstances. Prosumption may result in decreased energy expenses, while selling excess energy may result in money, particularly in light of recently high energy prices. Also, by producing their own energy, prosumers may be better protected from price fluctuations over time.

Their integration will depend on the costs of instalment, which has the potential to decrease over time. Based on a wave of high energy prices, which could continue in the coming years, the shift to prosumption brings medium- to long-term benefits on energy cost efficiency. Some open questions are: what will happen if energy prices fall—will there still be interest in becoming a prosumer? What would happen if the primary sources of PV panels today were unable to produce them in the next years? Legislative support and support/promotion mechanisms (public policies) are therefore needed.

In terms of risks, the electricity injected into the grid by prosumers could affect the energy system because their production is hard to predict, and without proper measurement systems or new technologies for demand-side control, it would create imbalances in the electricity grid. Also, the national electricity transmission system needs to be prepared to take additional energy from this type of source—this being another topic related to investments in the grid. Another risk is the behaviour of the household prosumers, who may become more careless with their consumption, knowing that they produce their own energy. But at the same time they may become more careful, based on the fact that investment in this technology is based on financial reasons and interest in reducing their energy costs.

In conclusion, in Europe and globally, the rising number of household prosumers will have a significant impact on residential consumption in the future. The emergence of prosumers is a significant trend in the energy sector, driven by the need to transition to renewable energy sources and reduce carbon dioxide emissions. Prosumers have the potential to reduce household energy consumption and contribute to grid stability, but their growth also presents potential risks for the system. As such, it is important to carefully monitor and regulate the trend towards prosumers to ensure a smooth transition to a more sustainable energy system, and this will depend on the number of prosumers in the future. Last but not least, further research is needed to better understand the exact impact of prosumers on household energy consumption and to develop policies that can maximise the benefits of this trend while minimising its potential risks.

References

Bódis, K., Kougias, I., Waldau, A.J., Taylor, N., Szabó, S.: A high-resolution geospatial assessment of the rooftop solar photovoltaic potential in the European Union. Renew. Sustain. Energy Rev. **114**, 109309 (2019)

BP Statistical Review of World Energy 2019. https://www.bp.com/content/dam/bp/business-sites/en/global/corporate/pdfs/energy-economics/statistical-review/bp-stats-review-2019-full-report.pdf. Accessed 05 Mar. 2023

Buragohain, T.: Impact of solar energy in Rural Development in India. Int. J. Environ. Sci. Dev. **3**(4), 334–338 (2012)

Deng, G., Newton, P.: Assessing the impact of solar PV on domestic electricity consumption: exploring the prospect of rebound effects. Energy Policy **110**, 313–324 (2017)

Doračić, B., Knoefel, J., Naber, N.: Prosumers for the Energy Union: mainstreaming active participation of citizens in the energy transition. PROSEU—Horizon 2020 (H2020-LCE-2017) Grant Agreement N°764056 (2020)

European Commission. https://ec.europa.eu/eurostat/statistics-explained/index.php?title=Electricity_production,_consumption_and_market_overview. Accessed 03 Mar. 2023

European Environment Agency. https://data.europa.eu/doi/10.2800/030218. Accessed 03 Mar. 2023

European Environment Agency. https://www.eea.europa.eu/publications/energy-prosumers-and-cities. Accessed 03 Mar. 2023

European Commission. https://sustainable-energy-week.ec.europa.eu/news/lets-boost-prosumerism-models-and-state-play-2022-09-02_en. Accessed 03 Mar. 2023
European Commission. https://ec.europa.eu/eurostat/statistics-explained/index.php?title=Electricity_production,_consumption_and_market_overview#Household_electricity_consumption. Accessed 03 Mar. 2023
European Environment Agency. https://www.eea.europa.eu/publications/trends-and-projections-in-europe-1. Accessed 03 Mar. 2023
International Energy Agency. https://www.iea.org/reports/world-energy-outlook-2022. Accessed 03 Mar. 2023
International Energy Agency. https://www.iea.org/reports/the-future-of-solar-pv. Accessed 28 Feb. 2023
International Energy Agency. https://www.iea.org/reports/clean-household-energy-consumption-in-kazakhstan-a-roadmap. Accessed 03 Mar. 2023
International Energy Agency. https://iea.blob.core.windows.net/assets/a72d8abf-de08-4385-8711-b8a062d6124a/WEO2020.pdf. Accessed 04 Mar. 2023
International Energy Agency. https://www.iea.org/news/global-electricity-demand-growth-is-slowing-weighed-down-by-economic-weakness-and-high-prices. Accessed 03 Mar. 2023
International Energy Agency. https://www.iea.org/reports/approximately-100-million-households-rely-on-rooftop-solar-pv-by-2030. Accessed 03 Mar. 2023
International Energy Agency. https://www.iea.org/reports/electricity-information-overview/electricity-consumption. Accessed 03 Mar. 2023
International Energy Agency. https://www.iea.org/reports/european-union-2020. Accessed 03 Mar. 2023
International Energy Agency, https://www.iea.org/reports/global-energy-co2-status-report-2019. Accessed 03 Mar. 2023
International Energy Agency. https://www.iea.org/reports/global-energy-review-2020. Accessed 03 Mar. 2023
International Energy Agency. https://www.iea.org/reports/net-zero-by-2050. Accessed 03 Mar. 2023
International Energy Agency. https://www.iea.org/reports/world-energy-outlook-2021. Accessed 03 Mar. 2023
InvesTenergy. https://www.investenergy.ro/george-andrei-tecusan-si-lucian-pamfile-fel-romania-pretul-energiei-electrice-in-viitor-vom-mai-reveni-vreodata-la-preturile-de-dinainte-de-pandemie/. Accessed 07 Mar. 2023
Kampman, B., Blommerde, J., Afman, M.: The potential of energy citizens in the European Union. Delft, CE Delft 16.3J00.75 (2016)
Kroposki, B., Margolis, R., Ton, D.: Harnessing the sun. IEEE Power Energy Mag. **7**(3), 22–32 (2009)
Marszal, A.J., Heiselberg, P., Bourrelle, J.S., Musall, E., Voss, K., Sartori, I., Napolitano, A.: Zero energy building—A review of definitions and calculation methodologies. Energy Build. **43**(4), 971–979 (2011)
McLaren, J., Davison, C., Miller, J., Bird, L.: Impact of rate design alternatives on residential solar customer bills: increased fixed charges, minimum bills and demand-based rates. Electr. J. **28**(8), 43–58 (2015)
Moser, R., Xia, C., Thema, J., Vondung, F.: Solar Prosumers in the German Energy Transition: a multi-level perspective analysis of the German 'Mieterstrom' Model. In: Østergaard, P., Calise, F., et al. (eds.) 15th Conference on Sustainable Development of Energy, Water and Environment Systems (SDEWES) 2020 (2021)
Our World in Data. https://ourworldindata.org/energy. Accessed 01 Mar. 2023
Qui, Y., Khan, M.E., Xing, B.: Quantifying the rebound effects of residential solar panel adoption. J. Environ. Econ. Manag. **96**, 310–341 (2019)
Romanian Energy Regulatory Authority. https://www.anre.ro/ro/legislatie/prosumatori/prosumatori-rapoarte. Accessed 08 Mar. 2023

Romanian Energy Regulatory Authority, https://www.anre.ro/ro/presa/comunicate/informare-de-presa-09-02-2023-anre-incurajeaza-producerea-de-energie-electrica-de-catre-prosumator. Accessed 08 Mar. 2023

Romanian Energy Regulatory Authority. https://www.anre.ro/ro/print?modul=stiri&link=comunicat-de-presa-06-01-2021. Accessed 08 Mar. 2023

Spearman, C.: The proof and measurement of association between two things. Am. J. Psychol. **15**(1), 72–101 (1904)

Verdolini, E., Galeotti, M.: At home and abroad: an empirical analysis of innovation and diffusion in energy technologies. J. Environ. Econ. Manag. **61**(2), 119–134 (2011)

Wang, Z., Wu, Z., Hu, Z., Mu, E., Fu, X., Hernández, J.O., Zhang, Z.Y., Jevric, M., Liu, Y., Wang, F., Li, T., Moth-Poulsen, K.: Chip-scale solar thermal electrical power generation. Cell Rep. Phys. Sci. **3**(3), 100789 (2022)

Measuring the Circular Economy Inside European Union, Using Sankey's Diagram of Material Flows and Fuzzy Clustering

Andreea Pernici, Stelian Stancu, Denisa Elena Bălă, and Monica-Ioana Vulpe

Abstract Recently, the circular economy concept has been almost seamlessly integrated into all key strategies, action plans, and policies across international institutions. An example in this case is represented by the European Commission's Circular Economy Action Plan, along with Sankey's Diagram of Material Flows, which will both describe in detail the potential and necessary transition to the new economic framework. Based on that, in the current paper, we have proposed a model that splits the diagram into inputs and outputs, with two correspondent datasets and two research objectives: identify the main producers and owners of materials, and highlight the main types of actors concerning the outflow systems. To reach these objectives, we have computed the Unsupervised Possibilistic Fuzzy C-Means Clustering (UFPC) algorithm. As a result, in the case of the inputs dataset, the computation allocated the 27 member states of the European Union into four clusters, as follows: high-scale, medium-scale, and low-scale resource owners and producers, and importers. When referring to the output data, we have also obtained four allocations, three of those based on the volume of emissions and the performance of the waste recovery system (high, medium, low), and one showing the outlier countries with the highest volumes of waste landfills. Therefore, the constructed model will be relevant for the current European context, while also filling in the gap in terms of using fuzzy clustering as a machine learning algorithm.

A. Pernici (✉) · D. E. Bălă · M.-I. Vulpe
The Bucharest University of Economic Studies, Bucharest, Romania
e-mail: andreea.pernici@csie.ase.ro

D. E. Bălă
e-mail: baladenisa16@stud.ase.ro

M.-I. Vulpe
e-mail: ioanarenard@gmail.com

S. Stancu
The Bucharest University of Economic Studies, Centre for Industrial and Services Economics, Romanian Academy, Bucharest, Romania
e-mail: stelian.stancu@csie.ase.ro

M. Busu (ed.), *Rethinking Business for Sustainable Leadership in a VUCA World*, Springer Proceedings in Business and Economics,
https://doi.org/10.1007/978-3-031-50208-8_17

Keywords Circular economy · Circular economy action plan · Sankey's diagram of material flows · Fuzzy clustering · European Union

1 Introduction

The circular economy is one of the most recurrent concepts of the last decade, being generated by the astringent need of reducing the dependence on certain materials and resources, while also eliminating waste and regenerating the production-nature relationship. As a direct consequence, the European Union has been focusing on the concept, developing action plans, and constructing tools for monitoring progress. Moreover, the latest considerations on the subject have been integrated into the broader *2030 Agenda for Sustainable Development* (UN General Assembly: Transforming our world: the 2030 Agenda for Sustainable Development 2015), as well as the *European Green Deal* (European Commission: A European Green Deal 2020) as a key priority for sustainable growth. Therefore, it will be crucial to analyze the concept from a literature perspective, since there will be multiple definitions, as well as critiques. In this sense, the first section of this paper will look at the circular economy concept and its theoretical background. Next, Sankey's Diagram of material flows will be described, as the main tool of visualization and measurement employed by the EU in their efforts to illustrate the whole process of transforming resources into materials and their eventual production journey. In the same tool, processes such as material use, accumulation, emission, or transfer (imports and exports) will be profiled, along with a definite loop for circular reuse, backfill, and recycling.

Having all the above in mind, Sankey's diagram will be used in the methodological construction of the datasets, splitting them into inputs and outputs. Followingly, we will compute a model composed of an unsupervised possibilistic fuzzy c-means clustering (UFPC) algorithm in order to answer the two research objectives: highlight the main producers and owners of materials in the EU conglomeration, and illustrate the regional distribution from an output point of view, with the accent on the circular economy potential. Therefore, in the results section, the two perspectives will be aggregated, along with a brief Romanian profile. Lastly, considering all the above, the model will be relevant in nowadays context, since it responds to a clear-set priority of the European Union: the immediate implementation of circularity and sustainability practices. At the same time, employing a dynamic algorithm will leave room for regional economic specificity, as well as replicability, an important characteristic that brings value to the overall research field.

2 Literature Review

2.1 Circular Economy Concept

Despite its recent popularity, the term has been around for more than forty years, since the general agreement is that the concept was first defined by British economists, Pearce and Turner (Pearce and Turner 1990). The authors included it alongside the sustainable development framework meant to describe the way the economy should be organized in the future. Since then, there have been multiple approaches that aimed to profile the concept, generated by very different stakeholders, from related literature experts to national agencies, international institutions, and private companies.

Therefore, starting with the continental vision, the European recently Union adopted the *Circular Economy Action Plan (CEAP)* (European Commission: The Circular Economy Action Plan 2020), being seen as a central part of the *European Green Deal* (European Commission: A European Green Deal 2020), the most updated continental agenda that describes in detail the way to proceed to the contemporary sustainability transition. In terms of definitions, the European Parliament (European Parliament: Circular economy: definition, importance, and benefits 2023) will describe the circular economy (CE) as a model of production and consumption that involves sharing, leasing, reusing, repairing, refurbishing, and recycling existing materials for as long as possible, with the ultimate goal of a life cycle extension. To add on that, in the last decades, production flows have been constructed on the *linear take-make-consume-throw-away* pattern, that relies on accessible and cheap materials, limited product lifespan, and economies of scale. Therefore, the goal of the new economic structure would be to redesign that pattern in a cyclic way, by replacing the throw-away stage with different processes, a concept known as the 3Rs: *reduce, reuse, and recycle.* Some authors have further developed this framework, reaching a number of 10Rs: *refuse, rethink, reduce, reuse, repair, refurbish, remanufacture, repurpose, recycle, and recover* (Potting et al. 2017; Morseletto 2020). With that in mind, we can now illustrate the main needs that the circular economy is supposed to answer: environmental protection, limitation of biodiversity loss, reduction of greenhouse gas emissions, creation of jobs and new industries, and most importantly, waste reduction (European Commission: The Circular Economy Action Plan 2020). From those, the key actionable items would be the productive switch to sustainable products and the transition of highly potential sectors (such as plastics, textiles, electronics, packaging, batteries, and vehicles) to the circularity principles.

Continuing the exploration of the concept with the related literature approaches, Geissdoerfer et al. (2017) defined the circular economy as a new sustainability paradigm, not being synonymous with the sustainability concept. The authors will identify the main differences between the notions, with a specific distinction in terms of goals and motivation: while sustainability has an open-ended, multitude of goals depending on the considered agents, the circular economy will aim to construct a closed loop, ideally eliminating all resource input into and leakage out of the system.

Therefore, the circular economy is defined with the potential to lead to sustainable development, the terms being in a causal or symbiotic relationship, rather than synonymous. In other words, the CE can be perceived as a change of paradigm in the way that human society is interrelated with nature, but not as a panacea of sustainability (Prieto-Sandoval et al. 2018). Moreover, Moraga et al. (2019) will also emphasize this theory, by stating that CE will be an umbrella concept that includes multiple sustainable aspects and practices, such as minimizing waste generation, lowering material input, and decoupling economic growth from natural resources.

Further on, being an economic structure, the ultimate goal will be economic prosperity, followed by environmental quality, as mentioned by Kirchherr et al. (2017). The authors have focused on 114 definitions of the term, reaching the conclusion that social equity or the concern for future generations will not be prevalent. However, there will be voices from the related literature that include the social function in the equation. For example, Murray et al. (2017) will develop a newer, revised definition of the circular economy, namely as an economic model where planning, resourcing, procurement, production, and reprocessing are designed and managed, as both processes and outputs, to maximize ecosystem functioning and human well-being. Next, when referring to the previously mentioned environmental quality, Stahel (2016) will provide a conceptualization of the circular economy, as being one that maximizes the services produced from the linear nature-society-nature material and energy. Therefore, in his view, this new flow will set a level of production that nature can tolerate, combining the ecosystem and economic cycles together and respecting the reproduction rates. The author will further develop this framework, by stressing the importance put on innovation and enriching the already existing materials (Prieto-Sandoval et al. 2018).

In regards to the practical terms, as expected, the concept has generated new business models, as well as best practices or implementation mappings. One of them focuses on the levels that it requires in order to inflict real change. In this sense, Yuan et al. (2016) will be the ones that described the three-layer approach. In their vision, the first level, the *micro* or individual one, is represented by consumers and enterprises and their respective actions in transitioning to cleaner production, industrial ecology, and ecological modernization. The second level will be the *meso* one, which will refer to a system of economic agents, that together can develop an eco-industrial network that will ultimately grow into the *macro* level. This last stage will imply the full development of eco-cities and eco-municipalities, a butterfly effect being observed since individual actions impact the flow up until administrations or governments. Moreover, when it comes to business models, Stahel (2016) will mention that there will be two frameworks based on the circular-economy concept: the one that focuses on the extension and reuse of services and products through constant repairs and upgrades, and the one that focuses on recycling old products and turning them into secondary materials. From another perspective, the OECD (Organisation for Economic Co-operation and Development: Business Models for the Circular Economy 2019) will also profile five types of business models that can lead to a circular economy. Shortly, these will be: the circular supply model, in which traditional materials are replaced with bio-based, renewable ones, the resource recovery model, which will focus on recycling waste into secondary raw materials, the product life extension model, which will aim to extend the use period of existing products, the sharing model, which will facilitate the sharing of under-utilized products, in contrast to demanding new ones, and lastly, the product service system, where the focus is being switched on services, rather than products.

Finally, since the concept has emerged and developed at an exponential speed, it undoubtedly met its critiques. One of the most frequent revolves around the lack of clarity on the theoretical background and its diffused limits, which will ultimately lead to implementation obstacles (Corvellec et al. 2022). On this note, the main contra-argument will be the fact that the circular economy neglects or ignores established knowledge, since any loop necessitates new materials and energy, consuming and creating waste and emissions in the process (Corvellec et al. 2022). Another discussed aspect will be the unclear ways of implementation and in this case, Hofmann (2019) will argue that companies' practices in terms of long-lasting products or repair services are not that frequent, being most of the time part of the niche, luxury industries.

However, whether the circular economy can be fully implemented as a way of redesigning society remains to be seen, but the concept in its theoretical form is a useful tool for describing the steps needed in order to increase or ensure future prosperity, sustainability, and overall quality of life, a fact confirmed by the European Union institutions.

2.2 *Sankey's Diagram of Material Flows*

Returning to the European Union, the main tool used to monitor and evaluate the circular economy will be the material flows diagram, created using Sankey's framework (Kennedy and Sankey 1898). This tool has been developed as a Eurostat statistical product (https://ec.europa.eu/eurostat/cache/sankey/circular_economy/sankey), with the main goal of graphically representing the flow of materials in the production process of an economy and how they are eventually discarded back into the environment as residuals. As an example, the visualization instrument can be seen in Fig. 1, showcasing the EU total quantities in 2021.

The diagram will be composed of multiple elements, starting from the material inputs (whether natural resources or imports), continuing with the processing of materials, and the afferent rate of material use and accumulation, and lastly illustrating the level of different outputs (emissions, dissipative flows, waste). The most relevant part of the circular economy concept will be the closed loop, which will represent the residuals that are not discharged but reused to produce secondary raw materials. This will be part of the waste treatment node, that integrates multiple processes such as incineration, waste landfills, or recovery (recycling, backfilling).

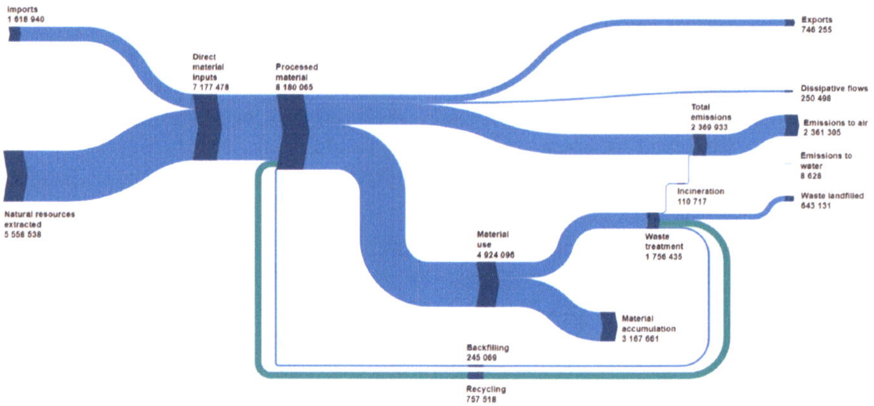

Fig. 1 Material flows diagram, European Union, 2021, thousand tonnes. *Source* Eurostat, Material flow diagram tool

2.3 *Cluster Analysis for Measuring the Circular Economy*

Lastly, before describing the model employed, we must illustrate some highlights from the related literature in terms of using cluster analysis. Therefore, we have found multiple papers that used clustering as a way of evaluating the level of progress and transition toward a circular and sustainable economy, inside the European Union. One example is profiled by Mazur-Wierzbicka (2021) who will compute multiple machine learning methods (principal components analysis, factor analysis, cluster analysis, ANOVA) in order to assess the advancement made by the EU member states. Another example will be constructed by Rodriquez-Anton et al. (2022) who will use cluster analysis to find out whether the performance of the member states in terms of CE transition is homogeneous in the region, eventually linking them to the Sustainable Development Goals existing in the 2030 Agenda. Lastly, Robaina et al. (2020) have used cluster analysis, along with several econometric estimation methods such as panel unit root tests, cointegration tests, or vector autoregression models to provide insights into the performance and evolution of the circular economy, in a sixteen-year timeframe. In all three sources, the main indicators were similar, collected from the Eurostat, while the algorithms were part of the hierarchical or k-means family. As a consequence, we have noticed a void in the related literature, that can be filled up by employing fuzzy clustering, a dynamic algorithm with multiple benefits.

Table 1 INPUTS dataset description

Indicator	Definition	UM	Eurostat
NAT_RES_EXT	Natural Resources (domestic) extraction	Thousand tonnes	ENV_AC_MFA
IMPORTS	Total Imports	Thousand tonnes	ENV_AC_MFA
IMP_WASTE	Imports of waste for recovery—recycling	Thousand tonnes	ENV_AC_SD
PROCESSED_MAT	Processed material—direct material inputs, recovery recycling and backfilling	Thousand tonnes	ENV_AC_SD
MAT_USE	Material use	Thousand tonnes	ENV_AC_SD
MAT_ACC	Material accumulation	Thousand tonnes	ENV_AC_SD

3 Methodology

3.1 Dataset

Having the diagram of material flows in mind, we will now construct our datasets based on the two main stages or components: INPUTS (Table 1) and OUTPUTS (Table 2). At the same time, as mentioned before, we will aim to achieve two research objectives. The first one will be to identify the main material producers and owners at the European level, with natural resources, imports, and processed materials quantities being included in the general outline. The second objective will refer to determining the main outputs resulting from the productive processes, starting from the total export emissions, dissipative flows, and concluding with waste treatments. Using these elements, we aim to eventually describe the continental poles of discharging, with a clear link to the circular economy potential and the performance of the waste management system.

As a methodological mention, all the indicators have been collected from the Eurostat portal, for all 27 member states of the European Union and the 2021 year.

Table 2 OUTPUTS dataset description

Indicator	Definition	UM	Eurostat
EXP_WASTE	Exports of waste for recovery—recycling	Thousand tonnes	ENV_ AC_SD
EMISSIONS	Total emissions	Thousand tonnes	ENV_ AC_SD
DISS_ FLOWS	Dissipative flows	Thousand tonnes	ENV_ AC_SD
CIRC_RATE	Circular use rate—Share of material recycled and fed back into the economy	%	ENV_ AC_SD
WASTE_ TREAT	Waste treatment	Thousand tonnes	ENV_ WASSD
LANDFILL	Waste landfills	Thousand tonnes	ENV_ WASSD
WASTE_REC	Waste recovery (recycling, backfilling)	Thousand tonnes	ENV_ WASSD

3.2 UPFC (Unsupervised Possibilistic Fuzzy C-Means)

In terms of applied algorithms, as mentioned before, we will use fuzzy clustering in the R Studio environment. We will compute the UPFC (Unsupervised Possibilistic Fuzzy C-Means) clustering technique (Yang and Wu 2006; Wu et al. 2010). In contrast to the *k-means* algorithm, in the fuzzy computation, each data point will be allocated to all the clusters with a certain degree of membership, between 0 and 1. The UPFC method will go a step further, producing both probabilistic and possibilistic membership simultaneously, being seen as a solution for the noise sensitivity of FCM (Fuzzy C-Means) or the generation of coincident clusters in PCA (Principal Components Analysis). Therefore, this method will provide certain advantages, such as flexibility, robustness, or interpretability, since it can handle outliers and noise more efficiently, identify overlapping objects, or allow a more detailed representation. Mathematically, the objective function is represented below (Yang and Wu 2006):

$$J_{UPFC}(\mathbf{X};\mathbf{V},\mathbf{U},\mathbf{T}) = \sum_{j=1}^{k}\sum_{i=1}^{n}(a\,u_{ij}^{m} + b\,t_{ij}^{\eta}\,d^2(\vec{x}_i,\vec{v}_j) + \frac{\beta}{n^2\sqrt{k}}\sum_{j=1}^{k}\sum_{i=1}^{n}(t_{ij}^{\eta}\,log\,t_{ij}^{\eta} - t_{ij}^{\eta}) \qquad (1)$$

Regarding the previous research in the field, we didn't find any papers that applied fuzzy clustering to assess the European evolution and identify the main poles of performance in terms of circular economy, so we can fill in this gap using the model described next.

4 Results and Discussions

4.1 INPUTS Dataset

In terms of results, we are going to present the computations separately for each of the two datasets. Starting with the INPUTS, the first step in the analysis is to describe the data from a statistical point of view, with several metrics being available in Table 3.

We can see that all variables will have a high amplitude, with the minimum and maximum values being extremely different. Next, looking at the distribution form, we can see that 3 variables will have a kurtosis coefficient bigger than 3, meaning that they have a leptokurtic distribution (NAT_RES_EXT, IMPORTS, and PROCESSED_MAT), that could signify the presence of multiple outliers. Those outliers will be handled through the fuzzy c-means algorithm computation. At the same time, the skewness coefficient will be positive for all the variables, meaning that the right tail will be longer than the left, so the small values tend to be prevalent. Lastly, in terms of economic interpretation, we can see that the level of material use will be higher than the one for material accumulation, which could mean an opportunity for replacing initial resources with secondary, recovered, or reprocessed ones, especially for stocks with long life periods, such as buildings and infrastructure. At the same time, we can see that the level of waste imports is low in absolute terms, meaning that there could be a potential for waste to be sent for treatment in more developed countries, for conversion into secondary raw materials.

The next step in the analysis would be to apply the described algorithm. Since all the indicators have the same unit of measurement, we have omitted the normalization step. The UPFC method is an unsupervised one, meaning that will require a number of clusters introduced apriori. In order to identify the optimal number, we will use the Fuzzy Silhouette Index. After testing with multiple variants, we obtained the highest value in the case of a 2-cluster allocation. Therefore, the results can be seen in Table 4.

Table 3 INPUTS dataset descriptive statistics, EU, 2021, thousand tonnes

Indicator	NAT_RES_EXT	IMPORTS	IMP_WASTE	PROCESSED_MAT	MAT_USE	MAT_ACC
Mean	201.539	121.344	4.928	361.076	175.687	109.558
Median	114.841	60.382	2.224	205.430	113.362	49.619
Std Dev	236.022	154.862	5.840	421.384	203.540	127.587
Kurtosis	3,13	4,61	1,77	5,32	1,66	0,67
Skewness	1,84	2,11	1,61	2,14	1,60	1,38
Minimum	2.128	4.985	16	9.618	7.676	2.661
Maximum	957.639	654.537	21.733	1.861.850	750.792	419.051

Table 4 EU member states allocation—INPUTS dataset

No	State	Degree C1 (%)	Degree C2 (%)	Allocation
1	Belgium	31	69	C2
2	Bulgaria	1	99	C3
3	Czechia	2	98	C3
4	Denmark	2	98	C3
5	Germany	81	19	C1
6	Estonia	1	99	C3
7	Ireland	0	100	C3
8	Greece	0	100	C3
9	Spain	74	26	C1
10	France	98	2	C1
11	Croatia	1	99	C3
12	Italy	95	5	C1
13	Cyprus	2	98	C3
14	Latvia	1	99	C3
15	Lithuania	0	100	C3
16	Luxembourg	2	98	C3
17	Hungary	1	99	C3
18	Malta	2	98	C3
19	Netherlands	46	54	C2
20	Austria	6	94	C3
21	Poland	91	9	C1
22	Portugal	4	96	C3
23	Romania	68	32	C1
24	Slovenia	1	99	C3
25	Slovakia	0	100	C3
26	Finland	2	98	C3
27	Sweden	12	88	C3

The initial distribution would be C1 (Germany, Spain, France, Italy, Poland, and Romania), C2 (Belgium and Netherlands, a cluster not clearly defined yet), and C3 (the rest of the countries). However, we need to be mindful of potential outliers, so next, we will compute the *detect.outliers()* function to discern if any aberrant values affect the analysis. The function calculates outliers using 4 approaches, described in detail in the package documentation. In Table 5 we can see the results for each approach. All the outliers' countries are part of the C1 or C2 cluster, so we will consider that allocation optimal and we will proceed with the exclusion of those from the initial INPUTS dataset.

Table 5 Outliers' detection results—INPUTS dataset

Approach number	Outliers detected
Approach 1	Belgium, Germany, Netherlands, Romania
Approach 2	Belgium, Germany, Spain, Netherlands, Romania
Approach 3	No outliers
Approach 4	Germany, Spain, France, Italy, Poland, Romania

Table 6 EU Member States Allocation—INPUTS dataset, without outliers

No	State	Degree C3 (%)	Degree C4 (%)	Allocation
1	Bulgaria	44	56	C4
2	Czechia	98	2	C3
3	Denmark	90	10	C3
4	Estonia	4	96	C4
5	Ireland	25	75	C4
6	Greece	34	66	C4
7	Croatia	1	99	C4
8	Cyprus	6	94	C4
9	Latvia	2	98	C4
10	Lithuania	4	96	C4
11	Luxembourg	14	86	C4
12	Hungary	90	10	C3
13	Malta	8	92	C4
14	Austria	84	16	C3
15	Portugal	94	6	C3
16	Slovenia	4	96	C4
17	Slovakia	9	91	C4
18	Finland	79	21	C3
19	Sweden	84	16	C3

Subsequently, we will apply again the UPFC algorithm, with the Silhouette coefficient registering the highest value for a bi-cluster aggregation. The results can be seen in Table 6, along with a C3 and C4 allocation based on the degree of membership.

The final repartition of states will be visible in Fig. 2 and Table 7. Each cluster has been named to capture the main characteristic of the group.

Therefore, the first cluster will be comprised of the countries with the highest natural resource extraction levels, as well as the highest processed materials volumes. Belgium and Netherlands will be the only states with a negative difference between imports and domestic extraction, so this will be the *Importers* cluster, highly dependent on international commerce. Next, the third cluster will aggregate all the countries

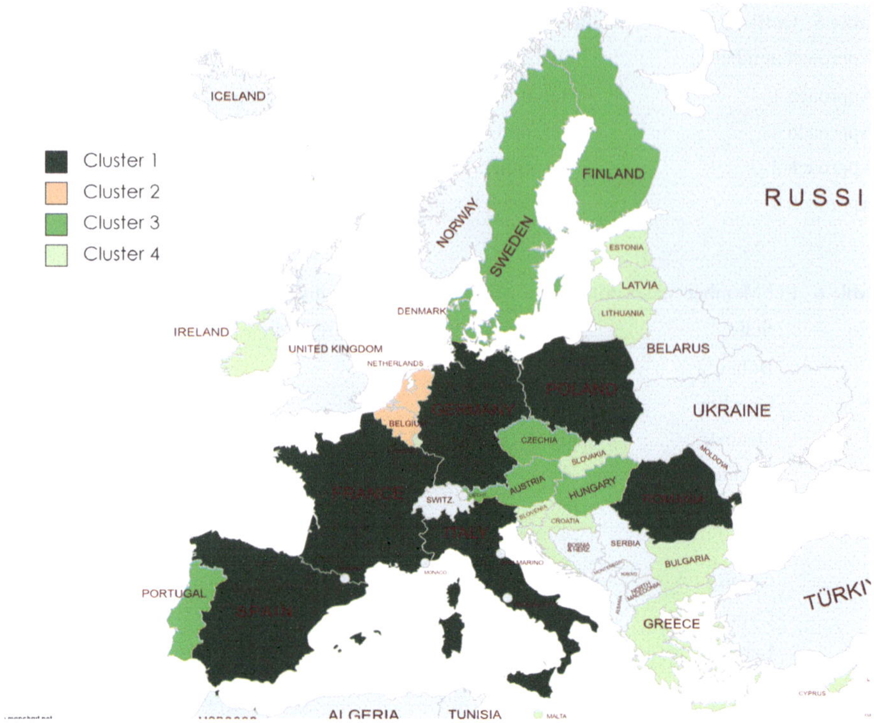

Fig. 2 Final cluster allocation—INPUTS dataset

Table 7 Final cluster allocation—INPUTS dataset

Cluster No	Cluster name	Countries
Cluster 1	High-scale resource owners & producers	Germany, Spain, France, Italy, Poland, Romania
Cluster 2	Importers	Belgium, Netherlands
Cluster 3	Medium-scale resource owners & producers	Czechia, Denmark, Hungary, Austria, Portugal, Finland, Sweden
Cluster 4	Low-scale resource owners & producers	Bulgaria, Estonia, Ireland, Greece, Croatia, Cyprus, Latvia, Lithuania, Luxembourg, Malta, Slovenia, Slovakia

that have considerable levels of natural resources, as well as a high level of material use—therefore they will be part of the medium-scale producers. Lastly, all the states with limited access to resources and relatively low levels of imports, have been grouped since eventually, they will register the lowest numbers in terms of processed materials. Therefore, the first research objective has been met, with a clear regional distribution in terms of natural resources and materials being profiled.

4.2 OUTPUTS Dataset

The same steps will be followed for the OUTPUTS dataset. In the descriptive stage, we will look at the statistics profiled in Table 8. There, we can see that the average level of emissions is very high, 92.209 tonnes (the maximum value being registered for Germany, while Malta will be the only country with zero emissions). The kurtosis is also registering the highest level, signifying a leptokurtic distribution and the existence of outliers.

When it comes to the Circular Use Rate (%), or the share of material recycled and fed back into the economy, we can see that the mean value is 9.32%, increasing versus last years, but still on an unsustainable level. The Netherlands will register the highest level (33.8%), at a considerable distance versus the next one, Belgium (21%). Regarding the waste treatment quantities, we can observe once again a very high amplitude. Moreover, when dividing them by the volume of processed material, we can see that Bulgaria, Finland, and Romania will be the countries with the highest shares of waste treatment and implicitly, landfills.

Next, we applied the UPFC algorithm. However, since for this dataset we had slightly different units of measurement, we have scaled the data as a preliminary step. Once again, the 2 cluster allocation will be optimal (Silhouette Index being the highest), with the first results illustrated in Table 9.

This time, the initial distribution is C1 (Germany, Spain, France, Italy, Netherlands, Poland), C2 (Bulgaria, Romania, Finland, and Sweden—a mention here is that Belgium was not integrated, since this will be the cluster characterized by the highest volumes of waste landfills), and C3 (rest of the countries). Since we have learned from the descriptive statistics that the dataset is susceptible to outliers, we will compute the same function as before, with multiple countries identified (Table 10). These will be the exact C1 and C2 allocations described, plus Belgium, so we can consider them optimal. Therefore, we will compute the algorithm once again with the removal of outliers. The updated results can be found in Table 11.

Table 8 OUTPUTS dataset descriptive statistics, EU, 2021, thousand tonnes*

Ind	EXP_ WASTE	EMISSIONS	DISS_ FLOWS	*CIRC_ RATE (%)	WASTE_ TREAT	LANDFILL	WASTE_ REC
Mean	4.735	92.209	9.332	9,32	66.129	20.130	38.193
Median	2.537	45.553	4.201	7,8	30.041	2.642	8.348
Std Dev	6.211	141.152	14.096	7,33	83.913	33.511	63.315
Kurtosis	5,01	11,82	8,50	3,55	3,00	5,26	5,11
Skewness	2,28	3,17	2,74	1,65	1,75	2,24	2,35
Minimum	76	0	55	1	1.340	74	514
Maximum	25.797	682.574	64.003	33,8	331.741	137.867	249.674

Table 9 EU member states allocation—OUTPUTS dataset

No	State	Degree C1 (%)	Degree C2 (%)	Allocation
1	Belgium	33	67	C3
2	Bulgaria	21	79	C2
3	Czechia	7	93	C3
4	Denmark	1	99	C3
5	Germany	76	24	C1
6	Estonia	7	93	C3
7	Ireland	7	93	C3
8	Greece	4	96	C3
9	Spain	53	47	C1
10	France	82	18	C1
11	Croatia	1	99	C3
12	Italy	78	22	C1
13	Cyprus	3	97	C3
14	Latvia	1	99	C3
15	Lithuania	2	98	C3
16	Luxembourg	2	98	C3
17	Hungary	0	100	C3
18	Malta	4	96	C3
19	Netherlands	59	41	C1
20	Austria	7	93	C3
21	Poland	88	12	C1
22	Portugal	3	97	C3
23	Romania	47	53	C2
24	Slovenia	2	98	C3
25	Slovakia	1	99	C3
26	Finland	30	70	C2
27	Sweden	25	75	C2

Table 10 Outliers' detection results—OUTPUTS dataset

Approach number	Outliers detected
Approach 1	Germany, France, Netherlands, Romania, Belgium, Finland
Approach 2	Germany, France, Netherlands, Romania, Belgium, Finland, Bulgaria, Sweden
Approach 3	No outliers
Approach 4	Germany, France, Poland, Italy, Spain, Netherlands

Table 11 EU member states allocation—OUTPUTS dataset, without outliers

No	State	Degree C3 (%)	Degree C4 (%)	Allocation
1	Belgium	78	22	C3
2	Czechia	96	4	C3
3	Denmark	44	56	C4
4	Estonia	33	67	C4
5	Ireland	64	36	C3
6	Greece	57	43	C2
7	Croatia	1	99	C4
8	Cyprus	6	94	C4
9	Latvia	3	97	C4
10	Lithuania	3	97	C4
11	Luxembourg	4	96	C4
12	Hungary	33	67	C4
13	Malta	9	91	C4
14	Austria	95	5	C3
15	Portugal	11	89	C4
16	Slovenia	7	93	C4
17	Slovakia	4	96	C4

Lastly, the final cluster allocation will be named considering first the total level of emissions and the exports of waste (so the final outputs that are flowing out of the economy), and second, the waste treatment, with an accent on the volume of landfills, waste recovery quantities, and circular use rate (Table 12).

Therefore, the first cluster will be composed of highly productive countries with a high quantity of emissions, but an overall performant waste recovery. The second cluster, as mentioned briefly before, will be composed of the outlier countries with the biggest volumes of waste landfills. Looking at the average values, Greece was also included in this cluster, since more than 78% of the residuals are disposed of in

Table 12 Final cluster allocation—OUTPUTS dataset

Cluster	Name	Countries
Cluster 1	High emissions & performant waste recovery system	Germany, Spain, France, Italy, Netherlands, Poland
Cluster 2	High volume of waste landfills	Bulgaria, Romania, Finland, Sweden, Greece
Cluster 3	Medium emissions & performant waste recovery system	Belgium, Czechia, Ireland, Austria
Cluster 4	Medium or low emissions & medium-performant waste recovery system	Denmark, Estonia, Croatia, Cyprus, Latvia, Lithuania, Luxembourg, Hungary, Malta, Portugal, Slovenia, Slovakia

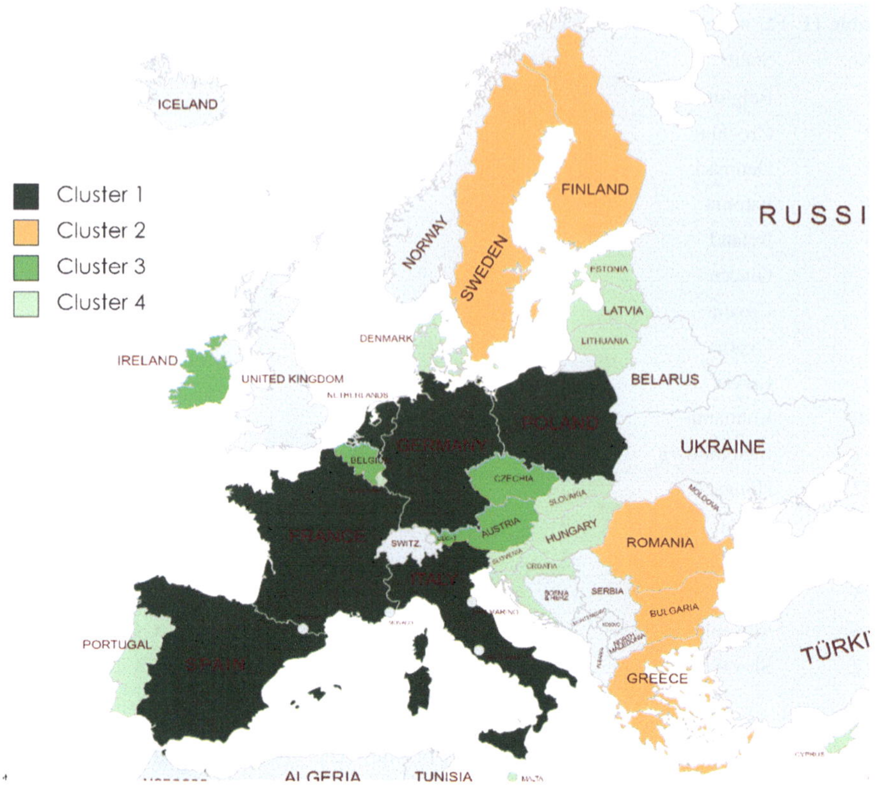

Fig. 3 Final cluster allocation—OUTPUTS dataset

landfills. Going further, the third cluster will comprise the countries with medium emissions, but a performant waste recovery, with large volumes being used for recycling purposes. Lastly, all the other countries have been grouped together, having the lowest levels of emissions in the European Union and an overall medium-performing waste recovery system.

Consequently, the mapping of the results can be seen in Fig. 3. We can now state that the second research objective has also been met, with clear profiling of the regional distribution in terms of outputs management, and circularity potential.

4.3 Aggregation of Datasets and Romanian Profile Considerations

Lastly, both findings have been aggregated in Table 13, with the characterization of each cluster and member state. Expectedly, the high-scale producers and resource owners will be the same states that will register high emissions, and most of the time

Table 13 Countries' profiles based on INPUTS and OUTPUTS

No	State	Resource owner and producer	Emissions & waste recovery system
1	Belgium	Importer	Medium & performant
2	Bulgaria	Low-scale	High volume of waste landfills
3	Czechia	Medium-scale	Medium & performant
4	Denmark	Medium-scale	Medium or low & medium-performant
5	Germany	High-scale	High & performant
6	Estonia	Low-scale	Medium or low & medium-performant
7	Ireland	Low-scale	Medium & performant
8	Greece	Low-scale	High volume of waste landfills
9	Spain	High-scale	High & performant
10	France	High-scale	High & performant
11	Croatia	Low-scale	Medium or low & medium-performant
12	Italy	High-scale	High & performant
13	Cyprus	Low-scale	Medium or low & medium-performant
14	Latvia	Low-scale	Medium or low & medium-performant
15	Lithuania	Low-scale	Medium or low & medium-performant
16	Luxembourg	Low-scale	Medium or low es & medium-performant
17	Hungary	Medium-scale	Medium or low & medium-performant
18	Malta	Low-scale	Medium or low & medium-performant
19	Netherlands	Importer	High & performant
20	Austria	Medium-scale	Medium & performant
21	Poland	High-scale	High & performant
22	Portugal	Medium-scale	Medium or low & medium-performant
23	Romania	High-scale	High volume of waste landfills
24	Slovenia	Low-scale	Medium or low & medium-performant

(continued)

Table 13 (continued)

No	State	Resource owner and producer	Emissions & waste recovery system
25	Slovakia	Low-scale	Medium or low & medium-performant
26	Finland	Medium-scale	High volume of waste landfills
27	Sweden	Medium-scale	High volume of waste landfills

will construct a performing waste recovery system (Germany, Spain, or France being some of the examples).

On another note, Romania has been classified as a high-scale resource owner, being the only country from the Eastern European region in that allocation. This is explained mainly by the large volumes of domestic extractions (ranked 4th in the EU) and the correspondent levels of material processing and accumulation. However, in contrast to that, based on the OUTPUTS data, Romania will be characterized by a very large distribution of waste landfills, which will ultimately translate into a poor waste management system. The Circular Use Rate confirms it, with Romania registering the smallest value in the EU, 1.4%, along with one of the smallest volumes of waste recovery. Therefore, we can conclude that there is a significant potential for circularity and sustainability practices, but they need to be backed up by broad legislative initiatives and strategies.

5 Conclusion

The circular economy has undoubtedly become a priority for the European Union, being included in the latest continental strategies such as The European Green Deal. Therefore, as a way of monitoring its potential, Sankey's diagram of material flows has been designed, to show the overall material journey inside an economy. Based on that, we have constructed the two research objectives presented in this paper, useful in the overall analysis of production and consumption dynamics on the continental stage. Each of the objectives had a correspondent dataset, on which we applied a fuzzy clustering algorithm (UFPC). As a result, when referring to the first objective, four clusters of states have been aggregated, from high-scale resource owners and producers to medium-scale, low-scale, or highly dependent on imports. On the other hand, when studying the outputs distribution and the circular economy potential (second research objective), we have also obtained four clusters, aggregated based on the total volume of emissions and the performance of the waste recovery system.

Therefore, we can conclude that the model helps in profiling both the individual member states and the regional distribution, based on the level of production, consumption, or the general management of materials and residuals. Moreover, by splitting the diagram into two key processes we have identified the centers that need further help in understanding the benefits and ways of working of the circular economy (for example the states that have been classified as having a medium performing waste recovery system or the ones that are described by the very high distribution of waste landfills, Romania included). At the same time, we can pinpoint the main regions that have already employed certain practices of circularity and that can provide their support for the developing countries, either by accepting higher volumes of waste imports or by sharing general know-how and best practices. Lastly, the model could help in providing a benchmark for the future and setting achievable goals regarding the national status quo in terms of inputs and outputs.

As a final consideration, the circular economy can be a feasible concept in the sustainability race of humanity, being seen not as a full transition, but mostly as an instrument of reducing environmental threats and responding to dangerous consumerist trends such as material accumulation or waste landfills.

Acknowledgements This paper was co-financed by The Bucharest University of Economic Studies during the Ph.D. program.

References

Corvellec, H., Stowell, A.F., Johansson, N.: Critiques of the circular economy. J. Ind. Ecol. **26**(2), 421–432 (2022)

European Commission: A European Green Deal (2020). Retrieved from: https://commission.europa.eu/strategy-and-policy/priorities-2019-2024/european-green-deal_en. Accessed: 10 Mar. 2023

European Commission: The Circular Economy Action Plan (2020). https://environment.ec.europa.eu/strategy/circular-economy-action-plan_en. Accessed: 10 Mar. 2023

European Parliament: Circular economy: definition, importance, and benefits (2023). https://www.europarl.europa.eu/news/en/headlines/economy/20151201STO05603/circular-economy-definition-importance-and-benefits. Accessed: 10 Mar. 2023

Eurostat: Material flow diagram tool. https://ec.europa.eu/eurostat/cache/sankey/circular_economy/sankey.html. Accessed: 11 Mar. 2023

Geissdoerfer, M., Savaget, P., Bocken, N.M., Hultink, E.J.: The circular economy—A new sustainability paradigm? J. Clean. Prod. **143**, 757–768 (2017)

Hofmann, F.: Circular business models: business approach as driver or obstructer of sustainability transitions? J. Clean. Prod. **224**, 361–374 (2019)

Kennedy, A.B.W., Sankey, H.R.: The thermal efficiency of steam engines. In: Minutes of the Proceedings of the Institution of Civil Engineers, pp. 278–312 (1898)

Kirchherr, J., Reike, D., Hekkert, M.: Conceptualizing the circular economy: an analysis of 114 definitions. Resour. Conserv. Recycl. **127**, 221–232 (2017)

Mazur-Wierzbicka, E.: Circular economy: advancement of European Union countries. Environ. Sci. Eur. **33**, 1–15 (2021)

Moraga, G., Huysveld, S., Mathieux, F., Blengini, G.A., Alaerts, L., Van Acker, K., Dewulf, J.: Circular economy indicators: what do they measure? Resour. Conserv. Recycl. **146**, 452–461 (2019)
Morseletto, P.: Targets for a circular economy. Resour. Conserv. Recycl. **153**, 104553 (2020)
Murray, A., Skene, K., Haynes, K.: The circular economy: an interdisciplinary exploration of the concept and application in a global context. J. Bus. Ethics **140**, 369–380 (2017)
Organisation for Economic Co-operation and Development: Business Models for the Circular Economy. Opportunities and Challenges for Policy (2019). https://www.oecd-ilibrary.org/sites/g2g9dd62en/index.html?itemId=/content/publication/g2g9dd62-en/. Accessed: 10 Mar. 2023
Pearce, D.W., Turner, R.K.: Economics of Natural Resources and the Environment. Johns Hopkins University Press, Baltimore (1990)
Potting, J., Hekkert, M.P., Worrell, E., Hanemaaijer, A.: Circular economy: measuring innovation in the product chain. Planbureau voor de Leefomgeving, vol. 2544 (2017)
Prieto-Sandoval, V., Jaca, C., Ormazabal, M.: Towards a consensus on the circular economy. J. Clean. Prod. **179**, 605–615 (2018)
Robaina, M., Villar, J., Pereira, E.T.: The determinants for a circular economy in Europe. Environ. Sci. Pollut. Res. **27**, 12566–12578 (2020)
Rodríguez-Antón, J.M., Rubio-Andrada, L., Celemín-Pedroche, M.S., Ruíz-Peñalver, S.M.: From the circular economy to the sustainable development goals in the European Union: An empirical comparison. Int. Environ. Agreem.: Politics Law Econ. **22**(1), 67–95 (2022)
Stahel, W.R.: The circular economy. Nature **531**(7595), 435–438 (2016)
UN General Assembly: Transforming our world: the 2030 Agenda for Sustainable Development (2015). https://www.refworld.org/docid/57b6e3e44.html. Accessed: 10 Mar. 2023
Wu, X., Wu, B., Sun, J., Fu, H.: Unsupervised possibilistic fuzzy clustering. J. Inf. Comput. Sci. **7**(5), 1075–1080 (2010)
Yang, M.S., Wu, K.L.: Unsupervised possibilistic clustering. Pattern Recogn. **39**(1), 5–21 (2006)
Yuan, Z., Bi, J., Moriguichi, Y.: The circular economy: a new development strategy in China. J. Ind. Ecol. **10**(1–2), 4–8 (2016)

Navigating the Impact of Artificial Intelligence on International Financial Reporting Standards (IFRS)

Nermin Sharbek

Abstract Due to the continually evolving nature of International Financial Reporting Standards (IFRS) and the technology itself, this paper analyzes the ability for companies to leverage artificial intelligence (AI) into complying with the international financial reporting standards (IFRS). The purpose of this research is to demonstrate the gap in existing literature pertaining to the intersection of financial reporting and artificial intelligence, while also showcasing the innovative methods that companies have developed to facilitate accurate and efficient preparation of financial reporting disclosures. This paper seeks to answer the questions related to (1) the considerations for incorporating AI into a company's financial reporting disclosures and (2) Which International Financial Reporting Standards may indeed be facilitated by artificial intelligence. The results of the study indicate that many companies are still in the early stages of integrating artificial intelligence into IFRS, and ongoing development is necessary to ensure the effectiveness and efficiency of financial reporting disclosures. The tools utilized by companies are relatively new and continue to evolve, propelled by the widespread recognition of AI applications across the globe. Nevertheless, uncertainty remains regarding the reliability and appropriateness of using artificial intelligence, considering the privacy and security challenges that accounting companies face.

Keywords International Financial Reporting Standards · Artificial Intelligence · Accounting · Financial Reporting · IFRS · Machine Learning

1 Introduction

Financial reporting provides the foundation for accounting, and as technology becomes increasingly important in accounting (Smith 2018), financial reporting will likely adapt to keep pace with the changes. The intersection of financial reporting

N. Sharbek (✉)
Bucharest Academy of Economic Studies, Bucharest, Romania
e-mail: falsesharbeknermin@gmail.com

M. Busu (ed.), *Rethinking Business for Sustainable Leadership in a VUCA World*, Springer Proceedings in Business and Economics,
https://doi.org/10.1007/978-3-031-50208-8_18

and artificial intelligence is indeed becoming more prevalent and significant, as Artificial Intelligence (AI) technology has the potential to improve the accuracy and efficiency of financial reporting processes. Considering the level of technological change currently being experienced in economics and finance, a human's imagination can only go so far in imagining what the future will be like as AI continues to establish a foothold in every profession, from the arts and crafts to engineering. No industry has been spared by artificial intelligence according to the AI Forum (EY 2022).

The purpose of this article is to investigate the artificial intelligence tools employed by businesses in light of the recent innovations made to assist with the compliance of international financial reporting standards. The financial sector is eager to utilize novel tech like artificial intelligence and machine learning to its maximum benefit. Accounting is associated with repetitive and burdensome duties by its very nature, and accountants are seeking innovative solutions to this problem, whether through digitalization or artificial intelligence.

AI has the ability to automate mundane processes, improve financial reporting by detecting abnormalities, and aiding in the interpretation and execution of IFRS standards. The use of AI into accounting and financial reporting procedures can increase the quality and reliability of financial reporting while decreasing the likelihood of errors and fraud. As AI technology continues to develop, we may anticipate even more inventive applications of AI in accounting and financial reporting, which will undoubtedly have a substantial impact on the future of IFRS.

This research aims to highlight the gap in current literature regarding the intersection of financial reporting and artificial intelligence. Additionally, it aims to showcase innovative methods that companies have developed to improve the accuracy and efficiency of financial reporting disclosures. Companies are utilizing relatively new tools that continue to evolve with the growing recognition of AI applications. However, there is still uncertainty about the reliability and appropriateness of using AI, particularly given the privacy and security challenges faced by accounting firms.

2 Literature Review

2.1 Financial Reporting

Financial reporting and disclosures are a significant source for presenting a company's performance. Companies were not always required to submit and declare their financial reports in accordance with a set of standardized accounting body requirements or even issue financial statements (RamMohan 2012) however, the increased use of financial information impacting the decisions of current stakeholders has led to the establishment of stricter regulations and the adoption of standardized accounting practices worldwide. The International Financial Reporting Standards (IFRS) outline the guidelines companies must adhere to when reporting their financial results (IASB). Due to the accounting breakthrough that occurred in 2001, IFRS became the major contributor of processes adopted by multinational corporations. The primary goal of financial reporting is to provide investors, creditors, and other external stakeholders with the information they need to make the best possible financial decisions (Lekovic and Arsenovic 2013). Financial reporting is a tool for fostering confidence and transparency in the financial situation and performance of a firm. The objective of the IFRS was to produce a set of high-quality accounting rules that could be accepted and enforced globally (Wieczynska 2016).

2.2 Artificial Intelligence

Artificial Intelligence is a revolutionized technology that is changing the financial sector (Mahalakshm et al. 2021). The field of computer science known as "artificial intelligence" (AI) seeks to create machines that are capable of cognitive activities often reserved for humans, such as voice recognition, decision making, and natural language comprehension. "The science and engineering of constructing intelligent machines," as John McCarthy (2007) famously put it, is what researchers imply when they talk about artificial intelligence (Russell and Norvig 2010).

This paper's primary objective was to provide insights on leveraging artificial intelligence (AI) for efficient and effective disclosures in financial reporting. The article aims to address two essential questions: (1) What are the considerations for incorporating AI into a company's financial reporting disclosures? (2) Which International Financial Reporting Standards may indeed be facilitated by artificial intelligence? This paper's topic was motivated by the emergence of innovative companies that have introduced AI-powered platforms for preparing IFRS 16 disclosures, leading to the question of whether the developments in IFRS standards will involve the use of AI to achieve the IASB's goal of continuously improving financial reporting standards. These objectives seek to meet the demands and interests of stakeholders by eliminating information gaps, fostering transparency, and, as a result, improving the reliability of financial statements.

The reason why the topic of Financial Reporting and Artificial Intelligence has not yet been extensively researched is because of the constant modifications to IFRS from year to year and the rapid evolution of the technology itself. These factors make it challenging for researchers to keep up with the latest developments and to study the impact of AI on financial reporting comprehensively. While there has been a significant amount of research on the impact of artificial intelligence (AI) on accounting, much of the existing literature has focused on the efficiency and decision-making benefits of AI, as well as the limitations of its adoption in service delivery due to privacy concerns (Noordin et al. 2022; Ashoka et al. 2019; Omoteso 2012; Hasan 2022) However, these same studies have also acknowledged a gap in the literature regarding the specific ways in which AI can both enhance and pose challenges to financial reporting. Specifically, there is a need to examine how cutting-edge technologies such as AI can help to improve financial reporting by mitigating bias and errors. Therefore, the present paper seeks to address this gap by exploring the methods in which AI can improve financial reporting and ensure compliance with required reporting standards. Additionally, other researchers have called for further investigation by professionals in the domain of financial reporting, lacking a conceptual framework to guide researchers into making a comprehensive assessment on the impact of AI in financial reporting (Saxena 2022). The importance of addressing this literature gap is evident. Financial reporting is a crucial element of corporate governance and any errors or biases in the reporting system can have severe repercussions if the Enron's scandal has taught us anything. In addition, the emergence of artificial intelligence presents significant concerns regarding the ethical and regulatory consequences of this technology, particularly in financial reporting. It is crucial to address these concerns for ensuring the integrity of financial reporting and for maintaining stakeholder confidence in the financial system.

However, there have been a few brief mentions of financial reporting in the context of implementing AI in accounting. For example, (Simian and Dutescu 2022) discussed the use of Cassa Digital Software platforms for various accounting activities, including reporting. (Leitner-Hanetseder and Lehner 2022) proposed a Framework for AI-powered Information and Big Data to determine the value of data in financial reporting. (Han et al. 2023) explored the use of Blockchain for regulatory complexity, while (Glancy and Yadav 2011) wrote about a computational fraud detection model (CFDM) for detecting fraud in financial reporting. Furthermore, (Pau and Gianotti 1990) introduced the concept of using Artificial Intelligence (AI) to achieve consistency within routine-based operations that involve distributed sources and require adherence to a common policy. They claimed that substituting paperwork, information consolidation, and onerous control procedures with AI would be preferable, but no developments were made in light of the concept itself.

This paper is expected to make a significant contribution by igniting a discussion as to whether artificial intelligence will be incorporated into future IFRS developments. Moreover, it seeks to explore the extent to which companies can utilize artificial intelligence to rapidly integrate IFRS. This is a crucial goal towards achieving harmonization in accounting to ensure that various users across the globe can understand the financial accounting presented in annual financial reports.

A bibliometric analysis was conducted by (Ayad and El Mezouari 2022) related to the use of artificial intelligence in accounting subfields, the study found that artificial intelligence in accounting subfields has increased significantly in recent years with a peak in 2019. The bibliometric analysis was done on articles published between 1990 and 2022, respectively until September 2022, when the article was received by the publishing body. The results related to financial reporting and artificial intelligence indicated that more research had to be done as to whether the two work together.

However, these results make sense as the emergence of innovative technology specific to the preparation of IFRS standards has been recent. Having platforms like ChatGPT accelerated the trend of AI usage towards the end of 2022 has spiked the interest of big tech companies to implement AI into their services, such as Google or Bing search engines adopting AI-powered chatbots.

The launch of OpenAI, the creator of ChatGPT, has sparked a rush for generative AI technology, which can produce new text, images, video, code, and music without human input. However, the accuracy of these outputs is dependent on the data used to train the system, which can lead to biases, errors, or falsities. ChatGPT has caused disruption in various fields, with concerns about cheating in education, the quality of AI-generated content in publishing, and the use of the tool to write news stories in media outlets.

3 Methodology

This study identified a gap in the literature regarding the methods through which artificial intelligence can be applied to financial reporting. To identify and address the gap, it was considered most appropriate to conduct a systematic review of sources, such as developer websites and research on accounting technological developments. The credibility and universal applicability of the study are supported by a systematic review because it is based on an extensive and representative sampling of the existing literature (O'Donnell 2011), which is deemed appropriate in the emergence of a topic discussed so vaguely in the existing literature. The identified studies were analyzed thematically, specifically on individual International Financial Accounting Standards (IFRS) and their applicability in light of technological developments. The findings were then synthesized to determine the potential uses of AI in financial reporting. The websites of prominent AI software developers and accounting-related research were indeed pertinent to the research questions posed and provided distinct insights into the approaches companies employ to build AI solutions for financial reporting. Using this methodology, common practices, and trends in the creation of artificial intelligence (AI) solutions for financial reporting were found, which might influence future research in this field and provide practical guidance for companies striving to establish and deploy such solutions.

4 What Are the Considerations for Incorporating AI into a Company's Financial Reporting Disclosures?

4.1 Developments of Technology and Financial Reporting in Recent years

The International Financial Reporting Standards (IFRS) are constantly evolving to meet the needs of various stakeholders and ensure transparency in financial reporting. With this primary goal in mind, artificial intelligence can play a major role in meeting these requirements without creating the additional burden of transitioning into new reporting standards. To comply with the increased level of detail required by IFRS, organizations need information technology systems that can track a wider range of data fields and capture more granular information in financial reports (Emekaponuzo et al. 2017).

Many regulations and regional or international laws must be observed in accounting. AI-enabled technologies facilitate audits and assure compliance by comparing documents to standards and legislations, flagging problematic ones (Marr 2023). Each year, fraud costs businesses billions of dollars, with financial services firms incurring $4.23 for every dollar of fraud (LexisNexis Risk Solutions 2022). Machine learning algorithms can rapidly filter through vast volumes of data to identify possible fraud concerns or suspicious behavior that people may have overlooked and flag it for further investigation.

In 2016, the Financial Reporting Council (FRC) conducted a study on potential future technologies that could affect corporate reporting. The study found that artificial intelligence could address various challenges such as the efficiency of recording and aggregating transactions across multiple entities and transforming the data into external communication. Furthermore, AI might make it easier to digest data and draw useful conclusions. There are often several manual procedures involved in creating corporate reports that restrict the potential for AI, however some use cases include the production and distribution of reporting data. In 2019, another study took place by the FRC (Financial Reporting Council) on Blockchain and corporate reporting, highlighting the understanding of the past and current state of data which can improve timeliness and cost efficiency, however implying that blockchain technology may not substantially increase trust between organizations. Businesses must ensure that their systems are reliable and provide appropriate forms of evidence for internal and external assurance standards, and all stakeholders involved in preparing, verifying, and using financial reports should collaborate to educate and find solutions related to assurance concerns (CFR 2018).

4.2 Judgement

Before AI can be used to legitimate judgment, it must earn the confidence of businesses, regulators, and other stakeholders, which may require each stakeholder to first increase their knowledge and experience. Financial statements, like any other source of information, should be deemed credible. Automation should be used by accountants since it decreases the chance of mistake and malicious attempts to portray accounts unfairly. Since O'Leary first proposed the idea of artificial intelligence in Accounting, Finance, and Management in 1995, this remark has sparked dispute throughout the years. In June of 2018, the CFA Institute and the IFRS Foundation co-hosted a panel debate that questioned whether conventional measures are still applicable in light of fast technology improvements. The argument emphasized certain beneficial and negative elements of artificial intelligence in preparing financial statements, such as the impossibility of humans to compete with algorithmic reflexes, which enables trading to be cheap and efficient, but that system faults make flash crashes far more probable. Another element was whether more data correlates to better data, and if human biases are embedded into the algorithms that filter it (Fuller 2018), which contradicts the concept that financial statements are prepared without prejudice.

Notwithstanding possible obstacles, many feel financial accounting is the best setting for the use of artificial intelligence. Individuals working in the area already possess the advanced mathematical skills and commercial acumen necessary to gain the advantages of artificial intelligence. Nevertheless, technology in finance is a well-known idea (Boyall 2021). To aid in decision-making, accountants need high-quality financial data and have used technology to aid in this process for decades. Thus, individuals are likely to grasp instinctively when and how to depend on such technology and when to rely on their own senses. In addition, (Lehner et al. 2022) highlight the necessity of a collaboration between humans and machines in accounting to avoid unethical decisions that could arise from AI having sole decision-making power. The authors call for additional research on the nature of this transformation in accounting, including critical voices and normative thinking. The article concludes by stating that AI, as an accounting apparatus with numerous embedded instruments of valuation, will inevitably shape both values and valuers.

4.3 Ethics

The conversation of moral principles, values and artificial intelligence is a complex one, it is a philosophical question of right and wrong, in an ongoing debate, as ethics is a multifaceted field that deals with many different opinions (Baker 2008).

AI systems are based on complex algorithms that derive information from data and self-evolve as more data is being fed (McGregor and Carpenter 2020). The use of complex algorithms to process and analyze financial reporting data may result in biased or inaccurate outcomes, which can have serious implications for companies and investors. It is important to ensure that AI systems are designed and implemented in a way that prioritizes the fundamental accounting principles related to ethics such as transparency, accuracy, and fairness.

Another important consideration is the protection of privacy and data security in the use of AI for accounting and financial reporting. Companies must take steps to ensure that sensitive financial information is safeguarded against theft or misuse, while also being transparent about how this information is being used and processed by AI systems. According to (Saxena 2022) the current accounting standards are not designed to address the ethical concerns related to lack of transparency, technological and economical inequity between companies preparing their financial statements with or without the use of AI. However, it can be argued that both IFRS and GAAP both have ethical considerations related to transparency and equity embedded in these standards and if anything, AI can assist in improving accuracy and objectivity in preparing financial statements in accordance with the necessary standards by reducing human bias and errors.

To initiate the implementation of artificial intelligence in financial reporting, companies must explore and experiment with AI, even if they are hesitant to allocate significant funds to AI solutions at first. However, it is important that financial statement preparers do not neglect ethical considerations in their rush to expedite the reporting process through AI solutions. Compliance with AI ethics policies is crucial in maintaining the right principles, and in case of uncertainty, preparers of financial statements should seek external opinions from reliable regulatory bodies to ensure adherence to ethical regulations.

5 Which International Financial Reporting Standards May Indeed Be Facilitated by Artificial Intelligence?

5.1 IFRS 16 Leases

> IFRS 16 introduces a single lessee accounting model and requires a lessee to recognize assets and liabilities for all leases with a term of more than 12 months unless the underlying asset is of low value. A lessee is required to recognize a right-of-use asset representing its right to use the underlying leased asset and a lease liability representing its obligation to make lease payments.

In January 2021, the new lease accounting standard IFRS 16/ASC 842 went into effect for private companies, mandating that everything from real estate leases to automobile and machinery leases be reviewed holistically, classified correctly, and presented on the balance sheet accordingly. The majority of businesses who struggle with a large number of leases contemplated using digital solutions to aid with the heavy lifting (MRI Software 2020). To accelerate the adoption of IFRS 16 by 2019 and enable efficient lease management within their organizations, certain companies needed to collaborate with tech companies, such as Leverton, that could aid them in swiftly and precisely extracting data from leasing contracts. The need for such a partnership varied depending on the nature of the business and the degree to which they utilized and depended on IFRS 16. Leverton is a platform powered by AI technology which proved to be immensely useful in evaluating the population, importing data, collecting vital information, verifying, and extracting relevant data (McGregor Boyall 2021). Another ground-breaking application that could be used for IFRS 16 application is Trullion, which does not only extract information from lengthy contracts, but also incorporates a reporting tool that enables the comparison of relevant accounting periods with the provisions of the lease agreement. Trullion employs continuous monitoring to document any alterations made to the contract. The edit history can be examined as well.

Leverton employs Optical Character Recognition (OCR) technology as shown in Step 1 of the diagram above, which aids in identifying the contract population by recognizing characters, irrespective of quality. Leverton then compares the identified contracts to the current list and identifies any missing or problematic contracts.

Problematic contracts meet certain criteria, such as containing outdated information or paragraphs no longer relevant to current laws or company policies. Step 2 of the process involves the collection of all pertinent contract details, such as landlord, start dates, and charges, which are easily accessible to the user and specifically linked to the corresponding section of the contract. Within Step 3, the Leverton platform is synced with the accounting system to verify the accuracy of the data, similar to an automated reconciliation and finally, in Step 4, the platform is continuously used for lease management and reporting, as the database is consolidated in a single location, simplifying navigation, and streamlining reporting processes. The data is presented in an organized and user-friendly manner on the platform.

Overall, it can be said that at the implementation stage, some aspects of IFRS 16 accounting require human judgement, particularly when classifying lease contracts according to their nature. To minimize the need for manual review and achieve accurate judgement for the IFRS 16 lease standard, some companies are turning to Reinforcement Learning (RL) (Choi 2021), a type of machine learning algorithm. RL is a powerful mathematical framework because it allows agents to have direct interactions with their environment. This type of autonomous learning is based on the agent's past experiences and relies on trial and error to improve its performance in the pursuit of maximization of the total reward. (Singh et al. 2022). In the context of IFRS 16, reinforcement learning can be used to train a computer program to make lease classification decisions based on a set of predetermined criteria.

5.2 IFRS 15

Accounting for revenue and cash flows resulting from contracts with customers is outlined in IFRS 15, which is made available to businesses as a reference for reporting purposes. In accordance with these principles, businesses should record sales in a manner that fairly reflects the delivery of products or services to customers and the consideration expected in return (International Accounting Standards Board).

According to PwC (2017), companies must ensure that revenue recognition aligns with the substance of the arrangement when looking at the new standards (ASC 606 and IFRS 15). The new Revenue Standards introduce complexity that can lead organizations to incorrectly apply them. Accounting firms can provide valuable advice to their clients to ensure accurate application of these standards. One solution may be to leverage artificial intelligence, and creators of AI-powered accounting platforms did not waste any time in developing tools that facilitate the adoption of these standards, such as Trullion or Kira.

To streamline the adoption of ASC 606 and IFRS 15, Revenue by Trullion provides a new approach by integrating customer relationship management (CRM), billing, and contract data into a unified platform for both internal and external stakeholders. With the help of AI, the workflow generates enterprise resource planning (ERP) entries, disclosure reports, and advanced reporting.

Kira, on the other hand, is an AI-powered contract analysis tool that may aid with revenue recognition by automating the extraction and assessment of essential phrases in contracts, which are significant to revenue recognition under ASC 606 and IFRS 15. Such essentials as performance obligations, the contract price, and the variable compensation may be included. Kira use machine learning algorithms to analyze contracts and extract useful information, allowing users to spot possible revenue recognition difficulties in a flash.

5.3 *IRFS 9*

The accounting regulations for financial instruments, introducing fresh guidelines for categorizing and assessing financial instruments, recognizing anticipated credit losses, and executing hedge accounting are presented in IFRS 9. The principal aim of the standard is to enhance the reporting and accounting practices of financial instruments in financial statements, to decorate more relevant information to investors and other financial statement users (IASB).

The insights shared by (Milojević and Redzepagic 2021) were helpful in evaluating the methods in which AI can be used for IFRS 9. For instance, the article shed light on how banks can leverage AI and ML to manage risks effectively by analyzing extensive data, enhancing credit scoring models, and detecting fraudulent activities. The predicted credit loss models mandated by IFRS 9 can benefit from the use of AI and ML because of their ability to filter through massive volumes of data in search of patterns and correlations that would be difficult for human beings to detect.

Moreover, (Martinelli et al. 2020) explored data mining techniques to determine the likelihood of default under IFRS 9, such as logistic regression, decision tree, random forest, and neural network. The study evaluated each technique's precision in predicting default probabilities by examining a dataset of small and medium-sized enterprises (SMEs) in Italy. The results of the study indicated that the neural network model was the most effective in forecasting the probability of default under IFRS 9 compared to the other techniques. Inspired by the structure and function of the human brain, neural networks, also known as artificial neural networks (ANNs) or simulated neural networks (SNNs), which are a type of machine learning algorithm that underpins deep learning techniques (IBM Cloud Education 2020).

6 Results and Discussions

The advantages of AI are substantial; nonetheless, not everyone is convinced of its usefulness just yet. At EmTech Digital conference held in association to EY, a survey was conducted in 2019, where 49% the respondents stated they did not trust artificial intelligence-generated financial data as much as data generated by humans and 30% stated that data privacy, security, and improper management are major causes for worry. Ethical and legal issues are certain to arise in relation to some AI benefits, such as the ability to get insights from previously unstructured data (such as emails written by workers). However, AI relies heavily on massive amounts of data to function properly, therefore smaller businesses would do well to proceed cautiously when considering whether or not to implement AI. Adopting AI is a long and expensive process that should not be attempted without first doing a thorough risk analysis.

Considering IFRS 16, IFRS 15, and even IFRS 17, considering its effective date in 2023, new regulations present difficulties for which firms must anticipate. Yet, meeting the requirements of the new standards might not need sacrificing consistency or precision. When businesses integrate solutions that make it easier for them to comply with financial regulations in a way that is accurate, secure, and responsive, everyone from the finance team to the CEO and the shareholders can rest tightly knowing that the financial information being released publicly is of the highest quality possible, however not without addressing ethical matters first.

There have been ongoing efforts to standardize IFRS globally due to its reliability. However, incorporating artificial intelligence in an industry like accounting can be challenging without established regulations and ethical guidelines. The objective of preparing financial statements under IFRS is to eliminate subjectivity, which cannot be guaranteed by artificial intelligence. This is because AI systems rely on data that reflects the biases and limitations of the humans who create and curate that data. AI systems learn from the patterns in the data they are fed, and if the data is biased or limited, the resulting AI system will also be biased or limited.

The potential for biases arises in the interpretation and application of IFRS standards by companies and their accountants, despite the fact that they are designed to present financial statements in a true and fair manner. Certain IFRS standards, such as IAS 37 Provisions, Contingent Liabilities and Contingent Assets, which deal with the recognition and measurement of provisions and contingent liabilities, may be more prone to biases due to their significant judgment and estimation requirements. The same can be said for IFRS 13, fair value measurement, and IFRS 2, share-based payment. Although artificial intelligence may support the consistent implementation and interpretation of IFRS standards, the complexity of certain standards and limited data available may lead to inaccurate interpretations. The use of artificial intelligence in certain IFRS standards can result in more informed decisions, such as the example of IAS 37, which may utilize historical legal cases to generate a probability-weighted estimate.

It is evident how blending human and AI intelligences would provide the most impressive outcome thus far. To ensure compliance, quality, and consistency when preparing IFRS disclosures in an accurate and fair manner, it is best to combine the expertise and judgement of a trained auditor with the precision offered by technology like Artificial Intelligence.

7 Conclusion

This paper discusses the impact of artificial intelligence (AI) on three main accounting standards: IFRS 16 Leases, IFRS 15 Revenue, and IFRS 9 Financial Instruments and explore the potential impact for standards requiring judgement. For IFRS 16, the adoption of the standard has led to the use of digital solutions to aid with the heavy lifting of reviewing and classifying leases, and AI-powered platforms such as Leverton and Trullion have been developed to extract data from leasing contracts and streamline lease management and reporting processes. For IFRS 15, the complexity of the standard has led to the development of AI-powered platforms such as Trullion and Kira to facilitate the adoption of the standard by automating the extraction and assessment of essential phrases in contracts significant to revenue recognition. Finally, for IFRS 9, banks can leverage AI and machine learning (ML) to classify financial instruments and recognize anticipated credit losses more accurately.

The paper concludes that AI can play a significant role in addressing stakeholders' expectations and manifesting transparency in financial reporting. Nevertheless, the elephant in the room must be addressed: artificial intelligence's present lack of reliability. Data is abundant, and systems for artificial intelligence are being offered more and more data. But, in a profession such as accounting, any external input must be deemed trustworthy and ethical in all aspects and forms, out of respect for clients and the regulations under which financial statements are prepared. However, the input of human intelligence is not indispensable, especially in areas for which IFRS requires judgement. For this reason, the most optimal outcomes may be accomplished through a combination of human intelligence and artificial intelligence.

The limitations of the study include the timing and scope of the paper's composition, as a larger study with investigative components on more IFRS standards must be done to gain a greater understanding of the extent to which AI can be employed in preparing IFRS disclosures.

This research can progress in a variety of directions, including the development of IFRS standards, the emergence of AI applications and/or software and their applicability in corporate reporting. More value might be given to this topic by conducting research on the benefits and challenges of the use of artificial intelligence in the preparation of IFRS-compliant financial statements.

Step 1 **Identification and data import**

- Identify the lease population for the transition project
- Leverton platform identifies problematic or missing contracts

Step 2 **Data collection**

- The platform extracts an appropriate IFRS 16 data set, which was co-developed between Leverton, technical accounting experts, and the client.
- The data set ensures that relevant IFRS 16 data points and those useful for ongoing lease management are accurately captured.

Step 3 **Data Extraction and Validation**

- Leverton uses AI and two layers of quality assurance to extract and validate the data accurately
- The accounting team performes quality reviews on high-risk and high-value assets to ensure an accurate, auditable data set.

Step 4 **Ongoing Management and Communication**

- Leverton is used to update lease changes, IFRS 16 data points and management needs
- The platform simplifies reporting and audit requirements, providing transparent, auditable data.

Fig. 1 Use case of Leverton within IFRS 16 (Source: Author's adaption of MRI Software (www.mrisoftware.com))

Acknowledgements This paper was co-financed by the Bucharest University of Economic Studies (ASE) during the PhD program.

References

Alles, M., Kogan, A.: Artificial intelligence in accounting and auditing: Volume, value, and future. 14(2), 1–14 (2017). https://doi.org/10.2308/jeta-52050

Ashoka, M.L., Abhishek, N., Divyashree, M.S.: Emerging Trends in Accounting: An Analysis of Impact of Robotics in Accounting, Reporting and Auditing of Business and Financial Information. International Journal of Business Analytics and Intelligence **7**(2), 28–34 (2019)

Ayad, M., El Mezouari, S.: Research on the implication of artificial intelligence in accounting subfields: current research trends from bibliometric analysis, and research directions. Int. J. Accounting, Finance, Auditing, Management and Economics, 3(5), 503–522 (2022)

Baker: Ethics and Social Justice: Fundamental concepts (2008)

Briink It's time to supercharge sustainable finance reporting with AI (2022). Retrieved February 26, 2023, from https://www.briink.com/post/supercharge-sustainable-finance-reporting-with-ai

CFR: Blockchain and the future of corporate reporting How does it measure up? (2018) https://www.frc.org.uk/getattachment/58866565-ab3b-44d3-93e1-1ef7158968d5/Blockchain-and-the-future-of-corporate-reporting-how-does-it-measure-up-(June-2018).pdf

Choi, B.: (2021, 1 5). Enhanced Audit Techniques Empowered by the Reinforcement Learning Pertaining to IFRS 16 Lease. Machine Learning. Retrieved from https://arxiv.org/abs/2101.05633

Emekaponuzo, D. E., Jeremiah, O. O., & Alfred, E. J. (2017). Information Technology Implications of IFRS Implementation in Nigeria: Challenges to Auditors. Journal of Investment and Management, 6(1).
Enhanced Audit Techniques Empowered by the Reinforcement Learning Pertaining to IFRS 16 Lease. (n.d.).
Fuller, J. (2018, June). Transforming the impact of financial information: the role of technology. Are traditional metrics still relevant? London: IFRS.org. Retrieved from https://www.ifrs.org/news-and-events/news/2018/06/transforming-the-impact-of-financial-information-the-role-of-technology/
Glancy, F.H., Yadav, S.B.: A computational model for financial reporting fraud detection. Decis. Support. Syst. **50**(3), 595–601 (2011)
Han, H., Shiwakoti, R. K., Jarvis, R., Mordi, C., & Botchie, D. (2023). Accounting and auditing with blockchain technology and artificial Intelligence: A literature revie. International Journal of Accounting Information Systems.
Hasan, A.R.: Artificial Intelligence (AI) in Accounting & Auditing: A Literature Review. J. Bus. Manag. **10**(1), 440–465 (2022)
IBM Cloud Education. (2020). Neural Networks. Retrieved from https://www.ibm.com/uk-en/cloud/learn/neural-networks
Kernbichler, S. (2020). Artificial Intelligence in Financial Reporting. Business Administration - Master Thesis.
Lehner, O. M., Ittonen, K., Silvola, H., Strom, E., & Wuhrleitner, A. (2022). Artificial intelligence-based decision-making in accounting and auditing: ethical challenges and normative thinking. Accounting, Auditing & Accountability Journal, 35(9).
Leitner-Hanetseder, S., & M. Lehner, O. (2022). AI-powered information and Big Data: current regulations and ways forward in IFRS reporting.
Lekovic, M., & Arsenovic, S. (2013, January). The role and importance of quality financial reporting. Researchgate. Retrieved from https://www.researchgate.net/publication/283637993_The_role_and_importance_of_quality_financial_reporting
LexisNexis Risk Solutions. (2022, November). Annual LexisNexis Risk Solutions Report Finds Fraud Costs up to 22.4% from Pre-Pandemic Levels Across U.S. and Canadian Financial Services Firms. CISION PR Newswire. Retrieved from https://www.prnewswire.com/news-releases/annual-lexisnexis-risk-solutions-report-finds-fraud-costs-up-to-22-4-from-pre-pandemic-levels-across-us-and-canadian-financial-services-firms-301675103.html#:~:text=Every%20%241%20lost%20to%20fraud,2020%20to%20%243
Mahalakshm, V., Kulkarni, N., Kumar, P. K., Kumar, S. K., Sree, N. D., & Durga, S. (2021). The Role of implementing Artificial Intelligence and Machine Learning Technologies in the financial services Industry for creating Competitive intelligence. Materials Today: Proceedings.
Marr, B. (2023). Artificial Intelligence in Accounting and Finance. Bernard Marr & Co. Retrieved from Bernard Marr & Co.
Martinelli, F., Mercaldo, F., Raucci, D., & Santone, A. (2020). Predicting Probability of Default Under IFRS 9 Through Data Mining Techniques. Web, Artificial Intelligence and Network Applications. Retrieved from https://link.springer.com/chapter/https://doi.org/10.1007/978-3-030-44038-1_87
McGregor Boyall. (2021, November 15). McGregor Boyall. Why financial accounting is the ideal environment for artificial intelligence. London. Retrieved from https://www.mcgregor-boyall.com/our-thinking-library/2021/11/why-financial-accounting-is-the-ideal-environment-for-artificial-intelligence/281
McGregor, D., Carpenter, R.: Potential threats for the auditing profession, audit firms and audit processes inherent in using emerging technology. Business and Management Review (2020). https://doi.org/10.24052/BMR/V11NU02/ART-06
Milojević, N., & Redzepagic, S. (2021). Prospects of Artificial Intelligence and Machine Learning Application in Banking Risk Management. Journal of Central Banking Theory and Practice, 41–57.

MRI Software. (2020, February 19). Artificial intelligence comes together with accounting expertise to simplify IFRS 16 transition. Artificial intelligence comes together with accounting expertise to simplify IFRS 16 transition. Retrieved from https://mricontractintelligence.com/blog/artificial-intelligence-comes-together-with-accounting-expertise-to-simplify-ifrs-16-transition/
Nitchman, M. (2016). The value of structured data: A review of information contained in XBRL filings. ournal of Business and Economic Perspectives.
Noordin, N. A., Hussainey, K., & Hayek, A. F. (2022). The Use of Artificial Intelligence and Audit Quality: An Analysis from the Perspectives of External Auditors in the UAE. Journal of Risk and Financial Management. Retrieved from https://doi.org/10.3390/jrfm15080339
O'Donnell, J.F.: What is a systematic review? Evid Based Nurs (2011). https://doi.org/10.1136/ebn.2011.0049
Omoteso, K. (2012). The application of artificial intelligence in auditing: Looking back to the future. Expert Systems with Applications, 8490–8495.
Pau, L. F., & Gianotti, C. (1990). Applications of Artificial Intelligence in banking, financial services, and economics. Economic and Financial Knowledge-Based Processing, (pp. 22–46). doi:1007/978–3–642–76002–0_4
RamMohan, Y. R. (2012, March). Incorporating International Financial Reporting Standards into The United States Financial Reporting System: Timeline and Implications. International Business & Economics Research Journal, II (3).
Russell, S. J., & Norvig, P. (2010). Artificial intelligence: a modern approach. Prentice Hall Press.
Saxena, R.: A Conceptual Framework for Assessing the Application of Artificial Intelligence for Financial Reporting. Universal Journal of Accounting and Finance (2022). https://doi.org/10.13189/ujaf.2022.100502
Simian, M. M., & Dutescu, A. (2022). Artificial Intelligence solutions for Romanian accounting companies. Sciendo.
Singh, V., Chen, S.-S., Singhania, M., Nanavati, B., Kumar kar, A., & Gupta, A. (2022, 11 2). How are reinforcement learning and deep learning algorithms used for big data-based decision making in financial industries–A review and research agenda. International Journal of Information Management Data Insights, 2(2). doi:https://doi.org/10.1016/j.jjimei.2022.100094
Smith, S. S. (2018). Digitization and Financial Reporting – How Technology Innovation May Drive the Shift toward Continuous Accounting. Accounting and Finance Research, 7(3). Retrieved from https://www.sciedu.ca/journal/index.php/afr/article/view/13762/8626
Van der Heijden, H. (2022). Predicting industry sectors from financial statements: An illustration of machine learning in accounting research. The British Accounting Review, 54(5).
Wieczynska, M.: The "Big" Consequences of IFRS: How and When Does the Adoption of IFRS Benefit Global Accounting Firms? The Accounting Review, 91(4), 1257–1283 (2016). Retrieved 6 3, 2021, from https://aaajournals.org/doi/full/https://doi.org/10.2308/accr-51340

How Demographics Influence Travel Choices? An Investigation on Sustainable Accommodation Booking Behavior

Neringa Vilkaite-Vaitone and Ilona Skackauskiene

Abstract Considering sustainable travel accommodation as a mechanism to increase the sustainability of the whole tourism industry, this paper explores whether demographic characteristics, such as gender, age, education, occupation, and income, influence tourists' green purchasing behavior. We adopted a quantitative approach to test the hypotheses concerning the relationship between consumer demographics and measures of green purchasing behavior (intention, actual behavior, and willingness to pay a price premium). The study surveyed 747 tourists who have visited the Canary Islands at least once. The research results indicate that gender plays a significant role in sustainable travel accommodation booking intention. Younger travelers tend to have a stronger intention and willingness to pay a premium price compared to older ones. Additionally, higher income is related to lower booking intention, actual booking behavior, and willingness to pay a premium price. These findings suggest that demographic variables can explain travelers' sustainable accommodation booking behavior to some extent.

Keywords Consumer demographics · Green purchasing behavior · Sustainable traveler accommodation

1 Introduction

The fast pace of economic and technological growth, coupled with the increasing global population, has resulted in environmental degradation, drastic climate changes, depletion of the ozone layer, air and water pollution. These environmental challenges have gained significant media exposure, and people are now more aware of their effects on humanity. As a result, marketers, academics, and policymakers are working to encourage consumers to choose green products over conventional options.

N. Vilkaite-Vaitone · I. Skackauskiene (✉)
Vilnius Gediminas Technical University, Sauletekio Av. 11, 10223 Vilnius, Lithuania
e-mail: ilona.skackauskiene@vilniustech.lt

M. Busu (ed.), *Rethinking Business for Sustainable Leadership in a VUCA World*, Springer Proceedings in Business and Economics,
https://doi.org/10.1007/978-3-031-50208-8_19

Despite the colossal endeavor to make tourism more sustainable during the last decade, the sector still lacks a balance between the environmental, social, and economic realms (Lukman et al. 2022). Rajaonson and Tanguay (2022) pointed out that tourism infrastructures, facilities, activities, and transportation negatively impact the environment. Therefore, tourism has become a significant research context for studying sustainable practices. In this sense, previous studies have focused on understanding consumer behavior and identifying possible differences among travelers in touristic contexts. Sustainable travelers' behavior has been widely studied, but surveys conducted globally have shown contradictory results, particularly concerning demographic characteristics. For instance, Holmes et al. (2021) found that sustainable travelers in Canada are mainly well-educated individuals aged between 25 and 44 with high incomes. On the other hand, Rodriguez et al. (2020) found that older tourists have a more assertive pro-environmental or circular attitude among tourists visiting Gran Canaria. These findings indicate a lack of consensus on the demographic profile of travelers who book travel accommodations while exhibiting sustainable behavior. To the best of the authors' knowledge, no previous studies have examined the combination of gender, age, education, occupation, and income to test the differences in sustainable travel accommodation booking behavior. Understanding consumer behavior towards more sustainable accommodation booking patterns is crucial for increasing the sustainability of the entire tourism industry. It is important to be aware of the different demographic profiles of travelers, as their degree of sustainability can vary significantly. Thus, this study aims to assess the role of demographic characteristics in promoting sustainability among travelers.

The paper is structured as follows. The paper proceeds with a literature review presenting a summary of research findings on the green purchasing phenomenon. Following an overview of the methodology, results are presented. The paper closes up with a discussion.

2 Literature Review

The research examines the phenomenon of green purchasing based on the Theory of Planned Behavior (TPB) (Ajzen 1991; Kwon and Ahn 2021). According to this theory, attitude, subjective norms, and perceived behavioral control all contribute to forming behavioral intentions. These intentions, in turn, lead to actual purchasing behavior. Previous studies (Chekima et al. 2016) have extensively tested and confirmed the theory's reliability. Therefore, sufficient confidence exists in its robustness and ability to explain sustainable travel accommodation booking behavior. Our study focuses on purchase intention and actual behavior, and while we do not cover attitude, subjective norms, and perceived behavioral control as latent factors, we recognize their existence and significance. The current research concentrates on the elements of purchase intention and actual behavior, confirming that green purchase intention is a proxy for actual purchasing behavior.

According to Chekima et al. (2016), green purchase intention can be defined as a specific environmentally friendly behavior that individuals engage in to express their concern for the environment. This intention is expected to result in an actual purchase (Holmes et al. 2021), which involves participating in green practices and consuming green products (Kwon and Ahn 2021). Consumer willingness to pay more refers to the willingness to pay a premium price for the eco-friendly version of the product (Wei et al. 2018).

Based on a review of existing literature, it has been established that demographic characteristics play a crucial role in explaining green purchasing intentions, actual behaviors and willingness to pay a price premium. Shahsavar et al. (2020) have even identified demographics as the most influential factors in pro-environmental behavior. Gender, age, education, occupation, and income are all key factors that significantly impact green purchasing behavior (Shahsavar et al. 2020; Diamantopoulos et al. 2003; Hojnik et al. 2021; Jaiswal et al. 2021; Saleem et al. 2018; Tasci et al. 2022; Pekerşen and Canöz 2022). Our study takes these characteristics into account.

Social behaviors are largely influenced by gender (Martínez and Bosque 2018). Various studies have examined gender in different research contexts, but empirical findings have been inconclusive. Rodriguez et al. (2020) found that females tend to engage in circular behavior more frequently than men in a sample of tourists who visited Gran Canaria. On the other hand, Saleem et al. (2018) argue that male enthusiasts are more likely to participate in environmentally conscious purchasing behavior. Diamantopoulos et al. (2003) confirmed that most empirical research suggests that men have a deeper knowledge of green issues than women, but women tend to participate more frequently in various types of green behavior. Taking into account previous literature and findings, this study investigates whether significant differences exist in green purchase behavior between men and women.

According to Holmes et al. (2021), age significantly predicts green purchasing behavior. They found that younger consumers have more favorable attitudes toward purchasing environmentally friendly products. However, a comprehensive analysis by Diamantopoulos et al. (2003) found older individuals to display higher levels of green behavior. Shahsavar et al. (2020) discovered that both Boomers and Generation X are willing to pay a higher price for eco-friendly furniture than Generation Y. Considering the conflicting findings of past studies, it is expected that differences in green behavior between younger and older consumers exist.

Research studies have consistently shown that formal education significantly promotes environmental concern that leads to green purchasing (Diamantopoulos et al. 2003). Saleem et al. (2018) found that consumers with a graduate degree are more likely to be eco-conscious enthusiasts and have a higher intention of adopting green behaviors. Holmes et al. (2021) also observed similar patterns in the sustainable travel behavior of Canadian tourists, where university graduates and undergraduates constitute a major share of sustainable tourists. Therefore, it can be inferred that the level of education of tourists can influence their green purchase behavior.

An individual's occupation is an important factor that influences their pro-environmental behaviors (Saleem et al. 2018). It also determines an individual's social class (Kaur et al. 2022). Regarding green purchasing, research indicates that individuals in middle and senior management or those in intellectual, scientific, and artistic occupations are more likely to engage in green purchasing behavior (Paço and Raposo 2010). However, a study on Indian millennials conducted by Kaur et al. (2022) found that an individual's occupation does not significantly affect the relationship between green marketing strategies and green purchasing intentions. Despite these contrasting findings, tourists' green purchase behavior is expected to differ based on their occupation.

It has been observed that green products are generally more expensive than conventional ones (Meng et al. 2022; Mukherjee and Carvalho 2020). Therefore, income is an essential factor to consider when examining green purchase behavior. A study by Hojnik et al. (2021) in the Slovenian renewable energy market confirmed that higher-income consumers are more likely to pay for green energy. Similarly, Holmes et al. (2021) found that higher income strongly predicts tourists' sustainability in a sample of Canadian travelers. Hence, it can be inferred that the willingness to purchase green products varies significantly depending on the tourists' income level.

Based on previous studies, demographic characteristics are expected to play a crucial role in defining green consumers based on their purchasing behavior. Thus, we predict that:

Hypothesis 1. Green purchase intention differs according to demographic variables such as gender, age, education, occupation, and income.

Hypothesis 2. Actual green purchase differs according to the demographic variables.

Hypothesis 3. Willingness to pay a price premium for green products differs according to the demographic variables.

Following the identification of the hypotheses related to travelers' demographic profiles for sustainable travel accommodation booking behavior, we present a theoretical model (see Fig. 1). The illustration connects the five demographic profile elements with the three elements of booking behavior: intention, actual behavior, and willingness to pay more.

3 Methods and Materials

Sample and procedure. We collected data through a survey conducted in the Canary Islands, Spain—one of Europe's most popular tourist destinations. The sample of tourists was chosen for convenience. Unfortunately, there were no available records for random selection of individuals. Therefore, it was necessary to directly contact individuals who have previously visited at least one of the Canary Islands. We can confirm that the selected sample is representative of the Canary Islands tourists and can provide relevant information.

12.3 million international tourists visited the Canary Islands in 2022 (Statista 2010), who made up the study population. The recruiting sample consisted of 747 individuals from 38 different countries. To take part in the survey, respondents were

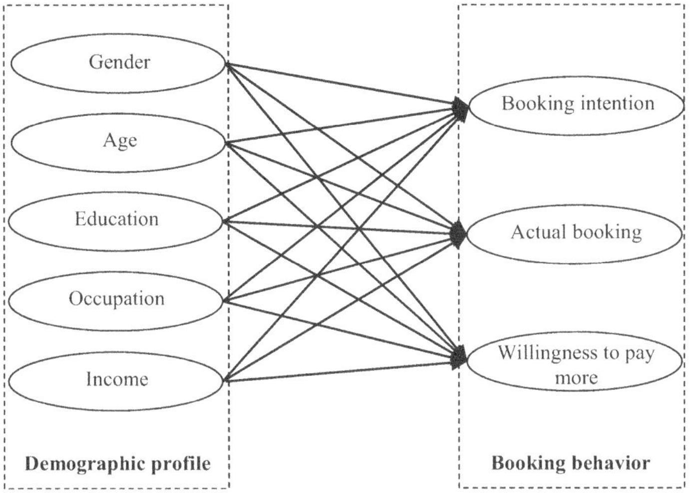

Fig. 1 Theoretical model

first asked if they had visited any of the Canary Islands, including Tenerife, Gran Canaria, Lanzarote, Fuerteventura, La Palma, La Gomera, El Hierro, and La Graciosa. Only those who had visited at least once were eligible for the survey.

The non-governmental organization 'Canary Green' team aided in data collection for the research study. The questionnaire was administered electronically, and participants were motivated to respond by an incentive - a complimentary week-long stay at a sustainable hotel in the Canary Islands. The data collection took place from December 2022 to February 2023. It took around 10–15 min to complete the questionnaire. We ensured that the confidentiality of the respondents was guaranteed.

Measures. To test the hypotheses, measures of green purchasing behavior were borrowed from the literature and adapted to the research context. All scales were measured with a five-point Likert (from 1 = strongly disagree to 5 = strongly agree).

The booking intention was measured on a scale consisting of five items. These items were proposed in the scholarship (Ku et al. 2012; Lee et al. 2012; Nekmahmud and Fekete-Farkas 2020; Nguyen-Viet 2022; Román-Augusto et al. 2022). An example of the item is 'From now on, I plan to book sustainable hotels'.

Actual booking behavior was assessed using three items to capture whether respondents choose sustainable accommodation in tourist destinations. The items were proposed by Riskos et al. (2021) and Shrum et al. (1995). An example of an item is 'I avoid booking accommodation that is not environmentally responsible'.

Willingness to pay a price premium consists of two statements that capture the tourists' willingness to pay more for environmentally friendly options. The items were adopted from Nguyen-Viet (2022) and Mehraj and Qureshi (2022). The sample item is 'I intend to pay more for green hotels'.

The internal consistency of the scales was sufficient (Cronbach's $\alpha > 0.70$) to capture both low and high levels of green purchasing behavior, as shown in Table 1.

Table 1 Summary statistics for green purchasing behavior

Measures	Number of items	Mean	SD	Possible range	α
Booking intention	5	3,79	0,73	1–5	0.861
Actual booking behavior	3	3,51	0,83	1–5	0.731
Willingness to pay a price premium	2	3,35	0,92	1–5	0.766

The survey questionnaire included questions about demographic information and travel habits. Before conducting the survey, a pre-test was done with a group of 20 people. The pilot study was created and conducted through a series of meetings with the 'Canary Green' team. Based on the feedback received, some minor adjustments were made to the final version of the questionnaire.

Data analysis. Data analysis was performed using IBM SPSS software. The full data set was subjected to *t*-tests, Spearman correlation tests, analysis of variance (ANOVA), and regression analysis.

4 Results

Table 2 provides information on the gender, age, education, occupation, and income of the respondents. Based on the demographic profile, it is evident that the majority of the participants in the study were female (60.37%). Furthermore, 57.43% of the respondents were between 18 and 29 years old.

The survey data found that most travelers (37.75%) had a graduate degree, while a significant portion (24.63%) had completed high school. It was also noted that half of the people surveyed were employed. The data further indicated that the Canary Islands have a strong tourism appeal, with Germany and the United Kingdom being the primary target markets.

According to our research, 40.70% of travelers visited the Canary Islands 2–5 times. Out of ten people, three were frequent visitors who have been to the islands six times or more. One-third of respondents had only visited one of the islands once. Tenerife, Gran Canaria, and Lanzarote were the most popular destinations, while La Graciosa and El Hierro were less popular.

Tourists were asked how long they had stayed on their last trip to the Canary Islands. 38.55% said they stayed for 4–7 nights, with the average stay being nine nights. Most people traveled with family (64.52%) or friends (32.93%), while some traveled alone.

Hotels were found to be the preferred accommodation choice. Seven out of ten people stayed in 1/2/3-star hotels on their last trip to the Canary Islands, while three out of ten opted for 4/5-star hotels. Self-catering lettings were chosen by 20.88% of travelers.

Table 2 Sample characteristics

Demographics		Frequency ($N = 747$)	Percentage (%)
Gender	Male	292	39.09
	Female	451	60.37
Age	18–29 years	429	57.43
	30–39 years	110	14.73
	40–49 years	74	9.91
	50–59 years	73	9.77
	60 + years	61	8.17
Education	High school	184	24.63
	Undergraduate	80	10.71
	Graduate	282	37.75
	Postgraduate	163	21.82
	PhD	25	3.35
	Other	13	1.74
Occupation	Student	160	21.42
	Employee	360	48.19
	Self-employed	137	18.34
	Retired	74	9.91
	Other	16	2.14
Country of origin	Germany	156	20.88
	United Kingdom	155	20.75
	Spain	97	12.99
	Sweden	51	6.83
	Other EU	226	30.25
	Other	62	8.30

Our research looked at sustainable accommodation options. This refers to overnight facilities that monitor environmental, social, and economic impacts and take responsible actions involving tourists to provide a better holiday experience. Half of the travelers chose sustainable or partly sustainable accommodation options. Italian, German, and Polish tourists were found to be the most likely to stay in sustainable accommodation.

We have analyzed research participants' sustainable accommodation booking behavior based on their demographic and touristic profile. The mean scores indicate only minor differences between males and females regarding their sustainable accommodation booking behavior. However, the t-test results are significant for booking intention (see Table 3). This means that there is enough evidence to suggest that female tourists have a stronger intention to book sustainable accommodation than their male counterparts.

Table 3 Impact of gender and occupation on sustainable travel accommodation booking behavior

Measures	Gender			Occupation		
	Male	Female	*t* value (2-tailed significance)	Employed	Others	*t* value (2-tailed significance)
	Mean (SD)	Mean (SD)		Mean (SD)	Mean (SD)	
Booking intention	3.72 (0.71)	3.84 (0.74)	2.24 (0.026)	3.79 (0.36)	3.79 (0.40)	–0.05 (0.963)
Actual booking behavior	3.51 (0.82)	3.50 (0.83)	–0.0202 (0.840)	3.51 (0.83)	3.50 (0.83)	0.26 (0.797)
Willingness to pay a price premium	3.37 (0.94)	3.34 (0.90)	–0.38 (0.706)	3.35 (0.89)	3.36 (0.94)	–0.20 (0.843)

In regards to occupation, there are no significant differences between employed individuals and others, contrary to previous hypotheses. Therefore, there is no clear evidence to suggest that being employed has a more substantial influence on sustainable accommodation booking intention, actual booking behavior, or willingness to pay a price premium.

Table 4 shows a significant negative correlation between age, booking intention, and willingness to pay more scores. However, the relationship is weak with Spearman's correlation coefficients of –0.113 and –0.087, respectively.

The research findings indicate a weak but meaningful connection between income and various measures of booking behavior. Although the correlation coefficients are not very high, they suggest that people with higher incomes tend to be less inclined to book, actually book less, and are less willing to pay a premium.

To investigate the impact of education level on booking behavior, one-way analysis of variance (ANOVA) tests were carried out, treating each education category as an independent sample (see Table 5). The results of the ANOVA tests showed that none of the measures of booking behavior varied significantly between different levels of education. This implies that the differences in booking behavior were not due to educational levels.

Three separate multiple regression analyses were conducted to determine the combined explanatory power of demographic characteristics in profiling tourists based on their sustainable accommodation booking behavior. The booking behavior

Table 4 Correlation between age and booking behavior measures

Measures	Age		Income	
	Correlation	Significance	Correlation	Significance
Booking intention	−0.113	0.002	−0.118	0.001
Actual booking behavior	−0.066	0.072	−0.086	0.019
Willingness to pay a price premium	−0.087	0.017	−0.093	0.011

Table 5 Impact of education on booking behavior

Measures	*F* ratio	Significance
Booking intention	1.918	0.089
Actual booking behavior	0.525	0.757
Willingness to pay a price premium	0.625	0.681

Table 6 Regression results

Measures	Multiple *R*	R^2	*F* value	Significance
Booking intention	0.185	0.034	5.208	0.000
Actual booking behavior	0.099	0.010	1.475	0.196
Willingness to pay a price premium	0.118	0.014	2.095	0.064

measures were used as the dependent variables, while the five demographic characteristics were utilized as predictors, as presented in Table 6.

The study indicates that only regression suggesting a link between booking intention and travelers' demographic characteristics is significant. However, demographic variables account for only a small fraction of the variance (3%). It has been found that demographic variables cannot predict actual booking behavior and the willingness to pay a price premium.

5 Discussion

The study found that the relationship between sustainable accommodation booking behavior and demographic characteristics is complex. It was initially anticipated that these relationships would remain consistent across all aspects of booking behavior, but this is not the case. For instance, while the assumption regarding gender's impact on booking intention was supported, gender does not affect actual booking behavior or willingness to pay a premium price. Additionally, education and occupation did not behave as expected concerning booking behavior. However, the research confirmed that age has an anticipated impact on booking intention and willingness to pay a premium price, consistent with previous literature (Shahsavar et al. 2020; Diamantopoulos et al. 2003).

The results of the multivariate analysis suggest that although demographic characteristics may be associated with booking intention, they have a weak explanatory power. This finding is consistent with a study by Diamantopoulos et al. (2003), which found that socio-demographic characteristics explain only 4% of the variance in purchasing behavior. A possible explanation for this could also be found in Akehurst et al.'s results (2012), who found that psychographic variables are more relevant than demographic variables in explaining green purchasing behavior. Therefore, we can conclude that while demographic variables can be used to some extent to profile tourists who visit the Canary Islands, an accurate profile of sustainable tourists cannot be created by considering only their demographic characteristics.

The current study contributes to the existing literature on sustainable tourism by examining the impact of demographic characteristics on the booking behavior of tourists regarding sustainable accommodations. The findings suggest that there is a relationship between sustainable accommodation booking behavior and specific demographic profiles, however, a more complete understanding of the sustainable traveler can only be obtained by examining other additional factors that influence sustainable accommodation booking decisions.

The results should be interpreted within the limitations of the study. The convenience sampling method and the non-probabilistic sample used in this study restrict the ability to generalize the findings beyond the Canary Islands. Future studies could expand the research to other tourist destinations to explain the differences in demographic profiles and accommodation booking behavior. Additionally, conducting longitudinal studies over different periods could provide further insights into this area of research.

References

Lukman, K.M., Uchiyama, Y., Quevedo, J.M.D., Kohsaka, R.: Tourism impacts on small island ecosystems: public perceptions from Karimunjawa Island. Indonesia. J Coast Conserv. **26**, 14 (2022)

Rajaonson, J., Tanguay, G.A.: An exploratory analysis of the negative environmental impacts of pandemic tourism on Canadian destinations. Annals of Tourism Research Empirical Insights. **3**, 100071 (2022)

Holmes, M.R., Dodds, R., Frochot, I.: At home or abroad, does our behavior change? Examining how everyday behavior influences sustainable travel behavior and tourist clusters. J. Travel Res. **60**, 102–116 (2021)

Rodríguez, C., Jacob, M., Florido, C.: Socioeconomic profile of tourists with a greater circular attitude and behaviour in hotels of a sun and beach destination. IJERPH. **17**, 9392 (2020). https://doi.org/10.3390/ijerph17249392

Ajzen, I.: The theory of planned behavior. Organ. Behav. Hum. Decis. Process. **50**, 179–211 (1991)

Kwon, J., Ahn, J.: Socio-demographic characteristics and green consumption behavior in developing countries: the case of Malaysia. SRJ. **17**, 1213–1231 (2021)

Chekima, B., Syed Khalid Wafa, S.A.W., Igau, O.A., Chekima, S., Sondoh, S.L.: Examining green consumerism motivational drivers: does premium price and demographics matter to green purchasing? Journal of Cleaner Production. 112, 3436–3450 (2016).

Wei, S., Ang, T., Jancenelle, V.E.: Willingness to pay more for green products: The interplay of consumer characteristics and customer participation. J. Retail. Consum. Serv. **45**, 230–238 (2018)

Shahsavar, T., Kubeš, V., Baran, D.: Willingness to pay for eco-friendly furniture based on demographic factors. J. Clean. Prod. **250**, 119466 (2020)

Diamantopoulos, A., Schlegelmilch, B.B., Sinkovics, R.R., Bohlen, G.M.: Can socio-demographics still play a role in profiling green consumers? A review of the evidence and an empirical investigation. J. Bus. Res. **56**, 465–480 (2003)

Hojnik, J., Ruzzier, M., Fabri, S., Klopčič, A.L.: What you give is what you get: Willingness to pay for green energy. Renewable Energy **174**, 733–746 (2021)

Jaiswal, D., Kaushal, V., Singh, P.K., Biswas, A.: Green market segmentation and consumer profiling: a cluster approach to an emerging consumer market. BIJ. **28**, 792–812 (2021)
Saleem, M.A., Eagle, L., Low, D.: Market segmentation based on eco-socially conscious consumers' behavioral intentions: Evidence from an emerging economy. J. Clean. Prod. **193**, 14–27 (2018)
Tasci, A.D.A., Fyall, A., Woosnam, K.M.: Sustainable tourism consumer: socio-demographic, psychographic and behavioral characteristics. TR. **77**, 341–375 (2022)
Pekerşen, Y., Canöz, F.: Tourists' attitudes toward green product buying behaviours: the role of demographic variables. TMS. **18**, 7–16 (2022)
Martínez, P., Bosque, I.R.: Determination of the role of consumers' demographic characteristics in the loyalty formation process based on socially responsible aspects. Cuadernos De Turismo. **42**, 643–646 (2018)
Kaur, B., Gangwar, V.P., Dash, G.: Green Marketing Strategies, Environmental Attitude, and Green Buying Intention: A Multi-Group Analysis in an Emerging Economy Context. Sustainability. **14**, 6107 (2022)
Finisterra Do Paço, A.M., Raposo, M.L.B.: Green consumer market segmentation: empirical findings from Portugal: Green consumer market segmentation. International Journal of Consumer Studies. 34, 429–436 (2010).
Meng, Z., Zhao, N., Shen, B., Zhai, C.: Optimal pricing strategy for green products under salience theory. Economic Research-Ekonomska Istraživanja. **35**, 2960–2983 (2022)
Mukherjee, A., Carvalho, M.: Pricing and Quality Investments in a Mixed Brown-Green Product Market. In: Lalla-Ruiz, E., Mes, M., and Voß, S. (eds.) Computational Logistics: 11th International Conference, ICCL 2020, Enschede, The Netherlands, September 28–30, 2020, Proceedings. Springer International Publishing, Cham (2020).
Statista: Number of international tourists in the Canary Islands, Spain from 2010 to 2022 (in millions) (2023), https://www.statista.com/statistics/449189/yearly-number-of-international-tourists-visiting-the-canary-islands/, last accessed 2023/08/12.
Ku, H.-H., Kuo, C.-C., Wu, C.-L., Wu, C.-Y.: Communicating green marketing appeals effectively: the role of consumers' motivational orientation to promotion versus prevention. J. Advert. **41**, 41–50 (2012)
Lee, N., Choi, Y.J., Youn, C., Lee, Y.: Does green fashion retailing make consumers more eco-friendly? The influence of green fashion products and campaigns on green consciousness and behavior. Cloth. Text. Res. J. **30**, 67–82 (2012)
Md., Nekmahmud, Fekete-Farkas, M.: Why not green marketing? Determinates of consumers' intention to green purchase decision in a new developing nation. Sujstainability. **12**, 7880 (2020)
Nguyen-Viet, B.: Understanding the influence of eco-label, and green advertising on green purchase intention: the mediating role of green brand equity. Journal of Food Products Marketing. **28**, 87–103 (2022)
Román-Augusto, J.A., Garrido-Lecca-Vera, C., Lodeiros-Zubiria, M.L., Mauricio-Andia, M.: Green marketing: drivers in the process of buying green products - the role of green satisfaction, green trust, green WOM and green perceived value. Sustainability. **14**, 10580 (2022)
Riskos, K., Dekoulou, P. (Evi), Mylonas, N., Tsourvakas, G.: Ecolabels and the attitude–behavior relationship towards green product purchase: a multiple mediation model. Sustainability. 13, 6867 (2021).
Shrum, L.J., McCarty, J.A., Lowrey, T.M.: Buyer characteristics of the green consumer and their implications for advertising strategy. J. Advert. **24**, 71–82 (1995)
Mehraj, D., Qureshi, I.H.: Does green brand positioning translate into green purchase intention?: A mediation–moderation model. Bus Strat Env. **31**, 3166–3181 (2022)
Akehurst, G., Afonso, C., Martins Gonçalves, H.: Re-examining green purchase behaviour and the green consumer profile: new evidences. Manag. Decis. **50**, 972–988 (2012)

Funding

This research was funded by the European Social Fund (project No 09.3.3-LMT-K-712–19–0077) under a grant agreement with the Research Council of Lithuania (LMTLT).

The Use of Intelligent Technologies in the Development of Digital Transformation as a Government Strategy

Marius Stefan

Abstract In the informational society, national and european, the modern history of Europe describes the industrial revolution as a factor that triggered the new economic order, through the scientific and technical progress made. While modernity as a term refers to a new social order, created as a result of the movement called Enlightenment. Through the totality and extent of the changes produced at the political and social level, summarized in the transfer of the individual from ancestral work to the activities corresponding to the urban industry, modernity became the first way of social organization that managed to dominate at the global level. Modernity is characterized by major transformations, which penetrated deeply into all levels of human life, in all economic, political and social spheres. The distinction between public and private life, the privileging of urban space, to the detriment of rural areas, specializations and qualifications on the labor market, increasing productivity and mass production, abandoning or at least minimizing conventional, traditional methods, and replacing them with new methods based on science, by reason, "on the new lights", all being legacies of modernity that laid the foundations for a present characterized by a new model of society, in which the purely industrial phase has been overcome, and production is replaced by services, knowledge and global information technologies, summarized under a term meant to describe this new structure, namely the informational society. The post-industrial society emphasizes, not on the production of goods, but on the preponderance of the use of services, depending to a large extent on the use of technology.

Keywords Intelligent technologies · e-Business · Digital transformation · Awareness and cybersecurity · Security of critical infrastructure of national interest · European funds · Development of the national economy

---

M. Stefan (✉)
Bucharest University of Economic Studies, Bucharest, Romania
e-mail: marius.stefan@mfe.gov.ro

M. Busu (ed.), *Rethinking Business for Sustainable Leadership in a VUCA World*, Springer Proceedings in Business and Economics,
https://doi.org/10.1007/978-3-031-50208-8_20

1 Introduction

The first stage of industrialization included a period in which scientific research, inventions and technological innovations took off. The progress achieved has determined a real advance in technology. The modernization of transport, the construction of the first railways and the appearance of the steam locomotive, generated notable developments in terms of trade. Distances were essentially reduced by electrical communication, thanks to the invention of the telegraph. The appearance of the first factories established a new work organization. All this produced, in the new industrial economic order, a demographic transfer from the rural to the urban environment, through the increase of the urban population and the number of cities. Production based on machines, thus replacing the manual work of the individual with the applications of new technological tools.

The second stage of modern industrialization being marked by the acceleration of urbanization, the emergence of corporations, socio-professional stratification, standardization and specialization, the modern educational system, the development of new industrial and technological branches, and until today in the economic era of financing from European funds.

E-business is the basis of the new economy, of an information society that catalyzes the future of an increasingly secure artificial intelligence for the development and use of new technologies of the future.

The technological infrastructure of the new economy, in constant need of ensuring all the principles related to cyber security, generates new e-business models, from e-commerce to the desideratum of the e-Government implementation strategy, through digitalization and computerization of public administration in Romania.

Electronic applications made because of the efficient management of European funds are a national interest, to achieve the goal of the positive evolution of the national economy, with critical values for achieving the balance of the rule of law and national security.

The field of European funds is constantly changing as well as the IT applications intended for their management, representing the first attempts at computerization and innovation in public administration. They are a new way of developing the national economy and e-business by promoting and pursuing European policies: distance learning (e-learning), remote work (e-working), electronic commerce (e-commerce), electronic banking services (e-banking), electronic government (e-government), electronic health services (e-health).

In an information society in which the quality of life, as well as the prospects for social change and economic development, depend to a greater extent on information and its exploitation, the institutional field of management of IT applications for European funds becomes a matter of national importance, with critical values for national security. Reinventing governance can be achieved through digitalization and government computerization, which involves modernizing the current IT infrastructure through specific external funding sources such as European funds, doubled and secured by advanced cyber protection and defence capabilities against possible vulnerabilities or cyber-attacks.

Knowledge and scientific information are of enormous importance in the global informational society, by: supporting innovation, promoting economic development, making decisions in an efficient and transparent way, at the governmental level and especially for the implementation and use of intelligent technologies in the development of the degree of digitization of public services through financing provided by European funds and the National Recovery and Resilience Plan.

In order to move on to building the knowledge society, it is necessary to reduce the digital gap, which accentuates disparities in development, excluding groups and even countries, from the benefit of information and knowledge. The limiting factor in development will be related to the human capacity to assimilate and develop these technologies, to use them in new fields of activity, for new products and services.

2 Intelligent Solutions in a Secure Cyber Environment

The use of intelligent technologies will result in the development of the degree of digital culture and cyber security, among civil servants, in the economic-social-political-post-pandemic context for e-Business.

The Ministry of European Investments and Projects, together with institutional partners, is the main developer for the IT exchange program between Romania as an EU member state, beneficiary of non-reimbursable European funds, and the European Commission, according to the provisions of the Regulation (EU) NO. 1303/2013 of the European Parliament and of the Council of 2013.

In Romania, the evolution of the cyber infrastructure intended for European funds is conditioned by inter-institutional cooperation, carried out in the form of strategies, harmonized with European legislation and materialized through specific projects to ensure cyber security, as well as awareness of the importance of the state of security at the level the entire state apparatus:

- project code SMIS 48723—Titeica 1—"The national system for the protection of IT&C infrastructures of national interest against threats from cyber space", financed by the Sectoral Operational Program for the Increase of Economic Competitiveness 2007—2013".
- project code My SMIS 127221—Titeica 2—"Updating and developing the national system for the protection of IT&C infrastructures with critical valences for national security against threats from cyber space", financed by the Competitiveness Operational Program 2014–2020 "Competitive together". (ICIN 54 MIPE—the project with national coverage, and implementation period according to the financing contract—23.08.2019–23.08.2022, with related maintenance services and support for applications and equipment until 23.08.2027).
- the Titeica 3 project—implemented through PNRR by the National Cyberint Center, intended for the development of the national cyber protection system,

included in Component C7, Digital Transformation, through the National Recovery and Resilience Plan—resulting in the expansion of the protection area of infrastructures of the type Informational Technology and Operational Technology, as a beneficiary entity of cyber security and protection, as well as participation in training programs organized in the field of cyber security.

The objectives of the projects in question being the updating and development of existing IT systems by including them in the system of new IT&C infrastructures with critical valences for national security in order to increase the capabilities to identify cyber-attacks, as well as to increase the level of national cyber security, subject to a common approach of the EU in cyber security.

Thus, through the expected results, the projects aim at increasing the security of IT and communications services at the national level, increasing the availability and level of security offered to institutions and entities of public interest and modernizing the security systems provided by the existing IT systems within the authorities/ institutions, by achieving the interoperability of the security systems to be implemented and integrated, regarding the corroboration of information, collaboration, analysis and reaction through the IT mechanism for rapid alerting and dissemination of information in real time.

All state institutions will be included in this national system of prevention and protection against cyber-attacks, through cyber defence activities, desirable and achievable with the considerable contribution generated by the cooperation between state institutions, specialized in ensuring cyber security, such as the National CYBERINT Center—the National Authority in the field of Cyber-Intelligence.

The faulty operation or in inappropriate parameters of the applications intended for European funds, integrated in the related cyber infrastructure, will generate the state of vulnerability constituted by blockages of the registration mechanisms and increase the absorption of European funds. The decrease in the absorption of European funds is and will be a real threat to Romania's national security, due to the following implications: economic; financial; social; political; as well as a result of the obligations assumed by Romania as a member state of the European Union.

National system of prevention and protection against cyber-attacks, through cyber defence activities, having a key role in the implementation of the new National Cyber Security Strategy and in ensuring Romania's compliance with the commitments assumed at the international level, including those related to the implementation of the Cyber Security Strategy EU Cybersecurity, the NIS Directive and NIS 2.0, as well as the European Cybersecurity Industrial, Technological and Research Competence Center (ECCC).

In any situation of blockage in this sensitive area of European funds, information for national security will be dominated by the need to capitalize by immediately informing the Minister of Investments and European Projects, as well as by adopting the necessary measures to remove the identified deficiencies.

Thus, avoiding the threats regarding the risks of disengagement, through an efficient management of the European funds, in the conditions of the desired good management of this objective of national strategic interest.

Touching the interests of the country, as well as the acts of destruction, degradation or bringing into disuse the structures necessary for the proper development of social-economic life—can constitute a threat—even by generating the state of blocking the absorption of European funds, a situation that falls under the provisions of Chapter 3 related to the National Defence Strategy 2015–2019/ The National Defence Strategy of the country for the period 2020–2024, and art 3, letter f, Law 51/91.

The identification and capitalization of these possible risks constitutes information for national security, which will be achieved gradually, due to an increased degree of persistence shown, as well as the perpetual lack of sufficient and efficient resources, endowed with the necessary specialization in the correct management of the IT system, in continuous development, placing more emphasis on results and efficiency and the creation of public values, including at the level of critical infrastructure—national cyber, in this newly developed branch of the economy—the field of European funds. A beneficial approach would be represented by the adoption of a cyber risk prevention strategy, by creating at organizational level in the Ministry of Investments and European Projects, a post-specialist, related to the cyber security component, by delegating an expert designated within Security Structures.

The poor functioning of the gear that is the basis of the attraction of funds, will bring damage both to the national budget and to the image at the European level, interoperability being a basic principle of the member states. Failure to fulfil the obligations assumed as a member state can cause economic disadvantages, the development of society and the increase of the quality of life depending on the evolution in the management of current financial resources, as well as for the future, the membership of the European Union and the financial exercises specific to the community space.

This data of national interest has an incidence in the current year—2023, when cyber-attacks become characterized by frequency and persistence, it is vital that organizations are armed with the most effective tools and security knowledge to prevent, detect and to respond to cyber threats. Vulnerabilities will always escalate into possible future threats and risks to national security, with awareness and prevention being the most effective approach.

In the context of the increase in the number of attacks, registered in the virtual environment, on public institutions considered to be targets, by cyber attackers, it is imperative to ensure the cyber protection of workstations and laptop devices within MIPE—through centralized administration of a unique antivirus solution management console, complementing the purchases related to the project carried out in the Cooperation Agreement with Cyberint: Titeica 2—"Updating and developing the national system for the protection of IT&C infrastructures with critical valences for national security against threats originating from cyber space".

The need for security is understandable in the context of the exponential increase in the number and complexity of cyber threats (malware/ransomware/social engineering in particular). Deficiencies encountered in development, hardware and software solutions, as well as the lack of an appropriate infrastructure dedicated to the MySmis2021 IT system, can cause malfunctions in the electronic services offered to beneficiaries/

the business environment, the IT system acquires as an area of involvement, importance even at the national level, but it can also slow down certain processes in the economy, such as even the annual drawing up of the national budget. Including interoperability requirements in the relationship with the European Commission, dictates a clear focus on functional and secure reporting processes, increasing processing of documents in electronic format, signed with digital certificates. The concentration of resources can only provide solutions under safe operating conditions, ensured only by a state of cyber security.

In terms of legislation, including harmonization with European provisions, it can constitute a vulnerability in the proper functioning of the activity, in the context of increasing rigors/requirements regarding interoperability from the EU (the request for expensive related services for the current technology used—EIF, part of Communication (COM (2017)134)).

Thus, it is necessary to ensure compliance with the requirements of the European NIS Directive2 and Law 362/2018—regarding the wave of digital information that must be managed and controlled by technical security measures, and Law no. 3652/2018, which transposes the European NIS Directive, regulates the necessary framework for the development the level of preparation of the EU states for dealing with potential incidents that may affect IT security. The NIS Directive (Network and Information Systems) EU 2016/1148 is an essential legal piece released at the EU level to increase the level of cyber security for critical infrastructure units, including entities in the field of utilities, transport, healthcare and digital services, as well as European funds.

The NIS Directive establishes a package of principles and rules by which it seeks to define, measure and improve cyber security. Considering the continuous increase in cyber-attacks, compliance with the requirements of the NIS Directive is imperative. A cyber security strategy based exclusively on prevention is not enough. Maximum involvement is needed, through quick detection and effective solutions.

The attack on a critical—cyber infrastructure of national interest, such as that of European funds, can occur as a result of security risks, not properly treated, having the effect of data leaks, through exfiltration, or by causing syncope in operation through unforeseen interruptions induced services (especially electricity and internet services).

The long-term trends show as the main threats that should be monitored, malware attacks especially in a government institution such as the Ministry of Investments and European Projects, which is a real area of interest for cyber-crime groups, for the purpose of cyber espionage or the theft of strategic information, such as that of the state.

Since 2016 and until now, the government sector has been at the forefront of data breaches. The methods are diversified, from the sale of credentials in the case of data encryption to the specific techniques of artificial intelligence and the tendency to use new IT technologies in any field, which allowed the increase in complexity of hacking attacks.

The awareness of the importance of the cyber security area, especially among the specialized staff and intended for ICT activities, implies the preparation to combat the risks that the public institution will face, starting from the software components maintained up to date from the point of view of security, and even avoiding periods of technological or governance changes, situations such as loss or lack of administration credentials/access to work environments created through electronic tools. Often political management changes in the institutional framework are also reflected at the level of specific activities of technical departments, by slowing down the decision-making process, which does not represent a good institutional practice, especially in the case of ensuring cyber security.

Failure to comply with the existence of access and security procedures to data and documents in the public institution, as well as the preservation of obsolete and morally worn equipment or technologies, creates a favorable environment for the incidence of security risks, including at the cyber level.

In order to align with the new trends in the field of cyber security and harmonization from a community legislative point of view, standards such as information security management (ISO 27001) and business continuity management (ISO/IEC 27,031:2011, 22,301:2012) will be considered. as well as the latest directions drawn by NIS—EU Directive 1148/2016.

Progress in the development of new technologies will establish the desire to align with the new standards of the future, new solutions such as the private or hybrid cloud will be adopted at the governmental level, from the point of view of budget efficiency, but especially for specific considerations cyber security.

Through clear and quick measures, possible cyber security breaches, identified in the organization, are remedied through the appropriate and responsible use of the vulnerability scanning IT component—with capabilities to propose corrective measures in ICT security. Analytical utilization of information as evidence of virtual activity in the organization is achieved through the management of technologies specific to the field of artificial intelligence, complementing specifically human cognition through additional learning methods generated by computing power.

Network traffic monitoring and security diagnosis based on behavioral analysis in technology, in order to avoid data leaks caused by possible cyber-attacks, should be carried out as an activity at the level of a public institution, by staff specialized in ICT and especially in Management of Information of importance to National Security. All correctly implemented measures, as well as the cooperation between the institutions involved, will considerably enhance the institutional capacity to react in the event of a cyber incident.

The Institutional Cyber Security Strategy needs a very good organization of the IT department, through a correct implementation according to the best practices in the field regarding the network department, the infrastructure department, the department responsible for IT applications.

The Cyber Security Department is a plus for any public institution in today's modern society. The best strategy involves the segmentation of security solutions, so that each component will make the work of attackers more difficult through the existence of security at every critical point of entry into the organization, more precisely at the level of critical areas of activity and institutional intervention: end-point, server, applications, switch, access-point, databases, website, etc.

A public institution should be directly interested in the protection against cyber-attacks and the prevention of data theft, as the main prevention measure being the adoption of the decision to establish a specific post-Cyber Security, within the Security Structure of the Minister of Investments and European Projects.

The incumbent of the established position will have to have the necessary specialization in the field, through certified resources in the field of Cyber Security with an emphasis on specific activities such as—National Security Information Management.

At the level of the Ministry of European Funds, this measure has not yet been fully implemented, the first paradigm-breaking measure, being carried out through the Participation and Cooperation Agreement, signed between MIPE and Cyberint, in a national cyber critical infrastructure project—ICIN 54 MIPE, intended for European funds and specific critical infrastructure, succeeding in the unification through cyber security solutions, of the majority of state institutions of national strategic importance.

The response to Cyber Security incidents, as an activity, requires the existence of its own personnel specifically intended for such situations, by nomination and inclusion within the Security Structure of MIPE, being extremely useful to the public institution, including in situations of Cooperation with the Authorities empowered in the field of cyber defence.

Cyber security awareness and training is also very useful, especially among users, or even using solutions to protect them from incidents related to their own password, which should meet certain security standards.

Raising awareness will be achieved by informing users, which is the first measure of protection against cyber-attacks. Human error can be avoided. Thus, paths to hacker attacks will be closed, through good periodic information, through constant emails, courses, trainings, eliminating the possibility of subsequent, much more serious problems, especially regarding information and data belonging to the state.

In the near future, the public administration will also evolve towards another approach to the use of technologies, being transposed into future strategies, the need to use solutions in cloud, on-premise, or hybrid cloud environments, depending on the available budgets and the advantages or disadvantages offered. In order to make the activity more efficient or in the situation of permanent blocking of new employment procedures in the public administration, the subcontracting of services that allow access to these technologies is an effective way to manage platforms, with a cost–benefit ratio in favor of the public institution, which will relieve you of the worry of using internal resources.

Security solutions are necessary, bringing real benefits to the institution when they are configured according to the needs of secure operation. Easy-to-use, intuitive management tools can optimize the time it takes to implement new policies, for proper monitoring and alerting. Also, the collaboration with the National Cyberint Center provides stability and access to the necessary knowledge in the activity of implementing these cyber security assurance systems.

Unified and integrated technologies offer a measurable advantage in results, benefiting from such consoles and tools adapted to the level of expertise, in accordance with the strategies built by the IT—Security department. Cooperation with other institutions such as CNC, participation in seminars and conferences in the field, is an excellent tool for improvement, and above all a guide of good practices, assimilated in order to adopt the best decisions for the public institution.

The budget allocated to innovation in public administration will create and maintain the stability so desirable, especially in critical areas, such as that of European funds. In industry and the economy, the role of robotics and process automation will increase considerably, with changes related to technology bringing both benefits and vulnerabilities, especially in cyber terms. A virtual parallel world will be created, where the existence of the state, with everything it represents, must be protected, so the environment will be safe and secure, including for the individual. The consequences of competition in innovation produce major transformations including in society, simplifying the complex life of modern man, in the information society. Thus constituting a national interest for the Governmental Strategy, areas such as attracting European funds and ensuring cyber security, with the aim of modernizing and computerizing the public administration in Romania.

Including the interoperability requirements in relation to the European Commission, dictates a clear focus on functional and reliable reporting processes, increasing processing of documents in electronic format, signed with digital certificates. Concentration of resources can only provide solutions in safe operating conditions, provided only by a state of cyber security.

Legislation, including harmonization with European provisions, may be a vulnerability in the proper functioning of the business, in the context of increasing EU interoperability requirements request for related services costly for current technology used—EIF, part of Communication (COM (2017) 134)).

A common case for public sector reasons, in terms of modernization strategies through public procurement, is that of equipment failure outside the support/warranty period, thus posing a threat to the network topology.

Expiration of security equipment support is a serious issue in the operation of the IT system parameters, as it leads to a serious breach in the security umbrella created by the specific equipment, continuously managed and which should contain the latest updates to be truly effective, in the cyber fight against increasingly complex and persistent attacks.

Poor technology, also manifested in uninteresting trends in continuous modernization, affects the limits of processing in the evolution of servers, especially in the case of increasing the need for balancing at the application level and/or saving and restoring data.

Although data generated and processed automatically is an important pillar of public sector activity, the focus on electronic data assurance policy still does not enjoy a high degree of importance in public institutions due to the perpetual intermediate stage, computerization of public administration.

Awareness of the importance of the Security area, especially among specialized personnel and intended for ICT activities, requires preparation for combating the risks that the public institution will face, starting from the software components maintained up to date in terms of security, and even avoiding in periods technological or governance changes, situations such as the loss or lack of management credentials/access to work environments created by electronic tools. Often the political changes of management in the institutional framework are also reflected in the specific activities of the technical departments, by slowing down the decision-making process, which is not a good institutional practice, especially in the case of ensuring cyber security.

Failure to comply with the procedures for access to and security of data and documents in the public institution, as well as the preservation of obsolete and morally worn equipment or technology, creates a favorable environment for the incidence of security risks, including at cyber level. To align with new trends in cybersecurity and harmonize Community legislation, standards such as information security management (ISO 27001) and business continuity management (ISO/IEC 27,031: 2011, 22,301: 2012) will be considered. as well as the latest directions drawn by NIS—EU Directive 1148/2016, NIS 2.0.

Progress in the development of new technologies will establish the desire to align with the new standards of the future, new solutions such as private or hybrid cloud will be adopted and at the governmental level, in terms of budget efficiency, but especially for specific reasons. cyber security. Clear and swift measures are used to address any cyber security breaches identified in the organization through the proper and responsible use of the vulnerability scanning IT component—with capabilities to propose corrective measures for ICT security. The analytical capitalization of information as evidence of virtual activity in the organization, is achieved through the management of technologies specific to the field of artificial intelligence, complementing the specific human cognition through additional learning methods generated by computing power.

Network traffic monitoring and security diagnostics based on behavioral analysis in technology, in order to avoid data leaks caused by possible cyber-attacks, should be carried out as an activity at the level of a public institution, by personnel specialized in ICT and especially in Management of Information of Importance for National Security. All the measures correctly implemented, as well as the cooperation between the institutions involved will considerably enhance the institutional capacity to react in case of a cyber incident.

The Institutional Cyber Security Strategy needs a very good organization of the IT department, through a correct implementation according to the good practices in the field in terms of the network department, the infrastructure department, the department responsible for IT applications.

The Department of Cyber Security is a plus for any public institution in today's modern society. The best strategy involves segmenting security solutions, so that each component will hinder the work of attackers by the existence of security at each critical point of entry into the organization, more precisely at the level of critical areas of activity and institutional intervention: endpoint, server, applications, switch, access point, databases, website, etc.

A public institution should be directly concerned with the protection against cyber-attacks and the prevention of data theft, with the main measure of prevention being the adoption of the decision to set up an IT Security department, headed by a CISO.

The department will have to have the necessary specialization in the field, through certified resources in the field of Cyber Security with emphasis on specific activities such as: implementing security policies and user access control, tracking management platforms, managing, and correlating logs from SIEM and DLP solutions.

At the level of the Ministry of European Investments and Projects, this measure has not yet been fully implemented, the first paradigm-breaking measure being implemented through the Participation and Cooperation Agreement, signed between MIPE and Cyber-int, in a national critical cyber infrastructure project—ICIN 54 MIPE, intended for European funds and specific critical infrastructure, managing the unification through cyber security solutions, of the majority of state institutions, of national strategic importance.

Responding to Security incidents as an activity could involve the existence of staff assigned to such situations, or by outsourcing, outsourcing this service would be very useful to the public institution. The hardware and software solutions that an IT Security department should manage are the following: Next Generation Firewall with traditional Anti-Virus modules, URL Filtering, Application Control, Firewall, IPS, Anti-Bot but also with modules for detecting unknown attacks through Sandboxing technologies. In addition to NGFW, there should be solutions capable of detecting End-Point attacks. These should contain traditional modules such as: Anti-Virus, URL Filtering, Firewall, Application Control, Anti-Bot, but to be able to deal with the latest cyber-attacks, advanced Anti-Ransomware protections are also needed.

In this way, SIEM and DLP solutions are urgently needed for such an organization, which wants to ensure a secure environment for creating excellence through effort and efficiency in results and protected information.

The two-way management of security solutions should be able to easily integrate as many of the existing solutions as possible to prevent or identify attacks both horizontally and vertically. The export of logs should be constant and active, towards SIEM solutions. In MIPE being used the Arc-sight solution, which manages to collect and transmit the data necessary for security analysis, being centralized used to prepare the counter-offensive for cyber-attacks in the Cyberint National Center.

A SIEM-type solution involves the existence of a strong log correlator to facilitate easy interpretation of the information provided by providing alerts, reports, trends, etc. The public administration should also be concerned with vulnerability patching solutions to ensure that the latest signature updates are available in the operating systems and applications used at the institutional level.

Perfection in computer systems does not yet exist, which is why it would be necessary, for the best results, for NGFW components, Sandboxing, to be doubled by a dual-solution approach, in order to ensure protection. enhanced with superior coverage to a single solution, given that there is no perfect IT Security manufacturer or one that offers 100% protection. At the MIPE level, two technology models are used, thus covering any possible security breach in the ever-changing virtual environment. The technologies used were evaluated before the acquisition to respond effectively to internal needs. ale MIPE. The Proof of concept (POC) scenarios as well as the establishment of certain POC Success Criteria for pursuing the objectives, were made at the level of the IT Department of MIPE.

Proper implementation is achieved using virtual media, using both hardware and software-based technologies. Optimized use of existing solutions will need to be implemented in a redundant cluster system. The configurations will be tested before deployment—live, in production, and they require backup copies, saved in case of disaster. The VMWARE virtualization solution is used in MIPE, and at the level of the MySMIS2021 computer application, of national interest, there are two activity environments, the test one and the production one. Access to applications, databases and more, must be optimized by the existence of a Disaster Recovery (DR) system, which must contain traffic load balancer elements—to avoid loading and ensure a constant flow for a quick response and easy access.

Public institutions can be targeted for cyber incidents, so protection against DDoS attacks must be ensured, which can be compromised on a voluminous basis, restricting users' access to exposed data and thus blocking the organization's activity.

The applications used by the organization, both those behind a site or portal, whether we are talking about employee access or the outside public, need to be protected by a dedicated solution—Web Application Firewall. It recognizes dedicated attacks at the application level and together with other modules protects the organization's website against other types of attacks at the web platform level. The application was provided within the ICIN 54 MIPE projects, currently being administered together with all other security solutions, by a designated MIPE expert, who will obtain for a better development of the cyber incident prevention activity, and the specific certification in national cyber security. Establishing the premises for the establishment of a special compartment for this activity of special importance, still not institutionally aware, as far as possible related impact, in case of disaster or cyber incident.

By implementing the use of intelligent technologies, advanced solutions will be used, either at the workstation level or at the network level, adding extra security against targeted APT (Advanced Persistent Threat) attacks that can exist for years infiltrated within institution, until the actual launch of the cyber-attack, which can be completed by exfiltrating data after exploiting a vulnerability in the process of penetration and bypassing the existing but insufficient security measures. By using algorithms based on behavior analysis—traps, for example honeypots, network anomalies that could represent a zero-day, signature-less attack can be identified. A critical cyber infrastructure can only be protected under conditions of maximum cyber security. The proactive use of specific technologies that can identify attacks at the level of files such as images, documents, sent in attachments, prevent unwanted situations such as identity theft, credentials, by redirecting users to the most used sites (phishing, smishing, etc.).

Cyber security awareness and training is also very helpful, especially among users, or even using solutions to protect them from their own incidents with passwords that should meet certain security standards.

Thus, avoiding the trend of password uniqueness, used between applications and IT systems, making a distinction between professional and personal. Public officials' access to the institution's resources will be achieved in the worst case, using a VPN platform, through the institutional internet service provider—the Special Telecommunications Service. Devices that store data, such as HDD, must be integrated into an encryption platform. By using a NAC system, the organization can have a unified endpoint security policy across network authentication.

User rights reflect the job description and are determined by specific network and endpoint permissions based on role status, need to know, and data access. This MIPE policy is provided through Active Directory, but requires migration to a higher solution as a production year. The password security policy is predetermined by this solution, according to the standards in force: change at a certain interval, a certain number of characters, numbers and letters. It is also necessary to implement a solution that provides protection against access environments, for example, managing access to USB ports.

2.1 Literature Review

In the context of the increase in the number of attacks, registered in the virtual environment, on public institutions considered to be targets, by cyber attackers, it is imperative to upgrade the current antivirus solutions, by using a superior solution such as the Endpoint Detection and Response.

Endpoint Detection and Response (EDR), also known as Endpoint Threat Detection and Response (ETDR), is an integrated endpoint security solution that combines real-time continuous monitoring and collection of endpoint data with automated rule-based response and analysis capabilities.

The main functions of an EDR security system are:

- Monitoring and collecting activity data from endpoints that might indicate a threat;
- Analyzing this data to identify threat patterns;
- Ability to automatically respond to identified threats to eliminate or contain them and notify cybersecurity personnel;
- Use of forensic and analysis tools to conduct research into identified threats and search for suspicious activities.

Thus, for a better visibility of advanced attacks and for quick response functionalities, it is necessary to preserve the current degree of cyber security ensured by Endpoint Detection and Response (EDR), benefiting from the automatic detection and response to possible cyber security incidents, thus eliminating the gaps created by the insufficient staff assigned in terms of human resource assigned to the implementing cyber security account at the MIPE level.

Using machine learning capabilities and automatic incident investigation, certain activities that should have been performed by a security incident response team will be performed automatically in conditions where MIPE does not currently benefit from an internal SOC (Security Operation Center) structure. Integrated and automated response flows will enable designated personnel to respond effectively by limiting lateral spread and stopping potential attacks.

Threat visualization features enable focus on specific aspects of investigations, helping to understand complex detections, and identify the root cause of attacks, thus maximizing immediate response capability. The EDR module provides automated alert prioritization with one-click remediation functions.

The EDR module will perform continuous analysis within the organization, using unique capabilities to identify risk based on hundreds of factors. Providing clear guidance for mitigating potential risks at the user, network and operating system levels.

EDR administration requires low maintenance effort, being easy to implement and integrate into the existing security architecture, compatible with the antivirus solution used at the MIPE level. The agent is resource-efficient, with low administrative costs in terms of disk space, memory, bandwidth, and CPU resources. The flexibility, scalability and upgradeability of the complete endpoint protection platform and managed detection and response (MDR) services are necessary in the process of ensuring the cybersecurity standard built at the MIPE level. By using cutting-edge threat detection technology, including files attacks, ransomware and other zero-day threats.

In threat analysis, the event logging feature continuously filters events produced on the endpoint, compiling a prioritized list of incidents for further investigation and response.

In the event recording process, continuous monitoring allows data to be passed to the threat analysis module to visualize the results generated by the events involved in an attack.

Relying on highly effective prevention, automatic threat detection and response technologies, the antivirus software product (the IT solution) greatly limits the number of incidents that require manual analysis, reducing the operational effort required to use an EDR solution.

A cybersecurity sandbox provides a safe environment for opening suspicious files, running untrusted programs, or downloading URLs without affecting the devices they're on. It can be used at any time, in any situation, to safely review a potentially malicious file or code before serving it on devices, while keeping it isolated from a computer and company network.

In cybersecurity, sandboxing is used as a resource to test software that might end up being classified as "secure" or "unsecure". As malware becomes more widespread and dangerous, there are malicious apps, links, and downloads that could gain endless access to a network's data if not first sandboxed. Sandboxing can be used as a tool to detect malware attacks and block them before they enter a network. The system allows IT to test code and understand exactly how it works before infecting a terminal device with malware or viruses; this provides IT teams with information and advice on what to look out for in other scenarios.

The Sandbox Analyzer component automatically executes suspicious payloads in a controlled virtual environment.

The threat analysis module then uses this analysis to make appropriate decisions about suspicious files, according to the automation achieved through the security policy implemented at the level of the single management console.

Cyber Security incident investigation and response processes will be automated through the IoC search capability, querying the event database to discover possible threats through ATT&CK techniques and indicators of compromise as well as updated information on discovered threats or other possible malware. Detecting in the pre-execution phase and eradicating malware (through customizable Machine Learning, real-time monitoring of processes, thus performing specific analysis in the sandbox-cloud).

To provide an additional level of protection for stations and servers, the solution includes a sandbox in the public cloud of its producer.

The Sandbox module will be able to automatically send files to the Sandbox from the manufacturer's cloud where they can be "detonated" for an in-depth analysis.

The Sandbox module includes two analysis options: only monitoring or blocking. In monitoring mode, the user will be able to access the desired file, while in blocking mode, the user will be blocked from running the file until the Sandbox in the manufacturer's cloud gives the verdict. The Sandbox module includes two types of remedial actions: default and safety. For the default action, it will be possible to set: only reporting, disinfection, deletion and quarantine. For the safety action, it will be possible to establish: deletion or quarantine.

The Sandbox module also includes the possibility of manually sending files to the Sandbox from the manufacturer's cloud. Thus, if the administrator suspects a file to be malicious, he can manually send it to the Sandbox to be "detonated" and find out the verdict. You will be able to send several files at once, with the possibility to specify whether they will be "detonated" individually or all at the same time.

The Sandbox module can support "detonation" of the following types of files: Batch, CHM, DLL, EML, Flash SWF, HTML, HTML/script, HTML (Unicode), JAR (archive), JS, LNK, MHTML (doc), MHTML (ppt), MHTML (xls), Microsoft Excel, Microsoft PowerPoint, Microsoft Word, MZ/PE files (executable), PDF, PEF (executable), PIF (executable), RTF, SCR, URL (binary), VBE, VBS, WSF, WSH, WSH-VBS, XHTML.

3 Methodology

The research was carried out at the level of the Ministry of Investments and European Projects, with the main aim of creating scientific and technological excellence, as well as obtaining advantages in the field of security and resilience of systems, services and critical infrastructure of national importance, as well as increasing the cyber security culture in the central public administration and among contractual users or civil servants, with the possibility of establishing within the institutional organization, at least 3 positions with specific tasks in the cyber field, in direct collaboration with the

Security Structure of the Ministry and in a cooperation agreement with the National Cyber Intelligence Center of the Romanian Information Service.

The period included in the analysis activity is between the years 2013–2023, including two programming periods of non-reimbursable financing from European funds, facilitated by the European Commission.

The three projects carried out by the Cyber-int National Center, to ensure cyber security at the national level, constituting a security umbrella, over the critical infrastructure of national interest, which will be reinvented through the digital transformation generated with the help of emerging technologies, have produced a considerable evolution in the governmental digital transformation.

Emerging technologies and the integration of machine learning or artificial intelligence functionalities, at the level of the Ministry of Investments and European Projects, as a development measure through innovation, having positive effects including on the development of the national economy by increasing the absorption of European funds in a secure cyber environment.

4 Results and Discussions

The first step in the use of emerging technologies through the integration of Machine Learning and Artificial Intelligence functionalities, at the level of the Ministry of Investments and European Projects, was made within the projects financed from non-reimbursable funds, as a measure of the development through innovation, of a critical infrastructure of national interest, through a cooperation agreement with the National Authority in the field of Cyber-intelligence—the National Cyberint Center.

The result is threat prevention, deep visibility, accurate incident detection and intelligent response to minimize exposure to infection and stop unauthorized access. As an integrated workstation protection package, the integrated device security management platform ensures a uniform level of security across the entire IT environment, so that attackers cannot find a weakly protected workstation to use as a starting point for dangerous actions against the organization.

The security equipment used, offering advanced management capabilities, to achieve the prevention, detection and investigation of cyber security incidents, by analyzing the risk generated by possible attacks, as well as the automatic remediation of threats.

Awareness raising will be achieved by informing users, which is the first measure of protection against cyber-attacks. Human error can be avoided. This will close the path of hacker attacks, through regular good information, constant emails, courses, trainings, eliminating the possibility of subsequent, much more serious problems, especially regarding the information and data belonging to the state.

In the near future, public administration will evolve towards a different approach to the use of technologies, being translated into future strategies, the need to use solutions in cloud, on-premises, or hybrid cloud environments, depending on budgets and available advantages or disadvantages. For the efficiency of the activity or in the situation of permanent blocking of the procedures of the new employees in the public administration, subcontracting the services that allow access to these technologies is an effective way of managing the platforms, with a cost–benefit ratio for the benefit of the public institution, it will relieve the worry of using internal resources.

Public administration services can be optimized using advanced information technology. The European Commission tries to set its own example in this sense, through the procedures and tools it uses in its day-to-day work, in its links with the administrations of the member states and with its own decentralized agencies, which are marked by advances in computerization. The aim is to facilitate citizens' access to public information through new IT applications, as well as to achieve better communication between all levels of public administration in the Union, thanks to the high-speed connection.

The development of the European information society requires a considerable and constantly growing financial effort, which cannot be fully assumed by the European Union and the governments of the member states. Experience has shown that the private sector is the most capable of taking risks in operating and developing new adaptable markets and has the necessary capital to make such investments necessary for the digital transformation strategy.

The integration of machine learning and artificial intelligence functionalities, at the level of the Ministry of Investments and European Projects, could be seen in Table 1 below, while the use of intelligent technologies such as Sandbox Analyzer and EDR—Endpoint Threat Detection and Response (ETDR) modules, can be seen in Figs. 1 and 2.

Table 1 Integrating Machine Learning and Artificial Intelligence functionalities, at the level of the Ministry of European Investments and Projects

Implementation period	Protected workstations	Increasing the degree of cyber protection	Automate responses to detected and remedied cyber attacks	Fixed vulnerabilities	Possible security risks
2014–2017	250 to 450	200 Endpoints	About 50%	75%	25%
2020–2023	450 to 1700	1250 Endpoints	About 75%	90%	10%
2023–2027	1700 to 3400	3400 Endpoints	About 95%	95%	5%

Source Author' own research

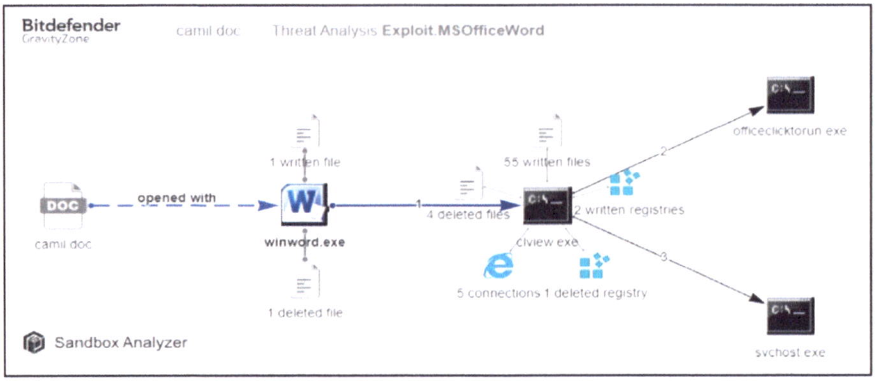

Fig. 1 Sandbox analyzer. *Source* www.bitdefender.com. *Source* www.bitdefender.com

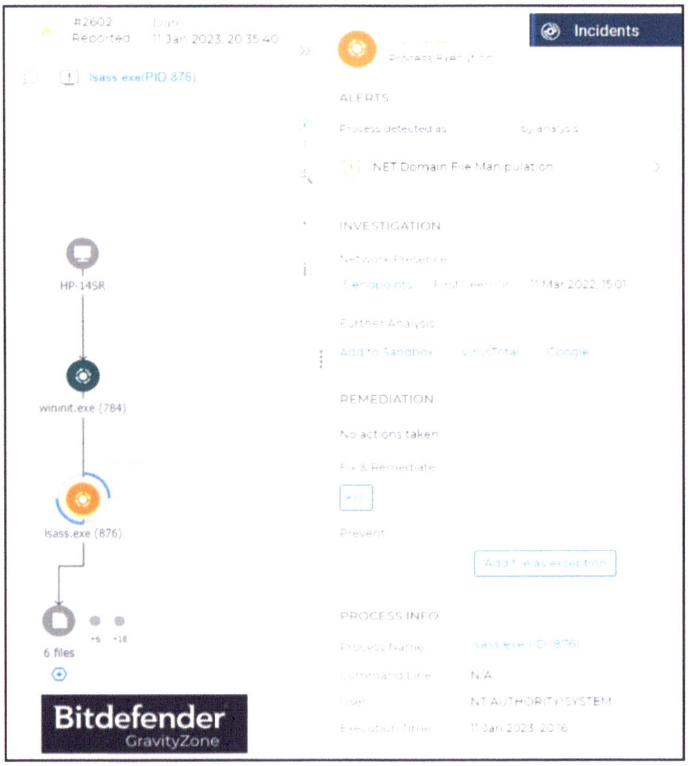

Fig. 2 Endpoint Threat Detection and Response (ETDR). *Source* www.bitdefender.com

5 Conclusions

It is necessary to use security solutions, resulting in the creation of real benefits to the institution once they are configured according to the needs of secure operation. Intuitive, easy-to-use management tools can optimize the time needed to implement new policies to achieve accurate monitoring and alerting. Also, the collaboration with the National Cyber Intelligence Center provides stability and access to the necessary knowledge in the activity of implementing these systems to ensure cyber security.

Unified and integrated technologies offer a measurable advantage in results, benefiting from consoles and tools adapted to the level of expertise, in accordance with the strategies built by the IT Security department. Cooperation with other institutions such as the National Center for Cyber Intelligence, participation in seminars and conferences in the field, represents an excellent tool for improvement, and especially a guide of best practices, assimilated to adopt the best decisions for the public institution.

The budget allocated to innovation in public administration will create and maintain such desirable stability, especially in critical areas, such as European funds. In industry and the economy, the role of robotics and process automation will grow considerably, with technology-related changes bringing both benefits and vulnerabilities, particularly in cyberspace. It will basically create a virtual parallel world, where the existence of the state, with all that it stands for, must be protected so that the environment is safe and secure even for the individual. The consequences of competition in innovation produce major transformations, including in society, simplifying the complex life of modern man in the information society. Thus, constituting a national interest for the Government Strategy, areas such as attracting European funds and ensuring cyber security, aiming at the modernization and computerization of public administration in Romania.

Thus, it is necessary to create a global framework of ICT security and trust, with an expansive trend towards process automation to achieve maximum efficiency. This strategic objective aims at creating scientific and technological excellence, as well as obtaining advantages in terms of the security and resilience of systems, services and critical infrastructure of national importance, as well as increasing the degree of cyber security culture in the central public administration, with the possibility of establishing within the institutional organization of at least 3 positions with specific attributions in the cyber field, in direct collaboration with the Security Structure of the Ministry in question and in a cooperation agreement with the National Authority in the field of Cyber-intelligence, following the designation by CSAT- the Supreme Council of National Defense—the Romanian Intelligence Service, through the National Cyber Intelligence Center.

In this way, an important stage in the inter-institutional collaboration will be achieved, for the achievement of the fundamental objectives of the country strategy, the field of funds becoming a critical infrastructure through the implications inherent in the national economy, affecting all the plans of society, from financial—to economic, social-educational, up to the political one, with all the necessary risks assumed. But most importantly, the efficient management of infrastructure and applications for European funds, having particular importance in the much-needed process of increasing the quality of life, representing the first step towards knowledge, innovation and development.

References

European Commission: Jobs and the economy during the COVID-19 pandemic (2022). https://ec.europa.eu/info/live-work-travel-eu/coronavirus-response/jobs-and-economy-during-coronavirus - pandemic.ro.

European Information Society (2005) - Publisher: Foundation for European Studies.

European Commission - Brussels, 3.3. (2021) One year since the outbreak of COVID-19: fiscal policy response https://ec.europa.eu/info/files/one-year-outbreak-covid-19-fiscal-policy-response_en.

Presidential Administration - Bucharest (2020) Romania - *National Strategy for National Defense for the period 2020–2024.* https://www.presidency.ro/files/userfiles/Documente/Strategia_Nationala_de_Aparare_a_Tarii_2020_2024.pdf

European Council - Council of the European Union - March (2010) - European Union Internal Security Strategy; https://www.consilium.europa.eu/ro/documents-publications/publications/internal-security-strategy-european-union-towards-european-security-model/.

Decision of the Official Gazette no. 677 (2020–August 14) - on the approval of the National Program for the digitization of micro, small and medium enterprises, financed under the Operational Program Competitiveness 2014–2020. http://legislatie.just.ro/Public/DetaliiDocument/229226 - OFFICIAL GAZETTE no. 756 of 19 August 2020.

EU Directive 1148 / (2016) - *Measures for a high level of security of networks and information systems in the Union.* https://cert.ro/pagini/ansrsi.

Regulation (EU) (2016) / 679 - *on the protection of individuals with regard to the processing of personal data and on the free movement of such data and repealing Directive 95/46 / EC* (General Data Protection Regulation).

The European Union Agency for Cybersecurity (ENISA), (2021) September 13 - *Methodology for a Sectoral Cybersecurity Assessment.* https://www.enisa.europa.eu/publications/methodology-for-a-sectoral-cybersecurity-assessment.

The European Union Agency for Cybersecurity (ENISA), (2020) April 15 - *Advancing Software Security in the EU* https://www.enisa.europa.eu/publications/advancing-software-security-through-the-eu-certification-framework

National Cybersecurity Directorate (DNSC) - (2021) September 30 - *European Cybersecurity Month – ECSM.* https://cert.ro/citeste/comunicat-luna-europeana-a-securitatii-cibernetice-2021

Oracle Romania: *Emerging technologies:* IoT, EoT, AI, Blockchain (2022). https://www.oracle.com/ro/emerging-technologies/.

Cloud Computing, Events - October 6, (2021 at 11:19 am) - Cloud Conference brings new technologies to the forefront - (clubitc). https://www.clubitc.ro/2021/10/06/conferinta-de-cloud-aduce-in-prim-plan-noile-tehnologii/.

Performance Evaluation of Green Ports via C-IFS AHP: A Case Study in Turkey

Mehmet Ali Taş and Esra Çakır

Abstract Seaports are an indispensable element of world maritime trade. All activities in the world that deal with problems such as global warming and the deteriorating ecological balance should prioritize the environment and be sustainable. The concept of the green port has inherently emerged in this direction. Various indicators can be used to assess the environmental performance of ports in a country. Seaports evaluated according to these indicators can obtain the green port title and certification. Seaports with the highest environmental sustainability performance can be scored and ranked according to the weights of the indicators. Fuzzy multi-criteria decision making (MCDM) methods are useful in calculating the importance of these indicators and ranking the alternative ports. Linguistic views and responses of experts can be placed as data for the proposed MCDM with a fuzzy approach. In this study, a novel methodology including circular intuitionistic fuzzy sets (C-IFS), which is newly introduced to the literature, and Analytic Hierarchy Process (AHP) approach is proposed. It is aimed that the criteria weights and alternative ranking obtained as a result of this study contribute to the appraisement of the seaports in terms of environmental sustainability and enable the emergence of the aspects that are required to be improved.

Keywords Circular intuitionistic fuzzy sets · Analytic hierarchy process · Green port · Fuzzy logic · Multi-criteria decision making

M. A. Taş (✉)
Turkish-German University, 34820 Beykoz, İstanbul, Turkey
e-mail: mehmetali.tas@tau.edu.tr

E. Çakır
Galatasaray University, 34349 Ortaköy, İstanbul, Turkey

M. Busu (ed.), *Rethinking Business for Sustainable Leadership in a VUCA World*,
Springer Proceedings in Business and Economics,
https://doi.org/10.1007/978-3-031-50208-8_21

1 Introduction

The importance of maritime transport in the global supply chain has increased with globalization (Bergqvist and Egels-Zandén 2012). Maritime transport offers an ascending cost advantage compared to railway, roadway, and airway, respectively. Thanks to its loading capacity, high tonnages can be transported at once. Moreover, because maritime transportation provides the lowest cost by distance, it is the most suitable option for long-distance transport that crosses the oceans (Behar et al. 2011). These reasons have caused the trade made by maritime transport to constitute the vast majority of the total international trade volume (Shi and Li 2017). Although maritime transport is considered the least harmful type of transport to nature, it is essential to strive for sustainability due to the size of the activities (Peris-Mora et al. 2005). In order to ensure sustainable development, all maritime transportation elements and activities should be managed appropriately.

A seaport can be defined as (Talley 2009): "a place at which the transfer of cargo and passengers to and from waterways and shores occurs". At least two seaports are used for goods transported by sea (Trujillo and Nombela xxxx). In the past, the main expectation of port managements focused on economic factors such as the transfer volumes at the port (Notteboom and Haralambides 2020). Over time, the increasing importance of sustainability around the world has pushed seaports to adapt in this regard. Thus, the concept of the sustainable seaport has emerged.

Sustainable ports serve sustainable development in economic, social, and environmental dimensions (Maritz et al. 2014). The main concern of sustainability is the limited resources that must be protected. It is necessary to focus on all three dimensions rather than making developments in only one. Nevertheless, improvements in environmental and social issues to ensure sustainability contribute positively to the economic result and customer retention (Lam and Dai 2015). The deteriorating ecological balance threatens all elements in the ecosystem (Bongaarts 1992). All activities in all businesses need to manage their waste and take control of resource consumption (Dwyer et al. 2009). Maritime transportation components and activities are expected to be similarly environmentally friendly. Green port (also called eco-port) is a port concept that expresses the adaptation of this environmental sustainability perspective to its processes. Ports consume high energy due to the operating volume they contain (Acciaro et al. 2014). There are negative impacts on the air and surface water quality of seaport activities. These effects can be observed and prevented by establishing or adopting environmental management plans (Gupta et al. 2005). Many countries including those in Europe and the Americas have embraced the green port concept and encouraged making management strategies in this regard (Lawer et al. 2019). Air quality, climate change, and energy efficiency are the top-three environmental priorities of European ports (Ecoports. 2021). While measuring the environmental performance of green ports, the establishment, management, and activities of the port, hinterland activities, and ships are taken into account (Chiu et al. 2014).

There are many indicators (factors) that are taken into account for evaluating green ports. Reflecting various perspectives makes the evaluation a complex problem (Chengpeng et al. 2018). Moreover, it may be necessary to select the green port with the best environmental performance based on these indicators. MCDM methods can be used to overcome these difficulties. MCDM methods are employed to make evaluations that contain conflicting criteria (Opricovic and Tzeng 2004). Numerous well-known MCDM methods have been used in the literature such as AHP (Saaty 1996), TOPSIS (Hwang and Yoon 1981), PROMETHEE (Brans et al. 1986), MOORA (Brauers and Zavadskas 2006) and others (Stojčić et al. 2019). In real-life applications, the methods can be performed individually or in combination with another one. These methods can also be combined with fuzzy logic. Thus, it is possible to utilize the uncertainty of subjective evaluations in solving the problem (Kahraman et al. 2004).

The aim of this study is to evaluate the green performance of seaports in Turkey. Four evaluation attributes and four alternative seaports are determined with the experts consulted. The fuzzy circular intuitionistic fuzzy AHP (C-IFS AHP) method has been proposed in order to evaluate the green performance of these ports. It is thought that the most important aspect of the study is its contribution to the environmental assessment literature in maritime transport. In this study, circular intuitionistic fuzzy AHP, which is a novel recent fuzzy MCDM method, is used for green port ordering. It has been observed that there is no green seaport selection practice where this novel approach is employed in Turkey.

2 Literature Review

In this section, the MCDM applications on issues related to the green port concept in the literature also were reviewed. Using factor analysis and fuzzy logic, Park and Yeo (Park and Yeo 2012) observed ports in Korea for their level of environmental friendliness. The most important criterion in the study, in which the authors evaluated five major ports with respect to five criteria, is "ease the environmental burden". Chiu et al. (Chiu et al. 2014) ranked three ports in Taiwan based on green port criteria. Thirteen factors were weighted by the fuzzy AHP method. As a result of the method in which triangular fuzzy numbers are used to express the assessments of decision makers, the most important factor in green port operations is "hazardous waste handling". Elzarka and Elgazzar (2014) determined the importance of key performance criteria of the ports in Egypt in becoming green ports with the fuzzy AHP method. As a result of this method, it was reached that the most important key performance indicator was "air pollution avoidance". Asgari et al. (2015) utilized AHP in an application that examine five ports in the UK in terms of sustainability. After the first two criteria in the economic dimension, "identify environmental objectives and priorities" in the environmental dimension of sustainability was determined to be the third most important criterion. Chengpeng et al. (2018) computed the weights of the criteria in the green port development evaluation of five ports in China using AHP and calculated the criterion with the highest weight as "gross domestic product". Munim et al. (2020) investigated governance models for green port management on the Indian Ocean Coast. The results of Best-Worst Method (BWM) and ANP were compared, and it was concluded that the BWM method was more applicable than the ANP method. Boljat (2021) employed another version of PROMETHEE (PROMETHEE II) to score energy-efficient mobility alternatives as a green port management activity for two ports in Croatia.

The studies in the literature in which C-IFS fuzzy sets and MCDM approaches are used together are also examined. Çakır et al. (2021a, b) benefited circular intuitionistic fuzzy MCDM methodology in order to select medical waste landfill sites and assess the enterprises in industrial symbiosis. AHP-VIKOR (Otay and Kahraman 2021), AHP (Çakır and Taş 2022) and TOPSIS (Kahraman and Alkan 2021) are also MCDM methods in that C-IFS sets are employed.

3 Methodology

After introducing intuitionistic fuzzy sets (Atanassov 1986), Atanassov (2020) expanded the definition and presented circular intuitionistic fuzzy sets (C-IFS). The steps of the proposed circular intuitionistic fuzzy AHP (C-IFS AHP) methodology are given in the flowchart in Fig. 1. The extended version of the method is detailed in article (Çakır and Taş 2022).

Turkey plays an important role in world maritime trade due to its location. Most of the country's foreign trade is transported by sea (Ateş et al. 2010). It is necessary to struggle with sea and air pollution caused by maritime activities. Efforts to improve

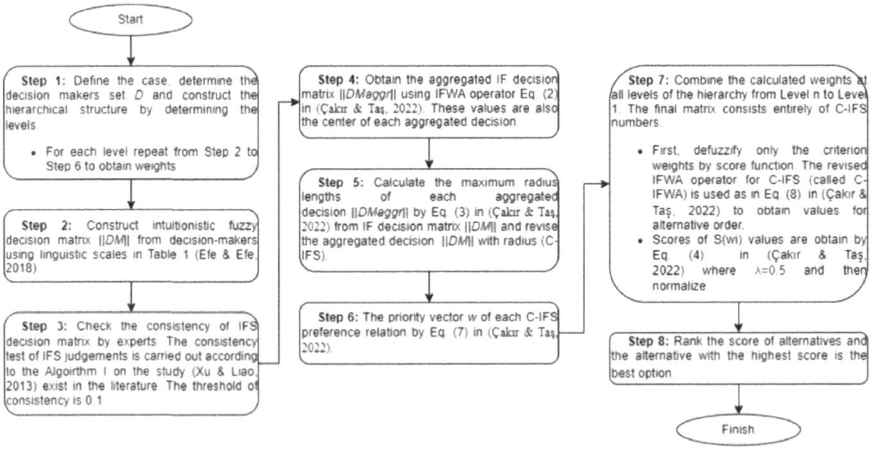

Fig. 1 The flowchart of proposed circular intuitionistic fuzzy AHP (C-IFS AHP) methodology.

Table 1 Intuitionistic fuzzy linguistic terms (Efe and Efe 2018).

Linguistic terms	Code	IFS
Very Low	VL	<0.05,0.95>
Low	L	<0.25,0.65>
Equal	E	<0.5,0.5>
Medium	M	<0.5,0.4>
High	H	<0.75,0.15>
Very high	VH	<0.95,0.05>

the environmental performance of ports' activities may prevent the harmful effects of increased maritime trade. Green port policies provide to prevent air and water pollution by controlling waste. Elements that threaten the life of the living creatures in the sea can be eliminated. There are environmental certifications related to green ports, such as the ISO14001 environmental management system. In addition, port executives regulate their activities in order to comply with international standards such as Port Environmental Review System (PERS) and Eco-Management and Audit Scheme (EMAS) (Puig et al. 2017). 105 ports in 25 countries are referred to as Eco-ports by the European Sea Ports Organization (Ecoports. 2021). Nevertheless, rewarding the highest performing businesses by evaluating the green performance of seaports can be an incentive for green port activities. MCDM methods can be used to achieve this goal.

As a case study, the green performance of seaports in Turkey is examined. P1, P2, P3 and P4 are the ports that evaluated their adaption on the indicators I1: Environmental management system, I2: Air pollution management, I3: Noise pollution, I4: Ecological monitoring in the port. The intuitionistic fuzzy number equivalents of the evaluations of the decision makers in the shipping industry in Turkey are given in Table 1.

After the execution of the proposed methodology step by step, the final table is constructed in Table 2.

Table 2 Final matrix of combined weights of levels.

Model	I1 $S(w_{c_1})=0.372$	I2 $S(w_{c_2})=0.256$	I3 $S(w_{c_3})=0.148$	I4 $S(w_{c_4})=0.224$	Ranking value	Score	Normalized score
P1	<0.31,0.51;0.21>	<0.19,0.42;0.24>	<0.51,0.25;0.12 >	<0.21,0.49;0.16 >	<0.296,0.433;0.240>	0.557	0.228
P2	<0.22,0.64;0.43>	<0.35,0.36;0.37>	<0.23,0.28;0.25 >	<0.39,0.31;0.24 >	<0.383,0.415;0.430>	0.638	0.261
P3	<0.43,0.34;0.22>	<0.23,0.21;0.22>	<0.65,0.33;0.13 >	<0.34,0.42;0.38 >	<0.429,0.314;0.380>	0.646	0.264
P4	<0.17,0.49;0.36>	<0.18,0.53;0.13>	<0.31,0.12;0.24 >	<0.37,0.52;0.21 >	<0.309,0.411;0.360>	0.603	0.247

4 Conclusion

Seaports are one of the most important components of the world trade. The increase in environmental initiatives all over the world has also affected the ports and created the concept of green ports. It may be possible to prioritize environmental investments by examining the green performance of ports. In addition, ports can be encouraged to show high environmental performances with various certification and award systems. MCDM methods can be used to evaluate green ports. In this study, four seaports in Turkey have been evaluated in terms of eco-friendliness. According to the results of the methodology, P3 has the best performance based on sustainable indicators. In future studies, neutrosophic fuzzy numbers or pythagorean fuzzy numbers can be used instead of C-IFS in order to compare the results.

Acknowledgement This work has been supported by the Scientific Research Projects Commission of Galatasaray University.

References

Acciaro, M., Ghiara, H., Cusano, M.I.: Energy management in seaports: A new role for port authorities. Energy Policy **71**, 4–12 (2014)

Asgari, N., Hassani, A., Jones, D., Nguye, H.H.: Sustainability ranking of the UK major ports: Methodology and case study. Transp. Res. Part E: Logist. Transp. Rev. **78**, 19–39 (2015)

Atanassov, K.T.: Intuitionistic fuzzy sets. Fuzzy Sets and Systems **20**(1), 87–96 (1986)

Atanassov, K.T.: Circular intuitionistic fuzzy sets. J. Intell. & Fuzzy Syst. **39**(5), 5981–5986 (2020)

Ateş, A., Karadeniz, Ş, Esmer, S.: Dünya Konteyner Taşımacılığı Pazarında Türkiye'nin Yeri. Dokuz Eylül Üniversitesi Denizcilik Fakültesi Derg. **2**(2), 83–98 (2010)

Behar, A., Venables, A.: Transport costs and international trade. In: de Palma, A., Lindsey, R., Quinet, E., Vickerman, R. (eds.) Handbook of transport economics, pp. 97–116. Edward Elga Publishing, Cheltenham (2011)

Bergqvist, R., Egels-Zandén, N.: Green port dues—The case of hinterland transport. Res. Transp. Bus. & Manag. **5**, 85–91 (2012)

Boljat, H.U., Vilke, S., Grubišić, N.: Maglić, L: Application of multi-criteria analysis for the introduction of green port management practices: an evaluation of energy efficient mobility in nautical ports. Zeszyty Naukowe Akademii Morskiej w Szczecinie **65**(137), 72–83 (2021)

Bongaarts, J.: Population growth and global warming. Popul. Dev. Rev. **18**(2), 299–319 (1992)

Brans, J.P., Vincke, P., Mareschal, B.: How to select and how to rank projects: The PROMETHEE method. Eur. J. Oper. Res. **24**(2), 228–238 (1986)

Brauers, W.K., Zavadskas, E.K.: The MOORA method and its application to privatization in a transition economy. Control. Cybern. **35**, 445–469 (2006)

Çakır, E., Taş, M. A.: Circular intuitionistic fuzzy analytic hierarchy process for remote working assessment in Covid-19. In: Intelligent and fuzzy systems: digital acceleration and the new normal-proceedings of the INFUS 2022 Conference, Volume 1, pp. 589–597. Springer International Publishing, Cham (2022)

Çakir, E., Taş, M. A., Ulukan, Z.: A new circular intuitionistic fuzzy MCDM: A case of Covid-19 medical waste landfill site evaluation. In: 2021 IEEE 21st international symposium on computational intelligence and informatics (CINTI), pp. 000143–000148. IEEE (2021)

Çakır E., Taş M. A., Ulukan Z.: Circular Intuitionistic Fuzzy Sets in Multi Criteria Decision Making. In: Aliev R. A., Kacprzyk J., Pedrycz W., Jamshidi M., Babanli M., Sadikoglu F. M. (eds.) 11th International Conference on Theory and Application of Soft Computing, Computing with Words and Perceptions and Artificial Intelligence–ICSCCW–2021. ICSCCW 2021. Lecture Notes in Networks and Systems, 362, 34–42. Springer, Cham (2022).

Chengpeng, W., Di, Z., Xinping, Y., Zaili, Y.: A novel model for the quantitative evaluation of green port development–A case study of major ports in China. Transp. Res. Part D: Transp. Environ. **61**, 431–443 (2018)

Chiu, R.H., Lin, L.H., Ting, S.C.: Evaluation of green port factors and performance: a fuzzy AHP analysis. Math. Probl. Eng. **2014**, 802976 (2014)

Dwyer, R., Lamond, D., Lee, K.H.: Why and how to adopt green management into business organizations? The case study of Korean SMEs in manufacturing industry. Manag. Decis. **47**(7), 1101–1121 (2009)

Ecoports.: (2021). https://www.ecoports.com. Last accessed 20 May 2021

Efe, B., Efe, Ö.F.: Intuitionistic fuzzy number based group decision making approach for personnel selection. Uludağ Univ. J. Fac. Eng. **23**(3), 11–26 (2018)

Elzarka, S., Elgazzar, S.: Green port performance index for sustainable ports in Egypt: a fuzzy AHP approach. In International Forum on Shipping, Ports and Airports (IFSPA) 2014. Sustainable Development in Shipping and Transport Logistics (2014)

Gupta, A.K., Gupta, S.K., Patil, R.S.: Environmental management plan for port and harbour projects. Clean Technol. Environ. Policy **7**(2), 133–141 (2005)

Hwang, C.L., Yoon, K.: Methods for multiple attribute decision making. Multiple attribute decision making, pp. 58–191. Springer, Berlin, Heidelberg (1981).

Kahraman, C., Alkan, N.: Circular intuitionistic fuzzy TOPSIS method with vague membership functions: Supplier selection application context. Notes Intuit.Istic Fuzzy Set **27**(1), 24–52 (2021)

Kahraman, C., Cebeci, U., Ruan, D.: Multi-attribute comparison of catering service companies using fuzzy AHP: The case of Turkey. Int. J. Prod. Econ. **87**(2), 171–184 (2004)

Lam, J.S.L., Dai, J.: Environmental sustainability of logistics service provider: an ANPQFD approach. Int. J. Logist. Manag. **26**(2), 313–333 (2015)

Lawer, E.T., Herbeck, J., Flitner, M.: Selective adoption: How port authorities in Europe and West Africa Engage with the globalizing 'Green Port' idea. Sustainability **11**(18), 5119 (2019)

Maritz, A., Shieh, C.J., Yeh, S.P.: Innovation and success factors in the construction of green ports. J. Environ. Prot. Ecol. **15**(3A), 1255–1263 (2014)

Munim, Z.H., Sornn-Friese, H., Dushenko, M.: Identifying the appropriate governance model for green port management: Applying Analytic Network Process and Best-Worst methods to ports in the Indian Ocean Rim. J. Clean. Prod. **268**, 122156 (2020)

Notteboom, T.E., Haralambides, H.E.: Port management and governance in a post- COVID-19 era: quo vadis? Marit. Econ. & Logist. **22**, 329–352 (2020)

Opricovic, S., Tzeng, G.H.: Compromise solution by MCDM methods: A comparative analysis of VIKOR and TOPSIS. Eur. J. Oper. Res. **156**(2), 445–455 (2004)

Otay, I., Kahraman, C.: A novel Circular Intuitionistic Fuzzy AHP&VIKOR methodology: An application to a multi-expert supplier evaluation problem. Pamukkale Univ. J. Eng. Sci. **28**(1), 194–207 (2021)

Park, J.Y., Yeo, G.T.: An evaluation of greenness of major Korean ports: A fuzzy set approach. Asian J. Shipp. Logist. **28**(1), 67–82 (2012)

Peris-Mora, E., Orejas, J.D., Subirats, A., Ibáñez, S., Alvarez, P.: Development of a system of indicators for sustainable port management. Mar. Pollut. Bull. **50**(12), 1649–1660 (2005)

Puig, M., Michail, A., Wooldridge, C., Darbra, R.M.: Benchmark dynamics in the environmental performance of ports. Mar. Pollut. Bull. **121**(1–2), 111–119 (2017)

Saaty, T. L.: The analytic hierarchy process, New York: McGraw Hill. International, Translated to Russian, Portuguese, and Chinese, Revised editions, Paperback (1996, 2000). RWS Publications, Pittsburgh (1980)

Shi, W., Li, K.X.: Themes and tools of maritime transport research during 2000–2014. Marit. Policy & Manag. **44**(2), 151–169 (2017)

Stojčić, M., Zavadskas, E.K., Pamučar, D., Stević, Ž, Mardani, A.: Application of MCDM methods in sustainability engineering: A literature review 2008–2018. Symmetry **11**(3), 350 (2019)

Talley, W.K.: Port economics. 2nd edn. Routledge, (2009)

Trujillo, L., Nombela, G.: Privatization and regulation of the seaport industry. The World Bank. https://documents1.worldbank.org/curated/en/973281468740398501/104504322_20041117155008/additional/multi-page.pdf. Last accessed 20 Oct 2022

Knowledge Vulnerabilities Scoring System and the Knowledge Economy

Vlad Mihai Ursache

Abstract The paper aims to study knowledge vulnerabilities correlations with the knowledge economy, focusing on identifying knowledge vulnerabilities considering new challenges in the knowledge economy. The purpose is to analyze and understand the knowledge vulnerabilities, focusing our research on identifying some vulnerabilities in the organizational knowledge dynamics. In order to achieve our goal, we propose KVSS method that may be used in the identification and analysis process of knowledge vulnerabilities. The process will be beneficial for a correct analysis of knowledge vulnerabilities and the impact in the knowledge economy, completing the research with some solutions that may reduce negative consequences for organizations that are an active component in the knowledge economy. In today knowledge economy were knowledge has become essential for every organization and disruptions are happening frequently which gives organizations less time to manage the change and the fact that they are almost constrained to develop new strategies to answer the challenges of the changing business environment, identifying knowledge vulnerabilities will increase visibility on organizational knowledge given the opportunity to threat weakness point identified and in the end to achieve the goal to strengthen the organization. The knowledge economy has evolved more than was predicted in a short time period, adding notable value to the knowledge economy. On other side, this fast evolution in the knowledge economy revealed that some components were not fully covered, leaving some gaps that may be exploited. Our research is focused on identifying these gaps in the knowledge economy, especially on knowledge vulnerabilities.

Keywords Knowledge · Knowledge economy · Knowledge vulnerabilities · Knowledge vulnerabilities scoring system · Knowledge management

---

V. M. Ursache (✉)
National University of Political Studies and Public Administration, Exposition Blvd., No. 30 A, Sector 1, Bucharest, Romania
e-mail: ursachevladmihai@yahoo.com

M. Busu (ed.), *Rethinking Business for Sustainable Leadership in a VUCA World*, Springer Proceedings in Business and Economics,
https://doi.org/10.1007/978-3-031-50208-8_22

1 Introduction

In a world of continuous change and marked by unpredictability, world pandemic crisis, military tensions between the United States of America and China, Turkiye and Greece, Israeli and Palestinian conflict, insurgencies in Africa and other parts of the world, adding energy crisis, global food crisis, financial crisis, warming climate crisis and cybersecurity war, "we ask how challenging all of this can be to knowledge management especially for knowledge risks management given the fact that organizations have less time to manage all the uncertainties and that they are constrained to continuously develop new strategies to assure healthy management of knowledge risks" (Ursache 2022b).

In a digitalized era where innovative technologies are on a daily basis and information is transferred at a very high speed, the number of research topics in the knowledge economy domain has become more and more diversified challenging researchers to explore unknown's areas and unexploited research fields of interest, leading knowledge economy domain to a larger academic scale than it is now. In this process of continuous change, knowledge has become more critical than ever, becoming the main component in the evolution process of organizations and societies and a key for organizations to achieve their success.

The goal of the current study is to gain a more relevant and objective understanding of the ideas of knowledge vulnerabilities in the field of the knowledge economy, starting from the SECI model that describes the dynamics of organizational knowledge elaborated by Ikujiro Nonaka and refined together with his colleagues, most notably with Hirohata Takeuchi (Nonaka and Takeuchi 1995). This model contains three main components: the SECI knowledge cycle, the dynamic context Ba, and the knowledge vision (Bratianu 2015). Nonakian dyad of tacit-explicit knowledge is completed with the research on the triple Helix of knowledge, the triad of cognitive-emotional-spiritual knowledge elaborated by Bratianu.

Consequently, the research is focused on the three conceptual pillars of our research, presented in Fig. 1 below, and presented in two progressive steps. First, the comprehensive literature review findings are thoroughly analyzed under the three conceptual pillars of the research, knowledge vulnerabilities, knowledge risks, and knowledge vulnerabilities scoring systems (KVSS). All three conceptual pillars of the research will be analyzed from the organizational dynamic's perspective. Second, we will focus our research on the proposed method Knowledge Vulnerabilities Scoring System to analyze and identify knowledge vulnerabilities.

The subject chosen to be researched, in this case, "Knowledge Vulnerabilities Scoring System and the Knowledge Economy", generates a complex set of future research directions, which may be clearly defined and specified in future research. The research objective is identifying, analyzing and proposing solutions that may be applied in the knowledge economy, which may reduce negative consequences within an organization increasing their knowledge in the economy area.

The expected results are to understand the primary sources of vulnerabilities and risks that may be generated for a knowledge economy system and to identify

Fig. 1 The pillars of the research (Author's research)

solutions that may be used to treat vulnerabilities and assess the risks in the knowledge economy.

Although there are some excellent papers focusing on knowledge (Nonaka and Takeuchi (2019); Bratianu and Vasilache, 2009; Bratianu, 2015) on knowledge risks (Bratianu et al. 2020; Durst and Wilhelm, 2013; Durst and Zieba 2017; Durst and Henschel 2020), there is no paper analyzing the vulnerabilities on knowledge economy from a dynamic organizational perspective. There are only very few papers on vulnerabilities (Timmerman 1981; Turner et al. 2003; Bejinaru 2022; Bratianu and Bejinaru 2022), but none of them analyze or identify and treat vulnerabilities purely from a knowledge perspective and only marginally about their relationship with the management domain from different perspectives.

Thus, there is a critical knowledge gap in the literature dedicated to analyzing, identifying, and understanding knowledge vulnerabilities in the knowledge economy, not to mention a method dedicated to scoring and metrics of knowledge vulnerabilities. The gap identified in the knowledge domain is critical for future improvements in knowledge economy research and organizational knowledge dynamics.

The present research aims to perform a systematic bibliometric study to identify some correlations of knowledge vulnerabilities with knowledge economy. Our research question can be formulated as follows:

RQ: *Are knowledge vulnerabilities treated in the knowledge economy domain?*

The research is qualitative and interpretive and is performed by using VOSviewer, specialized bibliometric software for massive literature reviews. To best fulfill the goals of the research, the introductory part will be followed by specific literature reviews with a specific focus on knowledge economy, knowledge vulnerabilities and knowledge risks. By monitoring the existing connections between the knowledge economy notion and other crucial concepts like knowledge vulnerabilities, this will enable us to depict the present state of study and demonstrate the most pertinent connections that we have found between them. Then, we will analyze the knowledge vulnerabilities challenges with data sources and the applied methodology that will be presented to conclude with results, study's limitations, and possible future research axes.

2 Literature Review

2.1 Knowledge Economy

From most famous work in the economy of Adam Smith (1723–1790), Wealth of Nations, published by Smith's in 1776 and considered the founding document of modern economics were it analyzes the division of labor, the role of self-interest in market economies, and the benefits of free trade and specialization to the Theory of Moral Sentiments, were Smith's earlier work on ethics and human behavior, published in 1759, explores the idea that people are naturally sympathetic to one another and that this empathy drives social cooperation.

Continued to the Principles of Economics of Alfred Marshall (1842–1924), considered to be a seminal work in the field of neoclassical economics. It introduced the concept of marginal utility and popularized the idea that supply and demand determine prices in markets. Marshall also developed the concept of consumer surplus and producer surplus, which are still used today to measure the welfare gains from trade.

To John Maynard Keynes (1883–1946) and the General Theory of Employment, Interest and Money, published in 1936, Keynes' book fundamentally challenged classical economic theory by arguing that market economies can experience prolonged periods of high unemployment and that government intervention can help to address this problem. Keynes being also known for his contributions to macroeconomics, including the idea of the "paradox of thrift" and the importance of aggregate demand in determining economic growth.

Reaching to Milton Friedman (1912–2006) and to the famous research, A Monetary History of the United States. This 1963 book, co-authored with Anna Schwartz, argues that monetary policy played a central role in the Great Depression and subsequent periods of inflation and deflation. It also advocates for a monetary rule to stabilize the economy. Friedman is known for his advocacy of monetarism, the idea that the money supply should be controlled to achieve stable economic growth, and for his support of free-market policies and deregulation.

To current research in the economy area where we have different needs and gaps, paying due attention to the knowledge vulnerabilities, will do nothing but finding solutions to various challenges and bringing that added value necessary for a better functioning of any organization and the economy.

It is important to understand that a knowledge-based economy has existed from the beginning of human civilization, and that it has evolved based on an ever-increasing body of information that has accumulated over time. Through the creation and distribution of commodities and services to satisfy socioeconomic demands, societies benefitted from knowledge.

Human capital, part of knowledge domain, was incorporated into the industrial process. The value of knowledge was not specifically mentioned in early economic literature. In the late nineteenth century, economists started to recognize its significance after Alfred Marshall proposed that "knowledge is our most powerful engine of production" and the organization facilitates the growth of knowledge. In the early twentieth century, Schumpeter considered the "new combination of knowledge" as an important element for innovation and entrepreneurship (Cader 2008).

In a positive economy, we discover that economics as a positive science is a body of tentatively accepted generalizations about economic phenomena that can be used to predict the consequences of changes in circumstances. Progress in expanding this body of generalizations, strengthening our confidence in their validity, and improving the accuracy of the predictions they yield is hindered not only by the limitations of human ability that impede all search for knowledge but also by obstacles that are especially important for the social sciences in general and economics in particular, though by no means peculiar to them (Friedman 1953).

In economies, the functions of knowledge are characterized by four important features: (a) knowledge ages rapidly and new knowledge is constantly replacing the old; (b) scientific (including social scientific) knowledge is highly valued, and the scale and economic penetration of scientific knowledge increases through subsequent economic development phases; (c) knowledge economies are especially characterized by the exploitation of new knowledge in order to create more new knowledge; and (d) knowledge is used in the production of goods and services, and to enhance the social welfare of its citizens (Cader 2008).

Knowledge economy is based on intangible resources which dominate now most of the companies in the well-developed economies. Powell and Snellman defined knowledge economy as "production and services based on knowledge-intensive activities that contribute to an accelerated pace of technical and scientific advances, as well as a rapid obsolescence". Knowledge has always been used in production and services but in the knowledge economy knowledge becomes dominant, such that knowledge management emerged as a necessary domain within the classical management (Massingham 2010).

The characterization and identification of knowledge is a complex process. There are kinds of know-ledge (know-what, know-why, know-how and know-who) which are important for knowledge-based economies. The stock or knowledge of these "kinds of knowledge" could vary from economy to economy, firm to firm, or region to region, and there is no clear understanding of what constitutes different kinds of knowledge (Cader 2008).

Understanding the knowledge economy means to understand first the concept of knowledge and its specific features. For instance, knowledge does not have a clearly delineated structure because its understanding is bounded by the metaphors used in getting its semantic field. Knowledge is intangible and nonlinear distinguishing this way clearly from the tangible resources like physical objects including monetary resources (Bratianu and Vasilache 2009; Nonaka and Takeuchi 1995).

Synthetically, OECD (2006) remarks on three basic features of the knowledge assets: "(i) they are sources of probable future economic profits; (ii) they lack physical substance; (iii) to some extent, they can be retained and traded by a firm" (p. 9). Knowledge is created by people and as a result of the knowledge creating spiral described by Nonaka and Takeuchi (2019), it amplifies and becomes organizational knowledge contributed significantly to the organization performance.

An excellent example of how knowledge powers firms could be unicorns. They are start-up firms that in a very short time reach the value of 1 billion dollars, by far surpassing traditional companies that already have experience in the market. For a unicorn, market experience, tradition and classic business models, doesn't even matter. All that matters is the knowledge they possess, how they used it and how they create value through knowledge in the new economy, where technological innovations and opportunities are everywhere, including risks (Ursache 2022a).

Another example will be on blockchain technologies and on crypto currencies, where the knowledge has reached an unimaginable well-being that is in a continuous growth, and research in these fields being almost non-existent.

2.2 *Knowledge Vulnerabilities*

Knowledge vulnerabilities is a new research field for the knowledge economy and this research area has become essential for any organization to understand how internal and external forces influence each other, how the organizational balance works, how important is the identification process of knowledge vulnerabilities to an organization, and how this process can make the difference between success and failure for an organization.

Aware that we live in an increasingly turbulent environment where uncertainty dominates, it is the knowledge of the world that one possesses. The capability to learn can make a difference (Bratianu and Bolisani 2017, p. 234).

Considering that knowledge is intangible and nonlinear, distinguishing this way clearly from tangible resources like physical objects, including monetary resources (Bratianu and Vasilache 2009; Nonaka and Takeuchi 1995).

Synthetically, OECD (2006) remarks three essential features of the knowledge assets: "(i) they are sources of probable future economic profits; (ii) they lack physical substance; (iii) to some extent, they can be retained and traded by a firm" (p. 9).

People create knowledge, and as a result of the knowledge-creating spiral described by Nonaka and Takeuchi (2019), it amplifies and becomes organizational knowledge that contributes significantly to the organization's performance.

Academic research on knowledge domain has been conducted in various fields, including management, economy, philosophy, and psychology, sociology, and computer science, but none of research focuses on knowledge vulnerabilities in the knowledge economy domain.

In academic research, vulnerabilities have been treated in disaster management, geography, climate change, economics, information security, and, more recently, cybersecurity. However, research has yet to be made so far in the knowledge domain, identifying and analyzing, in particular, the subject of knowledge vulnerabilities.

Having its roots in the Latin word *vulnus* and later on evolved into *vulnerabilis* and recently in vulnerability, the terminology has been used for centuries and from the beginning of its use until now, the same idea has been stated in numerous definitions. According to some studies, being vulnerable means having a weakness. However, it is incomplete because an organization may have some vulnerabilities, and as long as they are not exploitable, the organization does not have a weakness.

Our research is grounded on thoroughly examining literature to comprehend what has been explored about vulnerabilities. The aim is to identify different vulnerability classifications and analyze them to accommodate additional risks. Table 1 below outlines various vulnerability perspectives available in the literature. It will detail the definition, year, authors' name, journal name, and main findings.

On ISO 27005, we find vulnerability being defined as: a weakness of an asset or group of assets that can be exploited by one or more threats, where an asset is anything that has value to the organization, its business operations, and their continuity, including information resources that support the organization's mission; and on ISO/IEC 29,147:2018(en) Information technology, security techniques, vulnerability disclosure, define that a vulnerability can be thought of as a weakness or exposure that allows a security impact or consequence.

We agree more on the shape of the definitions given by the mentioned standards, at least for the mention that can be exploited by one or more threats, where an asset is anything that has value to the organization and allows a security impact or consequence.

From the above definitions, the term's weakness, threats, exposure, impact and consequence caught our attention for a deep analysis. Analyze-it together, each term can be associated with a process. Each process is part of a management system where vulnerabilities can be analyzed, identified, treated, and monitored and are directly connected with risks. Weakness, threats and exposure will not make sense without impact and consequence and vice-versa.

In all the processes of vulnerabilities identification and analysis, an important component is the risks calculation. The National Institute of Standards and Technology, NIST, defines risk as the net negative impact of the exercise of a vulnerability, considering both the probability and the impact of occurrence. The formula traditionally represents risk:

R (risk) = P (probability of occurrence) * C (consequence of occurrence either represented by some value or by a loss function).

Risk management is the process of identifying risk, assessing risk, and taking steps to manage risk by reducing risks to an acceptable level. Additionally, Smith

Table 1 Vulnerabilities definition available in the literature

Definition	Year	Author
Vulnerability is the degree to which a person, system or unit is likely to experience harm due to exposure to perturbations or stress	2002	Kasperson et al.
Vulnerability is defined as the extent to which a natural or social system is susceptible to sustaining damage from climate change. Vulnerability is a function of the sensitivity of a system to changes in climate and the ability to adapt to systems to changes in climate. Under this framework, a higly vulnerable system would be one that is highly sensitive to modest changes in climate	1996	Intergovernmental Panel on Climate Change (PCC). World Meteorological Organisation
Vulnerability is conceived as both a biophysical risk as well as a social response, but within a specific areal or geographic domain. This can be geographic space, where vulnerable people and places are located, or social space who in those is most vulnerable	1996	Cutter
By vulnerability we mean the characteristics of a person or group in terms of their capacity to anticipate, cope with, resist and recover from the impact of a natural hazard. It involves a combination of factors that determine the degree to which someone's life and livelihood are put at risk by a discrete and identifiable event in nature or in society	1994	Blaikie et al.
Vulnerability is defined in terms of exposure, capacity and potentiality. Accordingly, the prescriptive and normative response to vulnerability is to reduce exposure, enhance coping capacity, strengthen recovery potential and bolster damage control (i.e., minimize destructive consequences) via private and public means	1993	Watts and Bohle
Vulnerability is the differential capacity of groups and individuals to deal with hazards based on their positions within physical and social worlds	1992	Dow
Vulnerability is operationally defined as the inability to take effective measures to insure against losses. When applied to individual's vulnerability is a consequence of the impossibility or improbability of effective mitigation and is a function of our ability to detect the hazards	1989	Bogard
Vulnerability is the potential for loss	1989	Chambers
Vulnerability is the degree of loss to a given element or set of elements at risk resulting from the occurrence of a natural phenomenon of a given magnitude	1982	United Nations Disaster Relief Organization (UNDRO)
Vulnerability is the degree to which a system acts adversely to the occurrence of a hazardous event. The degree and quality of the adverse reaction are conditioned by a system's resilience (a measure of the system's capacity to absorb and recover from the event)	1981	Timmerman
Vulnerability is the threat (to hazardous materials) to which people are exposed (including chemical agents and the ecological situations of the communities and their level of emergency preparedness). Vulnerability is the risk context	1980	Gabor and Griffitch

Source Authors' own research

Table 2 Knowledge vulnerabilities identified

Authors	Year	Knowledge risks	Knowledge vulnerabilities
Jamieson & Loeng	2003	• Lack of effective knowledge base maintenance • Knowledge stealing • Risk of declining organizational creativity and innovation • Ineffective management	Knowledge stealing vulnerabilities Declining organizational creativity and innovation vulnerabilities Ineffective management vulnerabilities
Bayer &Maier	2006	• Knowledge transfer risks are concentrated at the level of operational business practices.	Knowledge transfer vulnerabilities
Perrot	2007	• Knowledge gap risk, which may hinder the company in fulfilling its objectives.	Knowledge gap vulnerabilities
Lambe	2013	• Knowledge articulation risks • Knowledge outsourcing risks • Knowledge acquisition risks • Knowledge continuity risks	• Knowledge articulation vulnerabilities • Knowledge outsourcing vulnerabilities • Knowledge acquisition vulnerabilities • Knowledge continuity vulnerabilities
Durst & Zieba	2017	• Knowledge risks due to unlearning • Knowledge risks due to forgetting	• Unlearning knowledge vulnerabilities • Forgetting knowledge vulnerabilities
Brătianu	2018	• Knowledge waste risk • Knowledge hoarding risk • Knowledge hiding risk • Knowledge attrition risk Knowledge obsolescence risk	• Knowledge waste vulnerabilities • Knowledge hoarding vulnerabilities • Knowledge hiding vulnerabilities • Knowledge attrition vulnerabilities • Knowledge obsolescence vulnerabilities

Source Authors' own research

(2012) and Aubert et al. (2001) agree that IS managers and researchers traditionally define risk in terms of negative consequences describing risk as the possibility of loss or damage and the possibility of suffering harm or loss. An alternative view by Billington (1997) points out that, when examined closely, 'risk' can actually lead to both positive and/or negative consequences.

Viewing risk as something more than a hazard is highly applicable to risk management in KM. This is consistent with the NIST risk definition so the NIST risk algorithm will be used as the basis for determining knowledge loss risk (Murray 2009).

Nowadays, we find at least one detailed definition of vulnerability in each discipline and field of research. For the knowledge field we did not find an explicit definition but starting from the fact that being vulnerable means to have a weakness, we may say that vulnerability is an abstract idea that encompasses the inner state of an internal environment of an individual or system that when interact with an unfriendly external environment loss may result and the internal environment may be affected.

Researchers from the global environmental risk area as Kasperson et al. (2001a, b) defined vulnerability as the degree to which a person, system or unit is likely to experience harm due to exposure to perturbations or stresses.

According to environmental change researcher Adger (2006), the concept of vulnerability has been a powerful analytical tool for describing states of susceptibility to harm, powerlessness, and marginality of both physical and social systems, and for guiding normative analysis of actions to enhance well-being through reduction of risk.

The key point when we discuss it further on vulnerabilities is that we have to analyze it according to their exposure and exploitation. In case the exploitation of vulnerabilities is not possible, we have to consider only a possible vulnerability that may occur in the near future and only when the actual action of exploitation takes place, an action that we may call it "0 Zero time", only starting from that moment we may consider to be indeed a vulnerability to the organization.

Starting from our above understanding of the essence of vulnerabilities, from the time when a possible vulnerability is identified, considered to be a weakness and to generate a treat, to the exposure and the exploitation process that will change a possibility into certainty, the whole process will be at the basis of our future research on knowledge vulnerabilities.

Vulnerabilities can be analyzed and identified separately from risks. Up to a certain point, this process is helpful for any organization, but if this process ends here and needs to be followed by an analysis, identification and risk assessment, will not only be complete, but it will also become a cost in the end for the organization.

On the other hand, risk analysis, risk identification, and risk assessment can be made without identifying vulnerabilities, but this approach can also have some shortages. A more practical method in correctly identifying knowledge risks is to include even the analysis of knowledge vulnerabilities.

The identification process of knowledge vulnerabilities is not easy at all and involves a very good understanding of knowledge concepts, risks, knowledge management, and many other constructs. Considering that vulnerabilities are essentially the weaknesses that allow threats to exploit an organization (Winkler et al. 2017, p. 164) and vulnerabilities enable risks, our research will analyze vulnerabilities, including the risks as also.

Observing that knowledge risks have been treated in various research and already have been identified by researchers some of the knowledge risks, we will continue our research analysis on knowledge vulnerabilities having for reference the existing studies.

Our research is grounded on a thorough examination of the literature to comprehend what risks have been identified in the research on knowledge risk and knowledge risk management so far. The aim is to identify knowledge vulnerabilities considering some identification of knowledge risks and to correlate risks with vulnerabilities, understanding the reason behind any risks and outlining from our perspective the knowledge vulnerabilities identified. It will detail the authors' name, journal name, identified knowledge risks and identified possible knowledge vulnerabilities.

The calculation part for each vulnerability identified through the knowledge Vulnerability Scoring System (KVSS), exposure part, and the exploitation point of each vulnerability will be treated in future research (Table 2).

3 Knowledge Vulnerability Scoring System (KVSS)

Being an intangible resource, because knowledge is not a tangible physical object and does not have a certain form and mass, it can be represented by a field (Cegarra-Navarro et al. 2023); knowledge is hard to quantify, measuring it and associate a value for this type of resource, not to mention giving it a score. Indeed, is hard, but it is not impossible.

Our research opens a new way in the research of knowledge vulnerabilities proposing an ambitious goal to set a system through which it can be analyzed, keep track of a vulnerability's key traits, capture them, and generate a numerical assessment of how serious it is. The qualitative depiction of the number score (such as low, medium, high, and critical) can then be used to assist organizations in correctly evaluating and prioritizing their vulnerability management activities. The scoring formula and principles will be treated in future research. In our present research paper, we propose a method that may be used in a very challenging and complex area, on the knowledge domain.

We know that knowledge is created by people, and as a result of the knowledge creating spiral described by Nonaka and Takeuchi (2019), it amplifies and become organizational knowledge contributed significantly to the organization performance.

In the process of assessing and prioritize the knowledge vulnerabilities, we propose a method by which several factors will be considered in the analysis process of a knowledge vulnerabilities.

KVSS will be composed of three metric groups: Base, Temporal, and Environmental. A referral system for our method will be the mechanisms conceived by Forum of Incident Response and Security Teams. In this method, a vulnerability's severity will be reflected by the Base based on its inherent traits, which remain consistent over time, and assumes the reasonable worst-case impact across various deployed scenarios. Based on variables that change over time, such as the availability of exploitation, taking into account the exposure surface, the Temporal metrics will modify the Base severity of a vulnerability. The Base and Temporal severities will be adjusted to a particular environment by the Environmental metrics. We'll take into account elements like the presence of mitigations in that setting. Exposure, exploitation and threats will be among the indicators that may be used to achieve a qualitative representation of the vulnerabilities impact and further, a score will be used to prioritize vulnerability.

In the KVSS method, a part mentioned indicators, the assessment and prioritization of knowledge vulnerabilities we should include a risk analysis and risks calculation, focusing first on principal characteristics of knowledge and thus using the KVSS method we will be able to identify correctly the existing vulnerabilities in the knowledge domain. In the knowledge economy or other related area of knowledge.

Method to establish a Knowledge Vulnerabilities Scoring System (KVSS) will use the notion of severity, which is common among indicators used in the cybersecurity field. We will use the severity measure based on a metric that will produce a score ranging from 0 to 10, with 0 being the lowest and 10 being the highest severity. Three categories of metrics—Base, Temporal, and Environmental—will be separated. The Temporal and Environmental metrics can be scored in order to change the Base metrics' score, which ranges from 0 to 10. The scores will be computed in sequence in such a way that the Base Score will be used to calculate the Temporal Score and the Temporal Score will be used to calculate the Environmental Score. Metrics, impact metrics, and equations will be detailed in future research.

A vector string, a compressed textual representation of the values may be used to calculate the score, also to express the KVSS technique. KVSS may be frequently used to prioritize knowledge vulnerabilities remediation efforts and to determine the severity of vulnerabilities found on a system.

KVSS method will not be a measure of risk, will be a method that may be used to supply a qualitative measure of severity of knowledge vulnerabilities that may be a very useful as measurement system for any type of organization, and to prepare them to anticipate threats and to calculate better the risks. In all this process the goal will be to help organizations to properly assess and prioritize the knowledge vulnerabilities in the knowledge management system and finally making the organization stronger, anticipating and preventing the exposure vulnerabilities faster, improving even the treatment process.

4 Methodology

Knowledge vulnerabilities in an unexplored area and finding no research that analyze and identify knowledge vulnerabilities, on our research we have performed a systematic bibliometric study to identify the main correlations of the existing vulnerabilities studies from various fields with the knowledge economy.

The present paper relies on bibliometric research or statistical bibliography, to answer the research question: *Are knowledge vulnerabilities treated in the knowledge economy domain?*

The research is qualitative and interpretive and is performed by using VOSviewer, specialized bibliometric software for massive literature reviews. To best fulfill the goals of the research, the introductory part will be followed by specific literature reviews with a specific focus on vulnerabilities, threats, exposure, exploitation, knowledge risks and knowledge management.

This will let us visualize the main correlations of the vulnerabilities research from various fields with the knowledge domain, observing the existing links between vulnerabilities research from various fields with the knowledge domain. In the next stage, we will analyze the vulnerabilities research with data sources and the applied methodology that will be presented to conclude with results, study's limitations, and possible future research axes.

In this regard, a complementary computer-aided analysis process was conducted, utilizing VOSviewer software. According to the software creators Van Eck and Waltman, VOSviewer can be used in academic research projects to define, explore, and visually illustrate network-based scientific maps by employing text mining analysis. Out of the available range of approaches, the author used the term co-occurrence analysis option in the present conceptual exploration. In the present study, the terms or the words represent the unit of analysis. The analysis outcome is an intellectual plan or a knowledge atlas of the studied topic (Iliescu 2021). The data was retrieved from Scopus and the retrieval model was through an advanced search function, and our search was within *article title*, *abstract* and *keywords*, the term *vulnerability*, while the retrieval period was: 2015–2024.

We have selected starting from 2015, considering that almost 10 years will reveal notable results for our study. The same advanced set of data will be used for the term's *threats, exposure, exploitation, knowledge risks* and *knowledge management*. The subject area was selected from Business Management and Accounting. Other available subject areas were considered not directly connected to knowledge management.

In the preparation phase and extract of data from Scopus, we have implemented settings for the data search, filtering, and extraction, to get the most conclusive results. In the first place, we have defined a topic category, and we have focused our study on titles, abstracts, author keywords and keywords plus field. In this way, we consider it more relevant to gather more accurate findings for our term co-occurrence analysis. In this regard, we divided our study in two parts.

In the first part, we set the search structure on "vulnerability" to identify relevant publications for this concept to view the connection with other relevant concepts and the connection strength with other concepts. In our study's case, the "vulnerability" search initially returned 231,283 research and after applying the filter on subject area Scopus returned 7,118 results meet the threshold and setting a minimum occurrence of a term to 5, for each of 7,118 terms, has been calculated a score.

Based on this score, the most relevant terms were selected, and the default choice selected by 60% for the most relevant terms. With this option, 502 out of the 10,190 keywords meet the threshold. The overall strength of the linkages between the co-occurrences of each of the 502 keywords was determined. Out of 502 keywords, the ones with the most total links were chosen. The result after clearing and filtering the data resulted in 480 terms identified, and 6 clusters were considered relevant for our study, where the term vulnerability was found in the same cluster with terms such as risk, management.

5 Results and Discussions

In this section, we will discuss in detail the connections established between the vulnerability and COVID-19 (cluster 2), network security (cluster 1), risk assessment and climate change (cluster 3). As illustrated in this section, the three clusters are gaining relevant meaning in the context of vulnerability correlation with other research fields, and this will also be reflected in the discussions below.

Table 3 lists the second cluster, "Vulnerability & COVID-19," which we deem pertinent to our study. The term that VOSviewer assigned to this cluster, together with the number of occurrences, links, and link strength values for each term. After performing the methodological data cleaning, the "Vulnerability" term appears 324 times across all of the analyzed publications. This value was obtained by implementing the full counting analysis option. The "Vulnerability" term registers the most significant values for all three parameters. The term "vulnerability" is related with phrases that are in all of our clusters, as indicated by the link's score of 390.

This suggests a direct connection between the research's focus and the other term co-occurrence analysis items, such as financial weaknesses. The link strength is a metric that always has a positive numerical value and can vary depending on how frequently a particular term appears in the examined texts. The first and most popular term in the first cluster has the largest incidence across the discovered documents, compared to all other terms, regardless of the cluster, according to the link strength value of 1352. We can also find the peak points in our study with the aid of the link strength values; the concepts in the literature that are most closely related to the knowledge concept are: "COVID-19" (196) with a link strength 590, "Human" (30) with a link strength 201, and "Pandemic" (33) with a link strength 153. At the same time, "Public Health" (17) with link strength 84, "COVID-19 Pandemic" (24) with the link strength 45 and "Financial vulnerability" (22) with link strength 28. From incidence of terms and connection, we may see that "Vulnerability" has a notable link strength and has been used in various research fields of interest.

From our study results, we may see that "knowledge vulnerabilities" is not treated directly, there are no studies so far, but the importance of the term "Vulnerability"

Table 3 Scopus cluster 2 analysis

Term	Cluster	Occurrences	Links	Link strength
Vulnerability	2—Vulnerability & COVID-19	324	390	1352
COVID-19		196	228	590
Human		30	98	201
Pandemic		33	93	153
Public Health		17	58	84
COVID-19 Pandemic		24	40	45
Financial vulnerability		22	26	28

Source Authors' own research

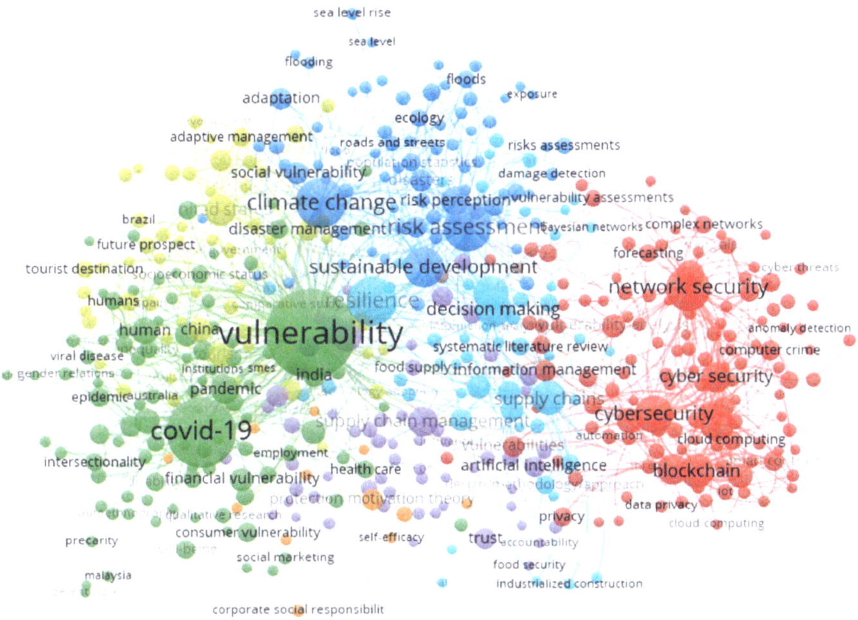

Fig. 2 Network clusters visualization by VOSviewer software version 1.6.18. *Source* Authors' own research

in other research is quite notable. As may be seen from Fig. 2, the term "Vulnerability" not only that is has connection with other terms such as "Risk assessment", "Resilience", "Sustainable development", "Climate change", "Network security", "Cybersecurity" and many others, but is has occurrence of 324, confirming us one more time the importance of our study in the knowledge vulnerabilities and risks area.

We provide the network visualization of all clusters in Fig. 2, although only three are significant: Vulnerability & COVID-19 (cluster 2), network security (cluster 1), risk assessment and climate change (cluster 3). According to Van Eck and Waltman, each term's sphere size and distance are visual representations of the link strength values listed in the tables.

In cluster 2, the term "Vulnerability" is more visible in the network together with the term COVID-19, and other terms as network security (cluster 1), risk assessment and climate change (cluster 3), as they have a notable presence in the network.

This means that there is one predominant concept in our analysis, and this is the "Vulnerability" concept, an essential term with strong connections in various fields. Another exciting network aspect relates to the fact that, even though directly connected specific items, cluster 2 "Vulnerability & COVID-19" and cluster 3 "risk assessment and climate change" and cluster 1 "network security", seem to be separated but still with strong connections between them. Understanding the insights gained through the systematic literature review, this visual distribution can be explained as follows: vulnerability is a fundamental concept for the knowledge economy and associated with risks this should be understood as a new challenge for our future research.

Figure 3 shows a representation of the density overview of the clusters, disseminating the most popular literary concepts associated with vulnerability. Each phrase, according to Van Eck and Waltman, has a corresponding sphere with a particular diameter and color density. As may be seen, there is a particular distance between each sphere in Fig. 3. The values of each item as shown in the cluster tables are directly related to these five factors. For example, "vulnerability", "COVID-19", "sustainable development", "climate change", "network security", and "cybersecurity" have the most obvious hollows on the map, and this is in line with their leading cluster positions and highest values in their cluster when it comes to the occurrences.

An interesting aspect is the fact that on the one hand, the "vulnerability" concept appears to be in closer relationship with "COVID-19", and on the other, "network security" is in a close relationship with "cybersecurity". This visual effect can be caused by the fact that closer items on the density map are part of the same article. It is also interesting to note the appropriation between items belonging to different clusters. The evident interest in vulnerability and risks is more than obvious.

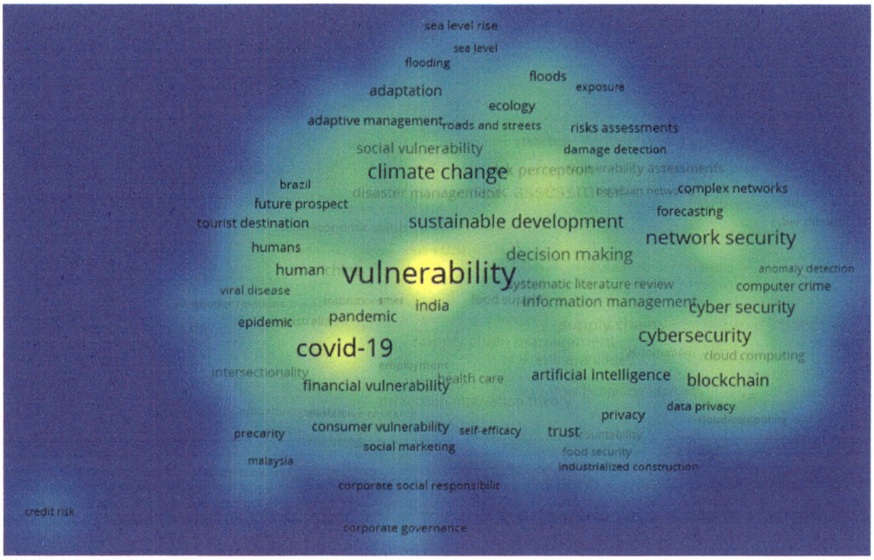

Fig. 3 Network density visualization by VOSviewer software version 1.6.18. *Source* Authors' own research

6 Conclusions and Limits of the Research

The purpose of present research was to analyze the knowledge vulnerabilities, concentrating our analysis on identifying some vulnerabilities in the organizational knowledge dynamics with some impacts in the knowledge economy. Through our research we register notable progress, particularly in the knowledge vulnerabilities area where some knowledge vulnerabilities were identified, which has not been done so far in previous research's and more than that in our research we conceived and proposed a new method for further research, named Knowledge Vulnerabilities Scoring System (KVSS).

The process will be beneficial for a correct analysis of knowledge vulnerabilities and the impact in the knowledge economy, completing the research with some solutions that may reduce negative consequences for organizations that are an active component in the knowledge economy. This was accomplished by first executing a thorough literature study, then using VOSviewer software for a text mining analysis. The bureaucratic system holds specific knowledge gaps, and research in this area would need more scientific attention, especially in the bureaucratic system analyzed as a risk generator for knowledge domain. Despite the fact that we were successful in identifying a set of research interests in knowledge management associated with the knowledge risk concept, we have also discovered that.

Not being a subject deeply analyzed, knowledge vulnerabilities have a high importance for the knowledge domain. Even if we didn't find specific research on the knowledge vulnerabilities in the knowledge economy, from the current research we identified that indeed there are gaps in literature and research are more than needed to fulfill this need. Research on this area will reveal much more conclusive results that will become a real support for further identification of vulnerabilities, comprehend in more details KVSS in a further research paper, finally identifying solutions to treat knowledge vulnerabilities.

In today knowledge economy were knowledge has become essential for every organization and disruptions are happening frequently which gives organizations less time to manage the change and the fact that they are almost constrained to develop new strategies to answer the challenges of the changing business environment, identifying knowledge vulnerabilities will increase visibility on organizational knowledge given the opportunity to threat weakness point identified and in the end to achieve the goal to strengthen the organization.

The knowledge economy has evolved more than was predicted in a short time period, adding notable value to the knowledge economy. On other side, this fast evolution in the knowledge economy revealed that some components were not fully covered, leaving some gaps that may be exploited. Our research identifies some of gaps in the knowledge economy especially on knowledge vulnerabilities. Given the unrestricted access to information and improvements in academic research, knowledge vulnerabilities will play a larger role in the current knowledge economy, and new techniques will be accessible to evaluate and prioritize the knowledge vulnerabilities.

References

Adger, W.: Vulnerability. Glob. Environ. Chang. **16**, 268–281 (2006). https://doi.org/10.1016/j.gloenvcha.2006.02.006

Aubert, B., Patry, M., Rivard, S.: Managing IT outsourcing risk: Lessons learned. (2001). https://doi.org/10.1007/978-3-662-04754-5_7

Bayer, F., Maier, R.: Knowledge risks in inter-organizational knowledge transfer. In: Proceedings of I-KNOW '06 Graz, Austria, pp. 200–208 (2006)

Blaikie, P., Cannon, Davis, L., Wisner, B.: At risk: Natural hazards, people's vulnerability and disasters, Routledge, London (1994)

Bogard, W.: Bringing social theory to hazards research conditions and consequences of the mitigation of environmental hazards. Sociological Perspective **31**, 147–168 (1989)

Bolisani, E., Bratianu, C.: Knowledge strategy planning: an integrated approach to manage uncertainty, turbulence, and dynamics. J. Knowl. Manag. **21**, 233253 (2017). https://doi.org/10.1108/JKM-02-2016-0071

Bratianu, C.: Organizational knowledge dynamics: managing knowledge creation, acquisition, sharing, and transformation. IGI Global (2015). https://doi.org/10.4018/978-1-4666-8318-1

Bratianu, C.: Knowledge dynamics. Manag. Dyn. Knowl. Econ. **4**, 323–337 (2016)

Bratianu, C.: A holistic approach to knowledge risk. Manag. Dyn. Knowl. Econ. **6**(4), 593–607 (2018)

Bratianu, C., Vasilache, S.: Evaluating linear-nonlinear thinking style for knowledge management education. Manag. & Mark. **4**(3), 3–18 (2009)

Bratianu, C., Nestian, A.S., Tita, S.M., Voda, A.I., Guta, A.L.: The impact of knowledge risk on sustainability firms. Amfiteatru Economic **22**(55), 639–652 (2020). https://doi.org/10.24818/EA/2020/55/639

Bratianu, C., Bejinaru, R.: Exploring vulnerabilities and risks related to knowledge management systems. (2022)

Cader, H.: The evolution of the knowledge economy. J. Reg.Nal Anal. Policy. **38**

Cegarra-Navarro, J.G., Bratianu, C., Martinez-Martinez, A., Vatamanescu, E.M., Dabija, D.C.: Creating civic and public engagement by a proper balance between emotional, rational, and spiritual knowledge. J. Knowl. Manag. (2023). https://doi.org/10.1108/JKM-07-2022-0532

Chambers, R.: Editorial introduction: Vulnerability, coping and policy. IDS Bulletin **20**(2)

Cutter, S.: Vulnerability to environmental hazards. Prog. Hum. Geogr. **20**(4), 529–539 (1996)

Dow, K.: Exploring differences in our common future(s): the meaning of vulnerability to global environmental change. Geoforum **23**, 417–436 (1992)

Durst, S., Henschel, T.: Knowledge risk management from theory to Praxis: From Theory to Praxis. https://doi.org/10.1007/978-3-030-35121-2

Durst, S., Wilhelm, S.: Do you know your knowledge at risk? Meas. Bus. Excell. **17**(3), 28–39 (2013). https://doi.org/10.1108/MBE-08-2012-0042

Durst, S., Zieba, M.: Knowledge risks—Towards a taxonomy. Int. J. Bus. Environ. **9**, 51 (2017). https://doi.org/10.1504/IJBE.2017.084705

Friedman, M.: The methodology of positive economics. The Philosophy of Economics: An Anthology. https://doi.org/10.1017/CBO9780511819025.010

Friedman, M.: A monetary history of the United States, 1867–1960. Princeton University Press (1963)

Gabor, T., Griffitch, T. K.: The assessment of community vulnerability to acute hazardous materials incidents. (1979). Retrieved from https://udspace.udel.edu/items/8f40bbe6-b9c9-49f7-86d7-93ecae197a84

Iliescu, A. N.: Conceptual atlas of the knowmad literature: Visual mapping with VOSviewer. Manag. Dyn. Knowl. Econ., **9**(3), 379–392 (2021). https://doi.org/10.2478/mdke-2021-0025

Intergovernmental Panel on Climate Change (PCC): Climate Change 1995: Impacts, adaptations and mitigation of climate change: summary for policy makers. World Meteorological Organisation, Geneva (1996)

Jamieson, R., Loeng, D.: An exploratory study of risks and issues in knowledge management. In: 14th Australasian conference on information systems, Perth, Western Australia (2003)
Kasperson, R.E., Kasperson, J.X., Dow, K.: Global environmental risk and society. Global Environmental Risk, Tokyo. pp. 1–48 (2001)
Kasperson, R.E., Kasperson, J.X., Dow, K.: Vulnerability, equity, and global environmental change. In: Global Environmental Risk. United Nations University Press and Eartscan, pp. 247–272 (2001b)
Keynes, J.M.: The general theory of employment, interest and money. Wordsworth Editions, (2017)
Lambe, P.: Four types of knowledge risk, pp. 1–3 [online]. Retrieved from www.greenchameleon.com/uploads/Four_Types_of_Knowledge_Risk.pdf
Marshall, A.: Principles of economics, 8th edn. Macmillan, London (1920)
Massingham, P.: Knowledge risk management: a framework. J. Knowl. Manag., **14**(3), 464–485 (2010). Q Emerald Group Publishing Limited, ISSN 1367–3270. https://doi.org/10.1108/13673271011050166
Nonaka, I., Takeuchi, H.: The wise company. How companies create Continuous innovation. Oxford: Oxford University Press. Organization for Economic Co-operation and Development (OECD) (2006). Creating value from intellectual assets. (2019). Retrieved from http://www.oecd.org/science/inno/36701575.pdf
Nonaka, I., Takeuchi, H.: The knowledge creating company: How Japanese companies create the dynamics of innovation. Oxford University Press, Oxford, UK (1995)
Perrot, B.E.: A strategic risk approach to knowledge management. Bus. Horiz. **50**, 523–533 (2007)
Schumpeter, J.A.: The theory of economic development. Harvard University Press, Cambridge, MA (Oxford University Press, New York, NY, 1961) (first published in German, 1912)
Smith, A.: Wealth of Nations. Wordsworth classics of world literature. Wordsworth Editions, Ware, England (2012)
Timmerman, P.: Vulnerability, resilience, and the collapse of society. Environmental Monograph No. 1. Institute of Environmental Studies, (1981)
Turner, B.L., II., Kasperson, R., Matson, P., Mccarthy, J., Corell, R., Christensen, L., Selin, N., Kasperson, J., Luers, A., Martello, M., Polsky, C., Pulsipher, A., Schiller, A.: A framework for vulnerability analysis in sustainability science. Proc. Natl. Acad. Sci. u.s.a. **100**, 8074–8079 (2003). https://doi.org/10.1073/pnas.1231335100
United Nations Disaster Relief Organization (UNDRO): Natural disasters and vulnerability analysis. Office of the United Nations Disaster Relief Co-ordinator, Geneva (1982)
Ursache, V.M.: Cybersecurity challenges in the knowledge economy. Proceedings of the International Conference on Business Excellence. **16**, 121–129 (2022). https://doi.org/10.2478/picbe-2022-0012
Ursache, V.M.: Knowledge risks management in a bureaucratic system. In: STRATEGICA international conference, October 21, 2022, Bucharest, Romania (2022b)
Watts, M.J., Bohle, H.: The space of vulnerability: the causal structure of hunger and famine. Prog. Hum. Geogr. **17**(1), 43–67 (1993)
Winkler, I., Gomes, A.: Advanced persistent. Security (2017). https://doi.org/10.1016/B978-0-12-809316-0.00023-3

SPRINGER NATURE

GPSR Compliance

The European Union's (EU) General Product Safety Regulation (GPSR) is a set of rules that requires consumer products to be safe and our obligations to ensure this.

If you have any concerns about our products, you can contact us on ProductSafety@springernature.com

In case Publisher is established outside the EU, the EU authorized representative is:

Springer Nature Customer Service Center GmbH
Europaplatz 3
69115 Heidelberg, Germany

The manufacturer's authorised representative in the EU is Springer Nature Customer Service Centre GmbH, Europaplatz 3, 69115 Heidelberg, Germany. If you have any concerns regarding our products, please contact ProductSafety@springernature.com

Printed and bound by CPI Group (UK) Ltd, Croydon, CR0 4YY
01/10/2026
02225899-0011